W9-AAT-842

Techniques and Applications of Hyperspectral Image Analysis

Techniques and Applications of Hyperspectral Image Analysis

Hans F. Grahn and Paul Geladi

John Wiley & Sons, Ltd

Other Wiley Editorial Offices

John Wiley & Sons Inc., 111 River Street, Hoboken, NJ 07030, USA

Jossey-Bass, 989 Market Street, San Francisco, CA 94103-1741, USA

Wiley-VCH Verlag GmbH, Boschstr. 12, D-69469 Weinheim, Germany

John Wiley & Sons Australia Ltd, 42 McDougall Street, Milton, Queensland 4064, Australia

John Wiley & Sons (Asia) Pte Ltd, 2 Clementi Loop #02-01, Jin Xing Distripark, Singapore 129809

John Wiley & Sons Canada Ltd, 6045 Freemont Blvd, Mississauga, Ontario, L5R 4J3, Canada

Wiley also publishes its books in a variety of electronic formats. Some content that appears in print may
not be available in electronic books.

Anniversary Logo Design: Richard J. Pacifico

Library of Congress Cataloging in Publication Data

Techniques and applications of hyperspectral image analysis / [edited by] Hans Grahn and Paul Geladi.
  p. cm.
Includes bibliographical references.
ISBN 978-0-470-01086-0 (cloth)
1. Image processing—Statistical methods. 2. Multivariate analysis. 3. Multispectral photography.
I. Grahn, Hans. II. Geladi, Paul.
TA1637.T42 2007
621.36'7—dc22   2007021097

British Library Cataloguing in Publication Data

A catalogue record for this book is available from the British Library

ISBN 978-0-470-01086-0

Typeset in 10/12pt Times by Integra Software Services Pvt. Ltd, Pondicherry, India
Printed and bound in Great Britain by TJ International, Padstow, Cornwall
This book is printed on acid-free paper responsibly manufactured from sustainable forestry in which
at least two trees are planted for each one used for paper production.

Contents

6   Hyperspectral Image Data Conditioning and Regression Analysis   127
James E. Burger and Paul L. M. Geladi

7   Principles of Image Cross-validation (ICV): Representative Segmentation of Image Data Structures   155
Kim H. Esbensen and Thorbjørn T. Lied

Preface

This book is about multivariate and hyperspectral imaging, not only on how to make the images but on how to clean, transform, analyze and present them. The emphasis is on visualization of images, models and statistical diagnostics, but some useful numbers and equations are given where needed. The idea to write this book originated at an Image Analysis Session at the Eastern Analytical Symposium (Somerset, NJ) in November 2002. At this session, the lectures were so inspiring that it was felt necessary to have something on paper for those not present.

An earlier book, also published by John Wiley & Sons, Ltd, came out in 1996. It was called *Multivariate Image Analysis* by Geladi and Grahn. This book contains a lot of the basic theory. The examples in this book are not very advanced because in the early 1990s it was not so easy to get 10 or more wavelength bands in an image. There has also been an evolution in theory and algorithms, requiring additions to the 1996 book, but a major difference is that image files are much larger in size and sizes are expected to keep on growing. This is a challenge to the data analysis methods and algorithms, but it is also an opportunity to get more detailed results with higher precision and accuracy.

It would have been possible to make a revised second edition of Geladi and Grahn, but it was considered more useful to include extra authors, thus creating a multi-authored book with chapters on hyperspectral imaging. The chapters would be written by groups or persons whom we felt would be able to contribute something meaningful. The book can roughly be divided into two parts. The earlier chapters are about definitions, nomenclature and data analytical and visualization aspects. The later chapters present examples from different fields of science, including extra data analytical aspects. The subdivision in theory and application parts is not ideal. Many attempts of putting the chapters in the correct order were tried and the final result is only one of them.

Chapter 1 is about the definition of multivariate and hyperspectral images and introduces nomenclature. It contains basic information that is applicable to all subsequent chapters. The basic ideas of multivariate interactive image analysis are explained with a simple color (three channels) photograph.

Chapters 2–5 give a good insight into factor and component modeling used on the spectral information in the images. This is called multivariate image analysis (MIA). Chapter 2 introduces interactive exploration of multivariate images in the scene and variable space in more detail using an eight channel optical image taken from an airplane. The role of visualization in this work is extremely important; something that Chapter 2 succeeds in highlighting. Chapter 3 gives a good overview of classification in optical images of agricultural products. The special topics of fuzzy clustering and clustering aided by spatial information are explained. In Chapter 4, the SIMPLISMA technique and its use on images are explained. This technique is an important alternative to those explained in Chapters 2 and 3. SIMPLISMA is not just exploratory, but tries to find deterministic pure component spectra by using spectroscopic constraints on the model. The examples are from Fourier transform infrared (FTIR) and time-of-flight - secondary ion mass spectrometry (TOF-SIMS) imaging. Chapter 5 is about even more factor analysis methods that can be applied to hyperspectral images. The special case of unsymmetrical noise distributions is emphasized.

Chapters 6–9 introduce the concepts and models for regression modeling on hyperspectral images: multivariate image regression (MIR). Chapter 6 is about regression on image data. This is the situation where the spectrum in each pixel is able to predict the value of an external variable, be it another image, an average property or something in between like localized information. Emphasis is also given on cleaning and preprocessing a hyperspectral image to make the spectral information suitable for regression model building. Chapter 7 takes up the important aspect of validation in classification and regression on images. The example of Chapter 2 is reused by defining one of the channels as a dependent variable. Also, a new example for the calibration of fat content in sausages is introduced. The advantage of image data is that many pixels (=spectra) are available, making testing on subsets a much easier task. Chapter 8 describes classical, extended and general least squares models for Raman images of aspirin/polyethylene mixtures. The theory part is extensive. Chapter 9 is about the need for expressing hyperspectral data in the proper SI and IUPAC units and about standards for multivariate and hyperspectral imaging. In particular, diffuse reflection, the most practical technique of imaging used in the laboratory, is in need of such standardization. Without the standards, reproducible image data would not be available and spectral model building and interpretation would be hampered severely.

The applied chapters do not give a complete overview of all possible applications, but they give a reasonable catalog of things that can be done with hyperspectral images using different types of variables. Chapter 10 is about multivariate movies in different variables, mainly optical, infrared, Raman and nuclear magnetic resonance (NMR). Multivariate movies represent huge amounts of data and efficient data reduction is needed. The applications are in polymer and pharmaceutical tablet dissolution. Chapter 11 describes the DECRA technique as it can be used on phantoms and brain images in magnetic resonance imaging. Chapter 12 gives an overview of agricultural and biological applications of optical multivariate and hyperspectral imaging. Chapter 13 is about brain studies using positron emission tomography (PET). The PET images are extremely noisy and require special care. Chapter 14 is about chemical imaging using near infrared spectroscopy. Pharmaceutical granulate mixtures are the examples used.

When writing a book one should always have students in mind. Books are ideal as course material and there is not much material available yet for learning about nonremote sensing hyperspectral imaging. Recommendations for newcomers are to read Chapters 1–9 together with Geladi and Grahn (Geladi and Grahn, 1996) in order to get the basics. Chapters 2–5 form the factor analysis block and Chapters 6–9 form the regression/calibration block. More advanced readers may review the basics quickly and plunge directly into the applied chapters (10–15). An alternative choice of reading would be Bhargava and Levin (Bhargava and Levin, 2005). There are also some interesting books from the remote sensing field (Chang, 2003; Varhsney and Arora, 2004).

REFERENCES

Bhargava, R. and Levin, I. (Eds) (2005) *Spectrochemical Analysis Using Infrared Multichannel Detectors*, Blackwell, Oxford.

Chang, C. (2003) *Hyperspectral Imaging: Techniques for Spectral Detection and Classification*, Kluwer Academic Publishers, Dordrecht.

Geladi, P. and Grahn, H. (1996) *Multivariate Image Analysis*, John Wiley & Sons, Ltd, Chichester.

Varshney, K. and Arora, M. (Eds) (2004) *Advanced Image Processing Techniques for Remotely Sensed Hyperspectral Data*, Springer, Berlin.

List of Contributors

Brian Antalek, Eastman Kodak Company, Research Laboratories B82, Rochester, NY 14650-2132, USA

Vincent Baeten, Département Qualité des Produits agricoles, Centre wallon de recherches agronomiques CRA-W, 24, Chaussée de Namur, B-5030 Gembloux, Belgium

Mats Bergström, Novartis Pharma AG, CH-4002, Basel, Switzerland

James E. Burger, Burger Metrics, Applied Hyperspectral Imaging for Automation and Research, Ladehammerveien 36, 7041 Trondheim, Norway

Pierre Dardenne, Département Qualité des Produits agricoles, Centre wallon de recherches agronomiques CRA-W, 24, Chaussée de Namur, B-5030 Gembloux, Belgium

Jane Dubois, Malvern Instruments, Analytical Imaging Systems, 3416 Olandwood Court #210, Olney, MD 20832, USA

Kim H. Esbensen, ACABS, Aalborg University Esbjerg (AAUE), Niels Bohrs Vej 8, DK-6700 Esbjerg, Denmark

Neal B. Gallagher, Eigenvector Research, Inc., 160 Gobblers Knob Lane, Manson, WA 98831, USA

Paul L. M. Geladi, NIRCE, The Unit of Biomass Technology and Chemistry SLU Röbäcksdalen, PO Box 4097, SE 90403 Umeå, Sweden

Hans F. Grahn, Division of Behavioral Neuroscience, Department of Neuroscience, Karolinska Institutet, S-17177, Stockholm, Sweden

Kenneth S. Haber, Malvern Instruments, Analytical Imaging Systems, 3416 Olandwood Court #210, Olney, MD 20832, USA

Joseph P. Hornak, Magnetic Resonance Laboratory, 54 Lomb Memorial Drive, Center for Imaging Science, Rochester Institute of Technology, Rochester, NY 14623-5604, USA

Sergei G. Kazarian, Department of Chemical Engineering, ACE Building 208A/210, Imperial College London, South Kensington Campus, London SW7 2AZ, UK

Michael R. Keenan, Sandia National Laboratories, Albuquerque, NM 87185-0886, USA

Linda H. Kidder, Malvern Instruments, Analytical Imaging Systems, 3416 Olandwood Court #210, Olney, MD 20832, USA

E. Neil Lewis, Malvern Instruments, Analytical Imaging Systems, 3416 Olandwood Court #210, Olney, MD 20832, USA

Sharon Markel, Eastman Kodak Company, Rochester, NY 14650-2132, USA

Jacco C. Noordam, TNO Defence, Security and Safety, PO Box 96864, 2509 JG The Hague, The Netherlands

Juan Antonio Fernández Pierna, Département Qualité des Produits agricoles, Centre wallon de recherches agronomiques CRA-W, 24, Chaussée de Namur, B-5030 Gembloux, Belgium

Pasha Razifar, Computerized Image Analysis & PET Uppsala Applied Science Lab (UASL) GEMS PET Systems AB, Husbyborg, 752 28 Uppsala, Sweden

P. M. Thompson, Eastman Kodak Company, Rochester, NY 14650-2132, USA

Thorbjørn Tønnesen Lied, Kongsberg Maritime AS, R&D, Hydrography & Hydroacustic Division, Horten, Norway

Willie H. A. M. van den Broek, TNO Defence, Security and Safety, PO Box 96864, 2509 JG The Hague, The Netherlands

Jaap van der Weerd, Department of Chemical Engineering, Imperial College London, South Kensington Campus, London SW7 2AZ, UK

Willem Windig, Eigenvector Research, Inc., 6 Olympia Drive, Rochester, NY 14615, USA

List of Abbreviations

2-D	Two-dimensional
3-D	Three-dimensional
ABES	Agricultural, biology and environmental sciences
ACD	Annihilation coincidence detection
AIS	Airborne imaging system
ALS	Alternating least squares
ANN	Artificial neural network
ARMA	Autoregressive moving average
AOTF	Acousto-optic tunable filter
API	Active pharmaceutical ingredient
ATR	Attenuated total reflection
AVHRR	Advanced very high resolution radiometer
AVIRIS	Airborne visible and infrared imaging spectrometer
CAS	Chemical Abstracts Service
CASI	Compact airborne spectrographic imager
CCD	Charge coupled device
CLS	Classical least squares
cFCM	Conditional Fuzzy C-Means
CSF	Cerebrospinal fluid
csiFCM	Cluster size insensitive Fuzzy C-Means
DECRA	Direct exponential curve resolution analysis
EDS	Energy dispersive spectrometer (mainly X-rays)
EMSC	Extended multiplicative scatter correction
ELS	Extended least squares
FA	Factor analysis
fALS	Factored alternating least squares
FCM	Fuzzy C-Means
FEMOS	Feedback Multivariate Model Selection

fNNMF	Factored non-negative matrix factorization
FBP	Filtered back projection
FOV	Field of view
FPA	Focal plane array
FT	Fourier transform
FTIR	Fourier transform infrared
FWHM	Full width at half maximum
GIS	Geographical information system
GLS	Generalized least squares
GOME	Global ozone monitoring experiment
GPS	Global positioning system
GRAM	Generalized rank annihilation method
HD	High density (for polymers)
HIA	Hyperspectral image analysis
HPLC	High pressure liquid chromatography
HPMC	Hydroxypropylmethylcellulose
ICV	Image cross-validation
ILS	Inverse least squares
ILS	Iterative partial least squares
IR	Infrared
IUPAC	International Union of Pure and Applied Chemistry
IWLS	Iteratively weighted least squares
LCTF	Liquid crystal tunable filter
LD	Low density (for polymers)
LDA	Linear discriminant analysis
LOR	Line of response
LUT	Look up table
MBM	Meat and bone meal
MCR	Multivariate curve resolution
MEIS	Multispectral electro-optical imaging spectrometer
MI	Multivariate image
MIA	Multivariate image analysis
MIR	Multivariate image regression
MIR	Mid infrared
MLPCA	Maximum likelihood principal component analysis
MNF	Minimum noise fraction
MRI	Magnetic resonance imaging
MSC	Multiplicative scatter correction
MVWPCA	Masked volume wise principal component analysis
NIR	Near infrared
NIRCI	Near infrared chemical imaging

NMR	Nuclear magnetic resonance
NNMF	Non-negative matrix factorization
NSLS	National Synchrotron Light Source
OSEM	Ordered subset expectation maximization
PAL	Phase alternating line
PAA	Polyacrylamide
PAMS	Poly (α-methyl styrene)
PAS	Photo-acoustic spectroscopy
PBMA	Poly (butyl methacrylate)
PC	Principal component
PCA	Principal component analysis
PCR	Principal component regression
PE	Polyethylene
PET	Poly (ethylene terephthalate)
PET	Positron emission tomography
PETT	Positron emisison transaxial tomography
PFA	Principal factor analysis
PGP	Prism–grating–prism
PGSE	Pulse gradient spin echo
PLS	Partial least squares
PMSC	Piecewise multiplicative scatter correction
PNNMF	Poisson non-negative matrix factorization
PVA	Poly (vinyl alcohol)
PVC	Poly (vinyl chloride)
PW	Pixel wise
PRESS	Prediction residual error sum of squares
QA	Quality assurance
QC	Quality control
RGB	Red green blue
RMSEC	Root mean square error of calibration
RMSECV	Root mean square error of cross-validation
RMSEP	Root mean square error of prediction
ROI	Region of interest
SECAM	Séquentielle couleur a mémoire
SECV	Standard error of cross-validation
SEM	Scanning electron microscope
SIA	Self-modeling image analysis
sgFCM	Spatially guided Fuzzy C-Means
SI	Système international
SLDA	Stepwise linear discriminant analysis
SIMCA	Soft independent modeling of class analogy

SIMPLISMA	Simple-to-use interactive self-modeling mixture analysis
SIMS	Secondary ion mass spectrometry
SNR	Signal to noise ratio
SNV	Standard normal variate
SPOT	Satellite pour l'observation du terre
SRM	Standard reference material
SS	Sum of squares
SSE	Sum of squared errors
SVD	Singular value decomposition
SVM	Support Vector Machine
SWPCA	Slice wise principal component analysis
TAC	Time–activity curve
TOF-SIMS	Time-of-flight-secondary ion mass spectrometry
TOMS	Total ozone mapping spectrometer
USP	United States Pharmacopeia
UV	Ultraviolet
VWPCA	Volume wise principal component analysis
WPCA	Weighted principal component analysis
WTFA	Window target factor analysis

1

Multivariate Images, Hyperspectral Imaging: Background and Equipment

Paul L. M. Geladi, Hans F. Grahn and James E. Burger

1.1   INTRODUCTION

This chapter introduces the concepts of digital image, multivariate image and hyperspectral image and gives an overview of some of the image generation techniques for producing multivariate and hyperspectral images. The emphasis is on imaging in the laboratory or hospital on a scale going from macroscopic to microscopic. Images describing very large scenes are not mentioned. Therefore, the specialized research fields of satellite and airborne imaging and also astronomy are left out. A color image is used to introduce the multivariate interactive visualization principles that will play a major role in further chapters.

1.2   DIGITAL IMAGES, MULTIVARIATE IMAGES AND HYPERSPECTRAL IMAGES

All scientific activity aims at gathering information and turning this information into conclusions, decisions or new questions. The information may be qualitative, but is often and preferably quantitative. This

Techniques and Applications of Hyperspectral Image Analysis   Edited by H. F. Grahn and P. Geladi
© 2007 John Wiley & Sons, Ltd

means that the information is a number or a set of numbers. Sometimes even a large set of numbers is not enough and an image is needed. Images have the dual property of both being large datasets and visually interpretable entities.

Freehand drawing and photography have been used extensively in the sciences to convey information that would be too complicated to be expressed in a text or in a few numbers. From the middle of the 1900s the TV camera and electronic image digitization have become available and images can be saved in digital format as files (Geladi and Grahn, 2000). A digital image is an array of I rows and J columns made of I × J greyvalues or intensities, also called pixels. A pixel is a greyvalue with an associated coordinate in the image. The image is also a data matrix of size I × J with the greyvalues as entries. (Pratt, 1978; Rosenfeld and Kak, 1982; Gonzalez and Woods, 1992; Schotton, 1993) For three-dimensional images, the array has I rows, J columns and H depth slices. The pixel becomes a voxel. For color imaging in TV, video and on computer screens, three images are needed to contain the red, green and blue information needed to give the illusion of color to the human eye (Callet, 1998; Johnson and Fairchild, 2004) For photocopying and printing the three primary colors are yellow, cyan and magenta. One may say that the pixels (or voxels) are not greyvalues anymore, but triplets of numbers (Figure 1.1).

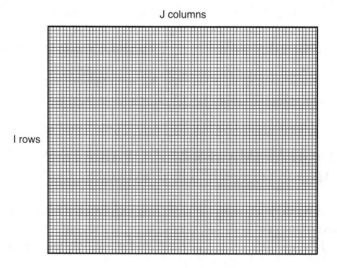

Figure 1.1 A digital image is an array of I rows and J columns. Each coordinate pair has a greyvalue and the small grey square (or rectangle) is called a pixel. For color images, the pixel becomes a red, green and blue triplet instead of a greyvalue

Because pixels are digitized greyvalues or intensities, they may be expressed as integers. Simple images may have a greyvalue range of 2^8 meaning that 0 is the blackest black and 255 is the whitest white. In more advanced systems, 2^{12} grey levels (0–4095), 2^{14} or 2^{16} greylevels are used. Some systems average images over a number of scans. In such a case, greylevels may have decimals and have to be expressed as double precision numbers.

The fact that a digitized image is a data matrix makes it easy to do calculations on it. The result of the calculations can be a number, a vector of numbers or a modified image. Some simple examples would be counting of particles (image to number), the calculation of intensity histograms (image to vector), image smoothing and edge enhancement (image to modified image). There are many books describing how this is done (Pratt, 1978; Rosenfeld and Kak, 1982; Low, 1991; Gonzalez and Woods, 1992).

Color images have three layers (or bands) that each have different information. It is possible to make even more layers by using smaller wavelength bands, say 20 nm wide between 400 nm and 800 nm. Then each pixel would be a spectrum of 21 wavelength bands. This is the multivariate image. The 21 wavelength bands in the example are called the image variables and in general there are K variables. An I × J image in K variables would form a three-way array of size I × J × K. (Figures 1.2–1.4).

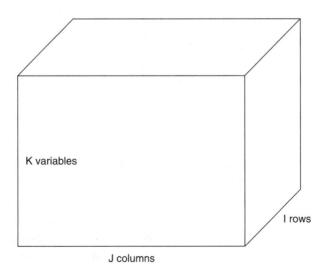

Figure 1.2 An I × J image in K variables is an I × J × K array of data

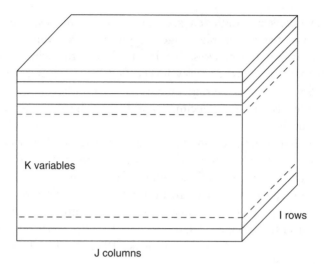

Figure 1.3 The I × J × K image can be presented as K slices where each slice is a greyvalue image

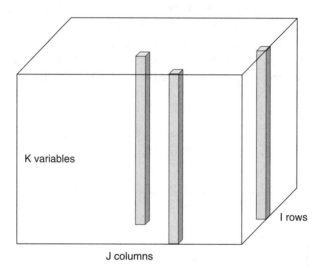

Figure 1.4 The I × J × K image can be presented as an image of vectors. In special cases. the vectors can be shown and interpreted as spectra

    The human eye only needs the three wavelength bands red, green and blue in order to see color. With more than three wavelength bands, simple color representation is not possible, but some artificial color images may be made by combining any three bands. In that case the colors are not real and are called pseudocolors. This technique makes no sense when more than three bands are combined because of the limitations of the human visual system.

Many imaging techniques make it possible to make multivariate images and their number is constantly growing. Also, the number of variables available is constantly growing. From about 100 variables upwards the name hyperspectral images was coined in the field of satellite and airborne imaging (Vane, 1988; Goetz and Curtiss, 1996), but hyperspectral imaging is also available in laboratories and hospitals. The following sections will introduce some multivariate and hyperspectral images and the physical variables used to make them with literature references.

Images as in Figures 1.2–1.4 with $K = 2$ or more are multivariate images. Hyperspectral images are those where each pixel forms an almost continuous spectrum. Multivariate images can also be mixed mode, e.g. $K = 3$ for an UV wavelength image, a near infrared (NIR) image and a polarization image in white light. In this case, the vector of three variables is not really a spectrum.

So what characterizes hyperspectral images? Two things:

- many wavelength or other variable bands, often more than 100;
- the possibility to express a pixel as a spectrum with spectral interpretation, spectral transformation, spectral data analysis, etc.

1.3 HYPERSPECTRAL IMAGE GENERATION

1.3.1 Introduction

Many principles from physics can be used to generate multivariate and hyperspectral images (Geladi and Grahn, 1996, 2000). Examples of making NIR optical images are used to illustrate some general principles.

A classical spectrophotometer consists of a light source, a monochromator or filter system to disperse the light into wavelength bands, a sample presentation unit and a detection system including both a detector and digitization/storage hardware and software (Siesler *et al.*, 2002). The most common sources for broad spectral NIR radiation are tungsten halogen or xenon gas plasma lamps. Light emitting diodes and tunable lasers may also be used for illumination with less broad wavelength bands. In this case, more diodes or more laser are needed to cover the whole NIR spectral range (780–2500 nm). For broad spectral sources, selection of wavelength bands can be based on specific bandpass filters based on simple interference filters, liquid crystal tunable filters (LCTFs), or acousto-optic tunable filters (AOTFs), or the spectral energy may be dispersed by a grating device or a prism–grating–prism (PGP) filter. Scanning interferometers can also be used to acquire NIR spectra from a single spot.

A spectrometer camera designed for hyperspectral imaging has the hardware components listed above for acquisition of spectral information plus additional hardware necessary for the acquisition of spatial information. The spatial information comes from measurement directly through the spectrometer optics or by controlled positioning of the sample, or by a combination of both. Three basic camera configurations are used based on the type of spatial information acquired; they are called point scan, line scan or plane scan.

1.3.2 Point Scanning Imaging

The point scanning camera configuration shown in Figure 1.5 can be used to measure a spectrum on a small spot. The sample is then repositioned before obtaining a new spectrum. By moving the sample

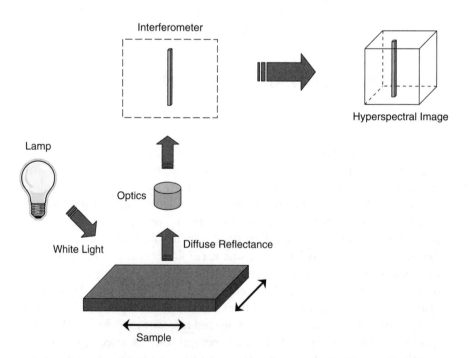

Figure 1.5 A scanning set-up measures a complete spectrum in many variables at a single small spot. An image is created by systematically scanning across the surface in two spatial dimensions

systematically in two spatial dimensions (pixel positions) a complete hyperspectral image can be acquired. Instrument calibration for wavelength and reflectance needs to be done only once before beginning the image acquisition. This system provides very stable high resolution spectra; however, the sample positioning is very time consuming and places high demands on repositioning hardware to ensure repeatability. The spatial size dimensions of the hyperspectral image are limited only by the sample positioning hardware. Continuous operation permits extended sample sizes or resampling at specific spatial locations (Sahlin and Peppas, 1997; Miseo and Wright, 2003).

1.3.3 Line Scanning Imaging

The second camera configuration approach indicated in Figure 1.6 is termed a line scan or push broom configuration and uses a two dimensional detector perpendicular to the surface of the sample. The

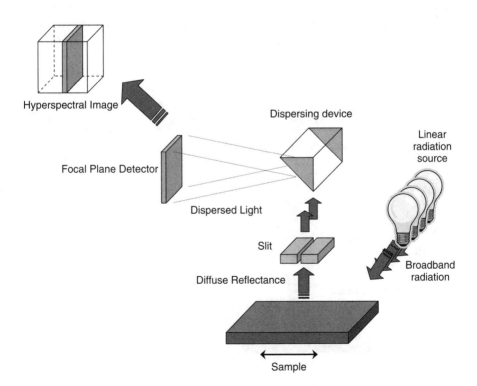

Hyperspectral Image

Dispersing device

Linear radiation source

Focal Plane Detector

Dispersed Light

Slit

Broadband radiation

Diffuse Reflectance

Sample

Figure 1.6 Illustration of how the capture mechanism of a linescan camera works. A two-dimensional detector captures all the spectra for one line in the image simultaneously. By scanning over the complete sample hyperspectral images are acquired

sample is imaged by a narrow line of radiation falling on the sample or by a narrow slit in the optical path leading to the detector. Hyperspectral images can easily be created by collecting sets of these matrices while moving the sample scan line. Since no filter change is necessary, the speed of image acquisition is limited only by camera read out speeds. Commercial instruments are available with frame rates of 90 Hz or higher with 256×320 pixel resolution InGaAs detectors. This speed allows images to be acquired in a matter of seconds. This configuration is also amenable to continuous operation for online monitoring of process streams (Aikio, 2001; Wold *et al.*, 2006).

1.3.4 Focal Plane Scanning Imaging

The focal plane configuration displayed in Figure 1.7 positions the detector in a plane parallel to the surface of the sample. The sample is imaged on the focal plane detector by a lens or objective. The spectrometer components including the sample remain fixed in position relative to the detector. If they are used, interference filters must be rotated into position for each image slice (Geladi *et al.*, 1992). For LCTFs (Gat,

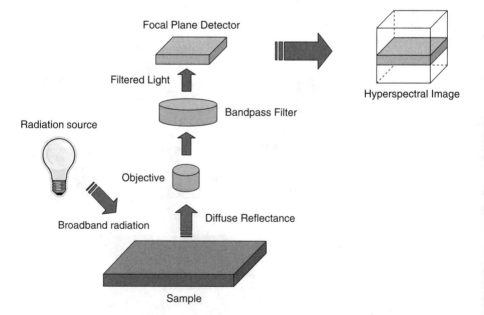

Figure 1.7 Using a bandpass filter in front of the camera a complete image scene in one wavelength is obtained. Many image slices are used to create a complete hyperspectral image while the bandpass is modified electronically. The sample does not move

2000) or AOTFs (Siesler *et al.*, 2002), the filter change is done electronically and more wavelengths can be used. Significant instrument settling times due to filter changes and movement within the sample itself will cause image registration problems. Lengthy image acquisition times can also be an issue for biological samples, which may be sensitive to heating caused by the continuous illumination from NIR source lamps.

1.4   ESSENTIALS OF IMAGE ANALYSIS CONNECTING SCENE AND VARIABLE SPACES

A simple but colorful color image is used to illustrate some of the principles of multivariate interactive image analysis. More details about the technique can be found in Chapter 2.

Figure 1.8   A still life of a flower arrangement in a vase was photographed digitally. The camera model was a Fujifilm FinePix A345 with the image size 1728 × 2304, 8 bits per color channel (see Plate 1)

Figure 1.9 The color image in Figure 1.8 is actually a stack of three grayscale images, that are used in the red, green and blue channels to create the illusion of color (see Plate 2)

Figure 1.8 shows a simple color image made with a digital camera of a flower bouquet. As is shown in Figure 1.9, this color image is made up of three planes representing intensities in red, green and blue. This is the typical way the human visual system perceives color images. The image is sometimes called a scene and doing operations on it is called working in scene space.

It is also possible to see each pixel in this image as a vector with three variable values. The values in these vectors may be transformed in a number of ways. What is used in this example is principal component analysis (PCA) (Esbensen and Geladi, 1989; Geladi *et al.*, 1989). This technique gives new axes in multivariate space. Figure 1.10 shows the results of PCA analysis. The three score images are combined in the red, green and blue channels and form a pseudocolor image. The score images can also be shown in a scatter plot called a score plot. This plot

Figure 1.10 A pseudocolor composite of the three principal component score images of Figure 1.9. Component 1 is in the red channel, component 2 in the green channel and component 3 in the blue channel (see Plate 3)

is shown in Figure 1.11. In this plot color coding is used to mark pixel density. One can clearly see single pixels and regions with high and low pixel density by observing the color coding.

The interactive analysis of the image is now done by comparing the variable space and the scene space. Clusters of pixels may be enclosed by a polygon in variable space and then the pixels under the polygon areas are projected to scene space. This projection is usually super-imposed on a black and white image. In Figure 1.12 the collection of pixels having the property red is enclosed in a polygon in vari-able space and then the matching pixels are highlighted in scene space. This can be done for all plant materials and the results are shown in Figure 1.13.

Figure 1.13 shows a number of important concepts of using the method as an analytical image analysis tool. Figure 1.13(a) shows the

Figure 1.11 A score plot of the second and third principal components. The colors are used to indicate pixel density. Most dense is white, then red, orange, yellow, dark green and olive. In such a plot, clustering, single pixels, gradients, outliers, etc., can be observed (see Plate 4)

scatter plot of components 2 and 3. A number of clusters are indicated by color polygons. The corresponding pixel values with matching colors in the scene space are shown in Figure 1.13(b). They are shown overlaid on a black and white picture so that parts of the scene space that were not used can be recognized. Figure 1.13(c) is the original picture used as a reference. Note the correspondence in color coding between the class colors and the true colors of the flower parts.

By discriminating pixel values from the variable space, the method can be used as a data reduction tool, where focus is pointed towards the relevant features of the problem studied:

- unnecessary or background pixels are not used (data reduction tool);
- the first component is not used. It mainly describes background (the wall, the table, the card, the dark vase);
- most colored plant material is detected;
- the plant material is classified according to color in the classes red, pink, green, yellow, light blue and dark blue;
- closeness in variable space means closeness in property/chemical composition;
- clusters and gradients can be detected;

(a)   (b)

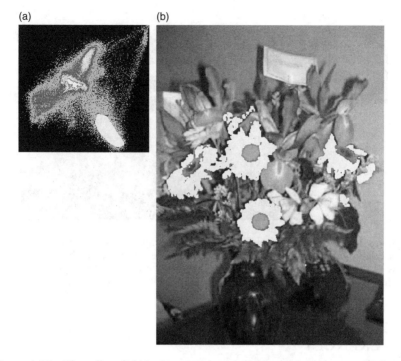

Figure 1.12 The yellow field in the scoreplot (a) illustrates the polygon selection of one dense class. By transferring this class to the scence space (b) an overlay is created indicating which pixel belongs to this class. In this case, red flowers (see Plate 5)

- single pixels with no neighbors are less interesting than dense clusters;
- mixed pixels and shadow pixels can be detected and studied.

One may wonder what PCA* analysis has done for this color image. Because there are three variables, it is only possible to calculate three principal components. The first component becomes the intensity in the image. The scatter plot of the second and third components is the color space and is not unlike the CIE color diagram.

---

* A multivariate or hyperspectral image of size I×J×K (K wavelengths) can always be rearranged into a data matrix **X** of size N×K where N=I×J. This data matrix can be decomposed as follows:

$$\mathbf{X} = \mathbf{TP}' + \mathbf{E}$$

where the A columns of **T** are orthogonal and contain score values, the A columns of **P** are orthogonal and contain loading values and **E** is a residual containing mainly noise. The vectors in **T** can also be reorganized back from size N×1 to size I×J whereby they become images again. These images are called score images. This is important knowledge useful in almost all the chapters.

(a)　　　　(b)　　　　　　　　　　(c)

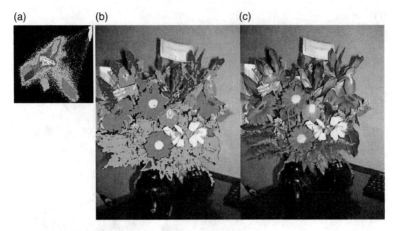

Figure 1.13 A more complete illustration using the same principle as in Figure 1.12. (a) In the score plot six different classes are indicated by colored polygons. (b) Overlays in matching colors highlight the positions of the different classes in scene space. (c) The original picture used as a reference (see Plate 6)

REFERENCES

Aikio, M. (2001) *Hyperspectral Prism–Grating–Prism Imaging Spectrograph*, PhD thesis University of Oulu, Publ. 435, VTT Publications, Espoo.

Callet, P. (1998) *Couleur-Lumière Couleur-Matière. Interaction Lumière-Matière et Synthèse d'Images*, Diderot, Paris.

Esbensen, K. and Geladi, P. (1989) Strategy of multivariate image analysis (MIA), *Chemometrics and Intelligent Laboratory Systems*, 7, 67–86.

Gat, N. (2000) Imaging spectroscopy using tunable filters: a review, *Proc. SPIE* , **4056**, 50–64.

Geladi, P. and Grahn, H. (1996) *Multivariate Image Analysis*, John Wiley & Sons, Ltd, Chichester.

Geladi, P. and Grahn, H. (2000) Multivariate image analysis, in R. Meyers, ed., *Encyclopedia of Analytical Chemistry*, John Wiley & Sons, Ltd, Chichester, pp. 13540–13562.

Geladi, P., Isaksson, H., Lindqvist, L., Wold, S. and Esbensen, K. (1989) Principal component analysis on multivariate images, *Chemometrics and Intelligent Laboratory Systems*, 5, 209–220.

Geladi, P., Grahn, H., Esbsensen, K. and Bengtsson, E. (1992) Image analysis in chemistry. Part 2: Multivariate image analysis, *Trends in Analytical Chemistry*, 11, 121–130.

Goetz, A. and Curtiss, B. (1996) Hyperspectral imaging of the earth: remote analytical chemistry in an uncontrolled environment, *Field Analytical Chemistry and Technology*, 1, 67–76.

Gonzalez, R. and Woods, R. (1992) *Digital Image Processing*, Addison-Wesley, Reading.

Johnson, G. and Fairchild, M. (2004) Visual psychophysics and color appearance, in Sharma G, ed., *Digital Color Imaging Handbook*, CRC Press, Boca Raton, FL, pp. 116–171.

Low, A. (1991) *Introductory Computer Vision and Image Processing*, McGraw-Hill, London.

Miseo, E. and Wright, N. (2003) Developing a chemical-imaging camera, *The Industrial Physicst*, **9**, 29–32.

Pratt, W. (1978) *Digital Image Processing*, John Wiley & Sons, Ltd, New York.

Rosenfeld, A. and Kak, A. (1982) *Digital Picture Processing*, 2nd Edn, Academic Press, New York.

Sahlin, J. and Peppas, N. (1997) Near infrared FTIR imaging: a technique for enhancing spatial resolution in FTIR microscopy, *Journal of Applied Polymer Science*, **63**, 103–110.

Schotton, D., ed. (1993) *Electronic Light Microscopy. Techniques in Modern Biomedical Microscopy*, John Wiley & Sons, Ltd, New York.

Siesler, H., Ozaki, Y., Kawata, S. and Heise, H., eds. (2002) *Near-infrared Spectroscopy: Principles, Instruments, Applications*, Wiley-VCH, Weinheim.

Vane, G., ed. (1988) *Proceedings of the Airborne Visible/Infrared Imaging Spectrometer (AVIRIS) Evaluation Workshop*, JPL 88-38, Jet Propulsion Laboratory, California Institute of Technology, Pasadena.

Wold, J.-P., Johanson, I.-R., Haugholt, K., Tschudi, J., Thielemann, J., Segtnan, V., Naruma, B. and Wold, E. (2006) Non-contact transflectance near infrared imaging for representative on-line sampling of dried salted coalfish (bacalao), *Journal of Near Infrared Spectroscopy*, **14**, 59–66.

2

Principles of Multivariate Image Analysis (MIA) in Remote Sensing, Technology and Industry

Kim H. Esbensen and Thorbjørn T. Lied
with contributions from Kim Lowell and Geoffrey Edwards

2.1 INTRODUCTION

Esbensen and Geladi (Esbensen and Geladi, 1989) and Lowell and Esbensen (Lowell and Esbensen, 1993) argued against what has been termed the traditionalist image analysis paradigm, the position of always starting image analysis in the scene space. The traditional approach is centred upon the concept of delineating scene space areas (objects, or part object) that are as homogeneous and/or spatially coherent as possible, in order to define representative training classes. It was argued that this is fraught with severe inconsistencies and necessarily must lead to suboptimal class representation. There is often a slight confusion in this image analysis tradition by not always specifying clearly in advance whether one is engaged in unsupervised (exploratory) or supervised (discrimination/classification) undertakings; for an indepth critique of this issue, see Lowell and Esbensen (Lowell and Esbensen, 1993).

Techniques and Applications of Hyperspectral Image Analysis Edited by H. F. Grahn and P. Geladi
© 2007 John Wiley & Sons, Ltd

2.1.1   MIA Approach: Synopsis

MIA takes its point of departure in feature space in general, in the score space in particular (Esbensen and Geladi, 1989; Geladi and Grahn, 1996). MIA can to a first delineation be understood as a truncated principal component (PC) modelling of the multi-channel image, producing sets of complementary score and loading plots. MIA's main thrust is that the score plot comprises the only valid starting point for image analysis, in that this is the only complete delineation of the covariance structure(s) of the total image pixel aggregation. The score plot visualizes the entire image inter-pixel dispositions (e.g. pixel classes, groupings, trends and outliers), while the complementary loading plot gives a graphic illustration of the underlying covariance/correlations responsible for these score dispositions. With MIA there is usually this mandatory order to the use of these spaces: the score space is usually the starting point of all image analysis but it is of course always possible to survey the scene space if so deemed necessary in special situations.

2.2   DATASET PRESENTATION

2.2.1   Master Dataset: Rationale

The comparison dataset to be used for all examples in this overview is an 8-channel digital image, recorded with the Canadian MEIS II airborne spectrometer, with a spatial pixel resolution of 70 cm, as described in McColl *et al.* (McColl *et al.*, 1984) and Esbensen *et al.* (Esbensen *et al.*, 1993). In satellite-based remotely sensed imagery all fine textural detail is usually lost at many of the prevailing resolutions. For airborne high resolution imaging the difficulty is high variability making traditional pixel classifiers ineffective. With high spatial resolution, individual image pixels (very) often tend to cover only minute fractions of the image objects and a pronounced smearing of discernible feature space classes is often observed. This is the well-known mixed pixel problem.

This particular dataset was chosen because it represents an easily manageable dimensionality, 8 channels. While clearly at the low(er) end of what is representative for modern remote sensing, or for technological/industrial imaging spectrometers, say 10 through 1024 channels or more (be they laboratory instruments or otherwise), this dimensionality still allows to present all principles and the full potential in the MIA approach. This dataset is characterized by its high-resolution overrepresentation of mixed pixels, but it will serve equally well in relation to less resolved imagery – and indeed for even more densely sampled image type, such as in current medical, tomographic and chemical imaging. It is in other words representative of all types of spatial and channel resolution.

At the outset it is emphasized however that this dataset primarily is a vehicle for presenting the general MIA approach. As such, the various examples of MIA analyses below should not be interpreted as specific for remote sensing type imagery. All examples and illustrations have been selected only because of relevance to the general MIA application potential.

2.2.2 Montmorency Forest, Quebec, Canada: Forestry Background

This forest scene was acquired in September 1986 over the Montmorency experimental forest belonging to Laval University, Quebec by the Canadian MEIS II airborne platform; MEIS II is described in detail in Kramer (Kramer, 1996). Table 2.1 lists the spectroscopic channel characteristics in the visible and near infrared.

The Montmorency Forest is located 80 km north of Quebec City near the southern edge of the northern boreal forest that dominates much of Canada. This partly heavily terrained forest contains mainly balsam fir, with minor white birch, white spruce and a small number of red spruce, trembling aspen and some other species.

Table 2.1 MEIS II channel characteristics

Band	Wavelength (nm)	Bandwidth (nm)
1	776	37.0
2	675	39.5
3	747	16.7
4	481	30.9
5	734	16.9
6	710	15.6
7	698	13.1
8	549	31.9

The site selected for this study contains a cutover which was cleared in 1975 and 1978, and which has both natural and planted conifer regeneration resulting from the forest experimental and observation campaigns in the ensuing period through 1986. Figure 2.1 shows the master Montmorency Forest scene, as depicted by MEIS-II channels 1:2:7 (R:G:B).

The cut clearing is characterized by balsam fir stands around the perimeter, one white birch stand in the central area, a small stream (bottom right), an east–west trending dirt road [bright green (false colours) in Figure 2.1].

The spatial and temporal regeneration history of this multiple cutover has been the subject of several Laval University Forestry Department studies and is accordingly very well understood. This scene is called the 'Clear-cut Study' in the illustrations that follow. Further in-depth scene description and full background forestry references were given by Esbensen *et al.* (Esbensen *et al.*, 1993). The specific scene history, as will transpire below, turned out to be a particularly illustrative context both

Figure 2.1 Composite scene display (1:2:7, R:G:B) of the master Montmorency Forest dataset (see Plate 7)

for illustrating the comprehensive MIA approach as well as allowing a powerful insight into the possibilities for spatio-temporal analysis of multivariate imagery characterized by emergent structures, i.e. temporally and/or spatially evolving/growing/changing structures. Thus, this example also extends the boundaries for static MIA applications.

2.3   TOOLS IN MIA

2.3.1   MIA Score Space Starting Point

Figures 2.2 and 2.3 show the most relevant PC cross-plots pertaining to the Montmorency Forest scene in the familiar MIA score plot rendition.

Note how the main MIA score cross-plot series specifically uses the same PC as the x-axis; this usually is PC1, although the image analyst may opt for any alternative (e.g. PC2 or PC3). In typical remote sensing imagery, and in many other comparative types of imagery, PC1 often represents an overall albedo/reflection/.... intensity or contrast measure, that either may, or may not, be well suited for this common x-axis role depending on the specific image analysis context (hence the alternatives).

Every scene has its own distinct score space layout, fingerprint, although many similarities and analogies eventually will be noted in building up one's own MIA experience. The integrity and individuality of each new multivariate image that is to be analysed cannot

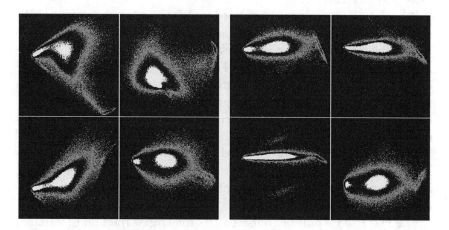

Figure 2.2 Standard MIA split-screen score plots (PC12, PC13, PC14, PC23) (PC15, PC16, PC17, PC18) (see Plate 8)

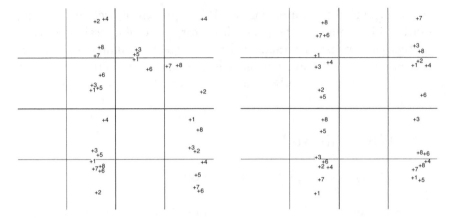

Figure 2.3   Standard MIA split-screen loading plots (cf. with Figure 2.2 for layout)

be overemphasized. The score space layout cannot be anticipated in advance, hence there is never any given a priori method (algorithm) for exploratory image analysis, and neither for classification nor for segmentation. This is the first main difference between traditional image analysis and MIA. This chapter will develop a completely general metaprinciples approach to MIA (which will include a novel topographic map analogy).

While Figure 2.2 showed the entire score space layout (eight PCs) of the Montmorency Forest scene in question (all 512×512 pixels are included in the PC analysis), the complementary loading plots are presented in Figure 2.3.

When a new image is to be analysed for the first time, the series of MIA score cross-plot is the only systematic, fully comprehensive approach to the objective data structures present in the feature domain, because absolutely all pixels are present. This score plot array will therefore display all there is to be learned from inspection of the spectral data structure of the image. The MIA approach is designed upon the central concept of having access to <u>all</u> image pixels, which is always an easy task with today's computer power. Without loss of generality we shall assume that all pixels are included in the analysis in the expositions below (but even when this cannot be achieved, in some specific hardware configuration cases, MIA's design philosophy allows for easy remedies, e.g. Esbensen and Geladi, 1989).

The PC12 score plot of course carries the largest fraction of variance modelled and is consequently usually assessed first. MIA analysts should make due note of the relative proportions of the total variance modelled by each component image. In the present case PC1 and PC2 explain (as this modelling parlance goes) 35 % and 27 %, respectively, totalling 62 % of the trace of overall variance in the spectral covariance X'X matrix. The image analyst should always take notice of the individual as well as the accumulated fractions pertaining to all score cross-plots inspected so far, lest interpretations accidentally be based on a too meagre residual variance. For the Montmorency Forest scene the decreasing variance fraction for all eight components breaks down as follows: 35, 27, 10, 9,... % (Figure 2.4), which by and large can be considered fairly typical of a large number of multivariate image types, but there are of course many individual exceptions to such a generalization, always as a distinct function of the specific image background.

The basic idea in MIA's score cross-plot set-up is to have one comparison axis with which to interrelate the whole series of cross-plots: PC12, PC13, PC14, etc. Adhering to such a standard set-up, one will need only a small number of cross-plots (one less the number of channels) in order to survey the gamut of all possible plots with, e.g. PC1 as the common x-axis. However, there is also an additional number of

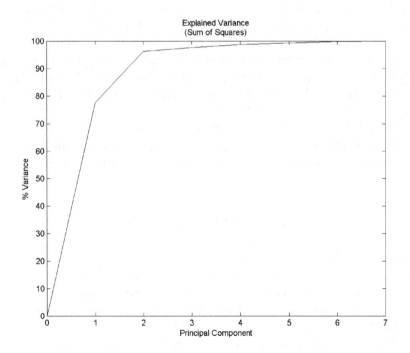

Figure 2.4 Variance modelled (%) per score image; general illustration

potential higher-order cross-plots available of the type PC23, PC24, . . . ,
PC34, PC35, . . . , through PC78 (in the present case). It is important,
especially for high(er) multi-channel work, not to be unnecessarily bewil-
dered by this overwhelming array of possible additional cross-plots
however. In principle everything there is to be known in score space
has been shown in the standard series with the one common anchor
axis. While it is indeed possible that certain higher-order cross-plots
sometimes may serve to depict (very) specific patterns, it is usually
advantageous to start out with this systematic approach. It may indeed
serve well specifically not to experiment with the higher-order option
without some reflection and experience. Thus the standard cross-plot
set-up includes only one higher-order plot, PC23 (Figure 2.2).

2.3.2   Colour-slice Contouring in Score Cross-plots: a 3-D Histogram

All MIA score cross-plots employ a colour-slicing technique for depicting
a 3-D histogram rendition of the relative number of pixels with identical score-pairs; for details see Esbensen and Geladi (Esbensen and
Geladi, 1989) and Geladi and Grahn (Geladi and Grahn, 1996). From
the peripheral black areas in the score cross-plots, the colour slicing
grades olive/dark green/green/yellow/orange/red/white, signifying that
$0<5<15<45, . . . ,>255$ image pixels have been plotted at the exact
same position in the pertinent PC cross-plot, i.e. at identical score-pair
coordinates. The exact numerical progression of the boundary values
of these bins is actually only of minor interest; it is the overall visual
impression of the relative patterns and trends revealed, which carries
the essential messages, very much in analogy with a topographic map.

2.3.3   Brushing: Relating Different Score Cross-plots

The choice of which score cross-plot is to serve as the starting point for
a MIA can be important. This is not always well understood, because of
the unavoidable next issue: What about the complementary cross-plots
in which the same classes of pixels can also be displayed? Brushing
comes to the fore.

   If no information to the contrary is present in a specific image
analysis situation, it can be assumed as a working hypothesis, that
the PC12 cross-plot carries the most dominating (variance/covariance)

information, because these two first PCs carry the largest and second largest fraction of the total spectral space variance. In specific situations, however, there is nothing against using any other, problem-specific combination of PC images as the starting score cross-plot. In the present remote sensing example, *if* it was decided that we are specifically not interested in the overall reflectance aspect of the original image, this could easily be compensated for simply by letting the analysis start out e.g. in the PC23 cross-plot.

We shall here make use of the standard PC12 cross-plot as the MIA starting plot. Figure 2.5 shows the technique of brushing, i.e. transferring a score space pixel class to the complementary other score cross-plots. In this example we have for illustration delineated a rather obvious class K. It will come as no surprise how the common PC1 anchor axis in the PC13 and PC14 plots strongly guides the brushed dispositions of

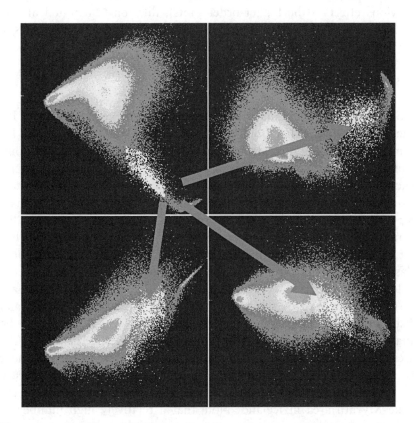

Figure 2.5 Brushing of MIA class K, delineated in upper-left quadrant in score plot PC12 (see Plate 9)

all pixels from the PC12 master class. One may perhaps appreciate the impression that in some of the plots the brushed class is floating above the main histogram. Note how the brushed class appears to dilute the complementary patterns in score space, as is quite possible, because the class was indeed defined in another score cross-plot dimension.

This is a fairly typical situation, in which one would then optionally decide to start the analysis with one of these higher-order cross-plots instead. With only a little MIA experience and careful consideration, the marked disposition in the PC23 plot might actually easily have been predicted directly from the PC13 plot.

With this powerful explorative brushing facility, it is possible to assess every tentative MIA class in the gamut of all other potential score cross-plots – indeed one should always do so! Features observed only by brushing include splitting (one apparently coherent class, actually splitting up into two, or more, classes in higher-order cross-plots); smearing (obvious effect in the higher-order plots), dilution (illustrated above) and others. We shall show several illustrations pertaining to these cases.

2.3.4   Joint Normal Distribution (or Not)

From extensive data analytical and statistical experience with PC analysis, it is clear that only truly ellipsoidal pixel clusters in all PC combination cross-plots can be said to meet the requirements of joint multi-normal distributions for all channels. It is thus very easy and uncomplicated to decide whether a particular class actually meets such requirements or not – and thus equally easy to find out whether unacceptable breaking of the premises of quite a number of traditionalist pattern recognition classifiers, etc., will take place or not. Alas, one has yet to see many good examples of truly joint multi-normally distributed classes in nearly all types of multivariate imagery from science, technology and industry! A very few have been noted, but these cases are vastly overwhelmed by the many other types of strikingly non-normal distributions (multi-modal distributions), all of which can easily be analysed with the standard MIA approach however. MIA allows the user to make allowances for any specific class shape in the pertinent boundary delineations. Figure 2.6 shows a relatively complex score aspect layout from this realm, which is a LANDSAT image from the Lake Myvatn area in Iceland. Note that a relatively high number of pertinent data classes in score space is absolutely no problem for the MIA approach, indeed there is no limit; full analysis of this scene (not presented here) showed there to be no less than 13 classes.

(a) (b)

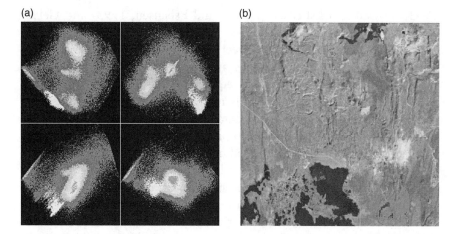

Figure 2.6 Example(s) of a complex, multi-modal layout in score space. Original class delineated in the PC12 score plot (a, upper-left quadrant) is brushed into the complementary score-plots, as well as into scene space (b), where it is seen to represent a particular volcanic unit (see text) (see Plate 10)

2.3.5 Local Models/Local Modelling: a Link to Classification

We are now in a position briefly to introduce a major theme of MIA, that of so-called local modelling.

The image analyst may for instance be interested in using just a subset of a scene, or of a score cross-plot aggregate, as a basis for a new, independent PC model. Subsets are often on the agenda when it is not the entire square image upon which we wish carry out a new PC decomposition in its own right. Esbensen and Geladi (Esbensen and Geladi, 1989), Esbensen and colleagues (Esbensen *et al.*, 1992) and Geladi and Grahn (Geladi and Grahn, 1996) develop the theme of local modelling in some detail.

Reasons for a local PC modelling facility are invariably closely related to the specific image analytical objectives, which will vary from image to image, and from case to case.

Image subclasses come in two distinct modes only:

(1) subclasses delineated in scene space (traditionalist fashion);
(2) subclasses delineated in score space as bona fide MIA classes.

The first mode was forcefully denounced by Esbensen and Geladi (Esbensen and Geladi, 1989), Esbensen and colleagues (Esbensen *et al.*,

1993), Lowell and Esbensen (Lowell and Esbensen, 1993) and Geladi and Grahn (Geladi and Grahn, 1996) as distinctly inferior.

It is a very simple matter to direct the MIA PC model to work only on a selected score space class as an alternative to the entire image. Based on such a local model it is equally simple to follow up and let MIA calculate scores for all pixels in the image, the said scores now corresponding to the covariance data structure of this local PC model only. This of course also applies to pixels in related scenes, images, etc.

This MIA concept of local modelling is very useful for more advanced work, but proper understanding and competence is dependent upon a thorough understanding and experience of the basic MIA PC modelling concepts. Once this has been mastered however, there is really only very little difference working with global or local models. The essential difference lies more with the specific reasons behind the need for local modelling. The main contribution of MIA here again resides with its design primacy of delineating appropriate (i.e. problem-dependent) local models in score space. Amongst other features, the local modelling feature can be shown to open up for the second-generation MIA concept of multivariate image regression (MIR), also to be presented in other chapters. The first introduction to the topic of MIR was by Geladi and Esbensen (Geladi and Esbensen, 1991), Esbensen and colleagues (Esbensen *et al.*, 1992) and Lied and colleagues (Lied *et al.*, 2000).

Any local model, selected and delineated on the basis of a pertinent problem-specific reason, may serve as a basis for a reclassification of the entire image, or of other images. It is important that any and all new images (or relevant parts thereof) may now be classified, or reclassified, in a completely analogous fashion to that of any MIA model classification. This feature opens up a complete range of discrimination/classification facilities of well-known features such as pattern recognition, SIMCA classification, etc., which is an entire topic for itself.

2.4 MIA ANALYSIS CONCEPT: MASTER DATASET ILLUSTRATIONS

2.4.1 A New Topographic Map Analogy

In the following we shall make use of topographic map terminology when discussing how to analyse a series of MIA score cross-plots. We shall use straightforward physical analogues: island, peninsula, peak, ridge, rise, flat, watershed, while also making use of virtual equivalents, e.g. brooks, or rivers supposed to follow virtual valley bottoms, etc.

The topographic analogy constitutes the core of the subject matter of the central MIA principles exposition below.

The colour-sliced score cross-plots, (Figure 2.2) are designed to be viewed, and interpreted, exactly like a topographic map. Thus, e.g. white areas, which invariably will be situated only in the centre of the topographic highs, signify the 3-D frequency histogram peaks, i.e. the highest densities of pixels with similar score-pairs. Unimodal and multi-modal pixel distributions are revealed with total unambiguity. In Figure 2.2, one thus observes three major topographic peaks (PC12 plot), more of which below. There is never any question about where and how such topographic peaks are to be found – and it is of no importance that not all peaks boast a pixel density which necessarily results in the white ("snow-capped") signature. It is the relative topographic expression of a peak which is important in this first interpretation stage (in fact there happens to be no snow-capped peaks in Figure 2.2, but see later).

Much more important, more subtle pixel groupings and trends are also clearly outlined, indeed such phenomena may be outlined in only the smallest of relative density terms, e.g. in the olive fringe areas only. A case in point in Figure 2.7 is the very prominent south-east trending ridge in the lower half of the PC12 quadrant, termed B. The relative proportion of all pixels encompassed by this ridge is actually less than 2 % of the total number of pixels in the original image, while its covariance trend occupies a much more significant part of the score plot. MIA was among other objectives designed towards the greatest possible sensitivity with respect to this type of subtle features in the score space domain.

In fact, MIA 3-D colour-sliced histogram score cross-plots often result in an inverse mapping of the frequency manifestation of the dominating scene data structures, so that these will be compressed into geometrically constricted peaks, etc., in score space. Any class of significant geometrical coherence in the score cross-plots by necessity must represent a valid image pixel class, irrespective of the corresponding spatial/geometrical disposition, or apparent size in the scene. Examples abound in which there is very little, or no correspondence, between well-defined score space pixel classes and the archetype, spatially coherent training classes defined in scene space. Note that this situation is what originally prompted the MIA's critique of the traditionalist image analysis scene space training paradigm (Esbensen and Geladi, 1989; Lowell and

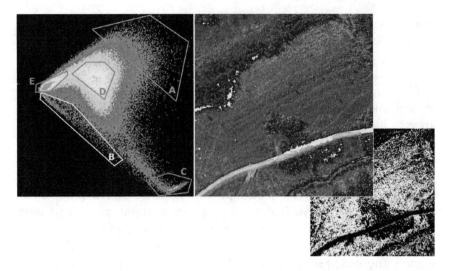

Figure 2.7 Maximum intensity/albedo contrast is represented by classes B and A, respectively. Magenta pixels in scene space correspond to class A and white pixels correspond to class B. The insert shows the scene-space distribution of class D, which is interpreted as undifferentiated regrowth (see Plate 11)

Esbensen, 1993; Geladi and Grahn, 1996), compared with which we here present the MIA alternative.

A(ny) major histogram peak in the score-plot(s) necessarily corresponds to a (very) large proportion of image pixels, but it is often not very illuminating to focus the attention of MIA on such prominent features; they are simply manifestations of the (by far) most dominating image structures, which are always very clearly observed by simple inspection of the scene space anyway. A case in point is shown in Figure 2.7, in which a MIA class of the clearly most dominating peak in the Montmorency Forest scene, the central peak of the island-like PC12 score cross-plot analogy, termed D, has been mapped back into the scene space. The forestry interpretation of this class is very clear: undifferentiated clear-cut regrowth.

This is the first general MIA rule: all dominating peaks in score space correspond to the dominating image space structures/segments. There would be nothing new about MIA were this the only MIA feature. In fact MIA merges with traditionalist image analysts when these first-order, most dominating image structures/objects/segments are the only items on the agenda.

This is also where MIA parts with the traditionalist image analysis concept in earnest: the traditionalist approach cannot – by definition – delineate subtle class features by starting out in scene space, with anything even remotely akin to the power of MIA. In this situation MIA rather presents itself as a most powerful complement that specifically only claims rights of true superiority when more and more subtle details in the image come to the fore. MIA comes on very strongly indeed for all weakly populated and/or subtly defined data structures in both scene space as well as score space. Exploratory MIA image analysis is especially aimed at finding and highlighting exactly these types of subtle, but well-structured and thus well-defined pixel aggregates (e.g. isolated islands, peninsulas, ridges), that otherwise run the risk of being swamped or drowned in the dominating structures and textures when delineated in scene space. Most of the remainder of this exposition is devoted to showing one or other aspect of this, much more difficult to analyse image analysis situation, the subtle class regimen. Several examples of the extraordinary power of MIA in this context will be demonstrated.

MIA will now be presented in a series of practical image analytical sessions. Both an exploratory image analysis mode as well as a pattern recognition (classification) mode shall be illustrated, as shall demonstrations of other related image analysis objectives, which lend themselves naturally in the MIA context. The totality of image analysis objectives and corresponding operations to be illustrated below need not all be put into use simultaneously, nor all be relevant for one particular image or scene.

2.4.2 MIA Topographic Score Space Delineation of Single Classes

MIA is so designed to allow the image analysts to focus on any interesting pixel cluster, prominent or subtle, by convex polygons, five of which are delineated in Figure 2.7. Thus class A is the class encompassing all pixels with highest reflectance in the entire scene. Delineation of pixel classes has been so designed to allow for a maximum freedom by the image analyst when outlining the enclosing perimeter of the (convex) polygons. Usually this type of convex polygon follows the topographic contours to a large extent, e.g. class D. By way of contrast, however, class B has a very different geometric layout, which suggests itself entirely by way of the covariance trend of the pixels involved.

All score space pixel class delineations should always be followed by immediate projection into the scene space, in order to facilitate interpretation of the meaning of the delineated class. Either a simple

binary mask is outlined with all designated pixels in white, or usually
MIA displays the original image together with the pertinent scene space
overlay (in a suitable monitor combination R:G:B). Figure 2.7 for
example shows the resulting spatial layout of projection of the three
pixel classes A, B and D. Upon inspection, class A turned out to represent
the pixels in this particular scene with the absolute highest radiometric
reflectance, while the opposite holds true of class B, interpreted as the
class composed of the absolute darkest pixels. This latter class represents
shadow – easily enough appreciated when the entire scene is viewed
with particular notice to the general sun illumination direction (from
NW) (cf. Figure 2.1).

The user is accorded complete freedom to iterate and refine this type
of (tentative) class definition procedure as often as needed, should the
first scene projection(s) indicate only a suboptimal class representation
when revealed in scene space, class D would be a case in point. Domain-
specific interpretation of the class masks in Figure 2.7 may for example
tell a forest expert, familiar with the imagery and the general features of
all the prevalent tree types in this particular context, that these resulting
scene boundaries are not optimally delineated yet – in which case only
an iteration of the score space class delineation is called for, etc.

This iterative score space/scene space interaction constitutes the most
important design principle of MIA in the explorative mode – an interactive interpretation stage: from score cross-plot pixel class delineation(s),
to projected scene space class outlines, complete with original image
underlay. It is up to the user, be it a domain specialist also versed
in image analysis or a two-person team covering both these fields, to
carry out this interactive procedure to as high an interpretation detail
as deemed necessary by the image analysis objective(s). MIA's on-screen
capabilities have been designed such that this interpretation work can
be as comprehensive and effective as needed (Lied, 1999).

Working systematically, MIA analysis of this particular forest scene
revealed five primary classes in Figures 2.1 and 2.7: a dirt road; areas
underlain by shadow; mature forest stands; undifferentiated regrowth;
and a somewhat unspecific type of class characterized by very high
albedo/reflectance. The primary classes are:

- A – areas composed of very high reflectance pixels;
- B – areas in shadow;
- C – road (dirt road, not metalled);
- D – regrowth (undifferentiated), mainly in the clear-cut areas;
- E – mature tree stands, also single crowns of old trees.

These primary class designations will play a central role when the next major image analytical features of MIA are to be developed, particularly the end-member mixing class concept.

The classifications revealed in Figure 2.7 happened to represent more or less equal areas in the score cross-plot. This characteristic should not lead to misunderstandings, however, when it is remembered that equal areas in the score plot may in fact represent very large differences with respect to the actual number of pixels in the 3-D histogram bins, with frequency differences as large as 5:255, or more. This inverse representation can be critically confusing if not well understood, but should not cause undue problems with experience.

The above illustrations show that careful, iterative pixel class perimeter delineation in score space is the main distinctive feature in MIA. The topographic analogy is natural, especially as regards peaks, ridges, etc. It is most likely the topographic expressions of peakedness that leads the human cognitive facility to form this type of spatial pattern cognition very easily. There is another, equally natural type of MIA class to be delineated – the concept of an end-member series or a mixing-class series.

2.4.3 MIA Delineation of End-member Mixing Classes

Figure 2.8 delineates three very prominent such mixing class series. Note how the direction of these class delineations are specifically related to the topographic ridge patterns, in particular what would be similar to watershed ridges. In the case of class series X and Y this terminology would appear obvious. Class series Y1–Y3 (numbering starting with the peripheral Y class) is made up of the mixing ridge between classes mature tree stands E and the regrowth class C (cf. Figure 2.7). Likewise class X1–X6 can be seen as representing a mixing series between end-member class C (regrowth) and the road class D (numbering again from outside inwards). We shall here also designate Z1–Z3 as a similar mixing class series, but now a mixing between the central end-members regrowth (C) and a virtual high reflectance end-member H. Observe how this similarity allows the image analyst to analyse all types of mixing classes by relying on only one common concept. Figures 2.9 and 2.10 show two examples of the gradual relationships displayed by these mixing classes, especially when followed from one end-member to its opposite in scene space.

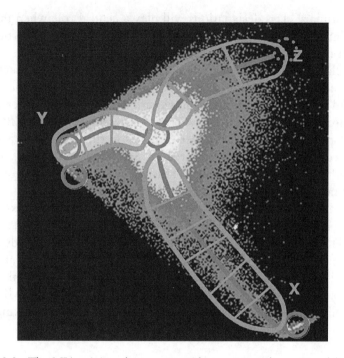

Figure 2.8 The MIA mixing class concept (three mixing class series delineated). Each mixing class series extends between two appropriate end-members. Numbering of mixing classes complies with the simple dictum: from outside inwards. Thus class Y_3 and X_6 meet at the central peak (class D, cf. Figure 2.7) (see Plate 12)

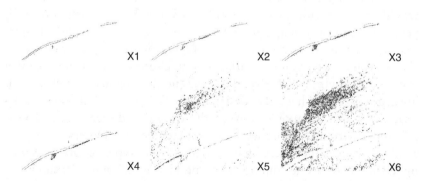

Figure 2.9 Scene space sequencing of successive mixing classes X1–X6 depicting the regrowth process

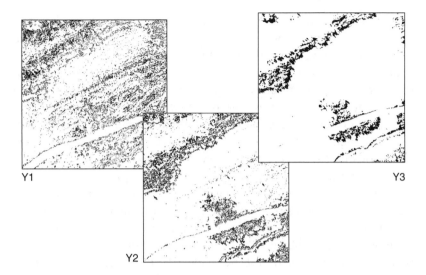

Figure 2.10 Scene space disposition of mixing classes Y1–Y3, continuing the forest growth process right through to the stage of mature trees (see text for details)

In this mixing class delineation we have at first only placed emphasis on identifying the end-members making up the series extreme end points, but there are two additional much more penetrating and powerful interpretative hidden information principles to be exploited. Starting with the road class, following the $X_{1\text{-}2\text{-}3\text{-}4\text{-}5\text{-}6}$ mixing series inwards to the central regrowth class D – continuing along mixing series $Y_{3\text{-}2\text{-}1}$, ending up with the mature tree stand class E – a specific forestry interpretation of the entire spatio-temporal sequence leads to the hypothesis which follows.

The dirt road was recently cleared for most/all regrowth saplings because of the regular traffic of heavy duty forest machinery involved in the overall clear cutting operations – consequently the scene space pattern class for X_1, the first in the direction of the central regrowth class (closely similar to the class dirt road) must represent the very first vestiges of whatever regrowth can be observed in the scene. In other words, the first mixing subclass immediately adjoining the road class must represent the absolute youngest regrowth saplings, with progressively older representatives forming a progressing sequence of the scene space rendition in the mixing class series X along the watershed route, ending up in the central regrowth class, D. This is graphically appreciated very well in the scene space rendition in Figure 2.9.

By similar reasoning, the grading mixing class series Y can be parsed in a corresponding fashion – in an identical spatio-temporal context – i.e. from D ending up with the oldest, most mature trees standings in the scene, class E, as is laid out in detail in Figure 2.10. As one progresses along the Y series (Y_{3-2-1}), the more mature the trees delineated become. Observe how one is actually able to follow this entire interpreted growth process progress in minute detail in this score space rendition (Figure 2.8) and immediately have access to its scene space dispositions in the sequenced imagery in Figures 2.9 and 2.10.

Thus, the ensemble of score space classes $X_{1-2-3-4-5-6} \rightarrow D \rightarrow Y_{3-2-1} \rightarrow E$ represent a spatio-temporal slicing of what could be interpreted (and termed) the regrowth process in the context of this particular scene. It is essential to appreciate that this interpretation takes place by starting out in score space, but it is only when the resulting MIA classes are displayed and interpreted in scene space, that full interpretation of their meaning is possible. Also, this interpretation is validated by reference to other, already interpreted or well-segmented features in the scene. In this particular dynamic, temporally interpreted scene, the biologic process of regrowth can be said to have been subjected to a kind of stroboscopic time-slicing, delineated by the sequencing of juxtaposed mixing classes.

By this development of the end-member/mixing class concept it has been possible to shed hitherto unavailable detailed light on the complex spatio-temporal forest growth process. Clearly it is the informed interpretation of the domain specialist – together with full command of MIA capabilities – which underlies this powerful image analysis.

MIA here allows a temporal-spatial decomposition which is unparalleled in traditionalist image analysis; there is simply no possible way to decompose the multivariate image in similar segments, were this to start out from the scene space, in which the would-be training classes are hopelessly far too disjunct and far too scantily distributed, as is dramatically shown in their intricate scene space context in Figures 2.9 and 2.10.

There is one more phenomenological identical mixing series present in the score space rendition of this scene, but with a distinctly different, nonbiologic interpretation, namely, the Z_{1-2-3} mixing series, situated in an almost perpendicular disposition with respect to the regrowth series. This series represents the ultimate span of the general low to high intensity (compare the above shadow–reflectance contrast phenomenon, Figure 2.7), but in the present context it can be seen as also tracking across the same central regrowth class. The physical interpretation of this new axis remains the same: the $Z_{1\text{-}2\text{-}3}$ vector must represent a presumably mostly physical reflectance direction (in score space), signifying gradation in the total reflectance recorded from within the confines of one ground trace pixel size (70 cm × 70 cm). The reflectance in this type of imagery is surely dependent on a complex set of factors, among which individual leaves, their colour, angle with respect to the sun illumination and degree of moisture coverage play important roles.

There are thus three general aspects of the above detailed MIA to be highlighted:

(1) It is observed how the two distinctly different mixing series arrays lie very close to an orthogonal disposition with respect to each other. This is no coincidence: one is dynamic, capturing a biological growth phenomenon; the other is distinctly physical, spanning an albedo–shadow factor. The orthogonality is an inheritance from the underlying PCA, the design purpose of which precisely is to decompose covariance trends according to forced orthogonal axes (PCs). In this particular case it is mainly PC1 which is interpreted as the dominating general intensity (albedo) axis, though the slightly oblique direction of the shadow–reflectance axis ($Z_{1\text{-}2\text{-}3}$) bears witness to a slight involvement of PC2 as well in delineating the direction of the overall increasing albedo.

(2) It is also observed how both these series meet, or cross over, at the scene's singular most dominant class, the central regrowth class D. This hammers home the message that class D is a huge mixed bag of many types of regrowth manifestations (with general average reflectance characteristics). It also explains why the simplest MIA class D is composed of an overwhelming number of scene space pixels, drowning out most, if not all, possibilities of making detailed interpretations other than that of the standard forestry undifferentiated regrowth category. In fact, similar crossover classes, showing up as analogous, centrally located, dominating peaks are often met with in MIA, certainly not only from (high-resolution) remote sensing, but from many other technological and industrial types of imagery as well. MIA constitutes the only image analysis tool with sufficient power for these kinds of complexities.

(3) Two, partly alternative, partly overlapping metaprinciples for MIA pixel class delineations have now been delineated: individual topographic peaks and mixing class series.

2.4.4   Which to Use? When? How?

One should not dismiss the above biologic growth analysis/
interpretation too quickly as a special, problem-specific event, even
though dynamic time-slicing might at first sight seem specific for
remote sensing in general and for scene change detection in particular. Very probably this type of change analysis constitutes a generic
type of interpretation, which can be modified and applied in several
other image type contexts as well. Certainly this is the case from
our experience based on a suite of different technological and
industry-related types of imagery. In any event, the other nondy-
namic mixing series reflects a more static, scene-dependent physical
phenomenon (albedo/intensity contrast/reflectance), the like(s) of which
will be present in almost any multivariate image from science, technology or industry in which (natural or artificial) illumination plays
a role.

In situations where this mixing class concept is not called for, a
return to the simple peak delineation of what will surely always be
representative, objective pixel classes will often then be quite sufficient.
The perhaps most interesting field here would be the interplay between
these two types of class delineation principles, but needless to say, this
problem will always be scene-specific to a very high degree.

In any event, the MIA prime directive rules supreme: each new multivariate image always should be analysed on its own accord.

However, experience with several types of multivariate images from a
broad range of different origins (e.g. remote sensing imagery, several types
of technological imagery and industrial imagery) reveals that the gamut
is indeed made up of peaks and end-member/mixing classes, to more than
95 %. So, when to use which approach: peaks or mixing series?

All mixing-class series can always be subjected to mixing-class analysis
as that presented above – and any end-member always constitutes a
legitimate single peak in its own right. By employing the concept of end-
member mixing/classes (directed along the watershed ridges, connecting
peaks in the topographic analogy setting) all types of connected peaks
are in principle always open to either type of analysis. Clearly it is the
scene space knowledge which determines whether it will be possible to
make meaningful interpretations of the sliced subclasses.

In this chapter two generic types of class definition have been
presented. It would probably be simplistic, were it to be suggested that
this is all there is; that still other types of analogous analysis axes
will not be found to be associated with other type(s) of imagery in

future applications. Actually there has not yet been found the need for additional MIA metaprinciples. Perhaps the suggested concepts of end-member/mixing class versus standard peak MIA will be of very general, perhaps even universal, applicability? Only extensive use of MIA will be able to tell.

2.4.5   Scene-space Sampling in Score Space

Often an image analyst may need to subsample an entire image for the specific meaning of a representative subsample of the pixels making up an image. This may be on the agenda for many different reasons, e.g. for forest inventory purposes, where a forester would like specifically to sample all known forest and growth classes present. Here one specifically needs to be certain that all classification classes indeed are equally represented on a spatial basis. Many procedures and sampling schemes have been developed over the years for this and related purposes, all of which operate exclusively in the 2-D scene space of the original image, or equivalently, on the map. There are many parallels to this subsampling situation from other types of imagery as well.

Interestingly MIA may also here offer at least an alternative to this scene space tradition. Again, we illustrate using the clear-cut study imagery. Figure 2.11 shows an extremely thin class delineated in the lower left quadrant – in fact this class is only one pixel wide. This class is now brushed into the three complementary PC cross-plots.

It is of the highest significance that we here have made use of the PC23 cross-plot for the class delineating purpose, for reasons that will become clear immediately. Note first how this PC23 class covers all the major classes present in score space – here we actually take advantage of the fact that most of the major classes are nonresolved (i.e. quasi-spherical) in the higher-order, e.g. PC23, plots. It is precisely because of this judicious use of the PC23 cross-plot that we have been able to acquire a representative, complete, equal-density sampling closely corresponding to the total covariance data structure in feature space by using the simplest of class delineations: a line, as is amply substantiated in the accompanying PC12, PC13 and PC14 cross-plots. It is especially gratifying to observe the inherent splitting in the PC13 (upper right) plot. This one-pixel thin sampling class has done a remarkable job sampling over all classes indeed!

Figure 2.11 also shows the corresponding spatial projection in scene space. Indeed a uniform spatial disposition of potential inventory

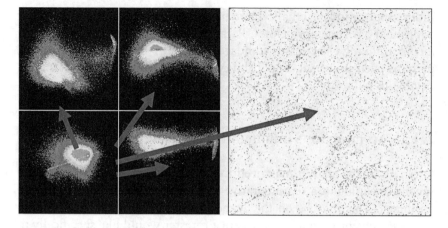

Figure 2.11 Generic MIA pixel subsampling concept. From score space to scene space: equal density sampling with respect to the image covariance structure(s), *not* with respect to scene-space area density (see Plate 13)

localizations has been achieved, complete with a number of denser structures present (major tree stands, etc.). It is now a simple matter to overlay this display with e.g. a road map or the like and to proceed with a logistical planning for the forest inventory, in which a further weeding out of surplus sampling sites no doubt will form an integral part. It is really not a problem worth mentioning weeding down an already acceptable spatial template relative to the opposite case. Thus, for such sampling purposes one might advantageously seek out the least structured score cross-plot.

There are other variations on this pixel sampling feature of MIA, all invariably related to the specific image analysis problems at hand and their special objectives, and we leave it to the reader to associate freely from this generic example.

Mortensen and Esbensen (Mortensen and Esbensen, 2005) illustrated other aspects of this interesting comparison between specific 2-D scene space sampling schemes and informed use of various chemometric data spaces, which may also be of interest in the above context.

2.5   CONCLUSIONS

MIA was presented as a set of interdependent image analysis unit procedures, encompassing both an explorative and pattern recognition mode, as well as classification-related facilities (local modelling). A synoptic overview of the most fundamental elements in MIA was presented by a series of different applications on a master dataset. We presented extensive justifications for the specific MIA approach versus the traditionalist image processing mode.

MIA is a type of problem-dependent, interdependent explorative analysis and a classification tool to be used by the informed analyst, according to the specific image analysis objectives. This review furnished a generic overview of the principles of MIA needed to compose one's own flexible, problem-specific strategy: mandatory pixel class delineation in feature space (score space) by a topographic analogy and a dual end-member/mixing class versus peak class delineation concept, of universal applicability also in other image modes.

MIA constitutes a most powerful image analytical concept for dealing with any degree of complex imagery: the design and analysis principles of MIA are invariant with respect to the number of channels present. With MIA it is not necessary to invoke massive, parallel computer approaches in order to deal with even the most complex imagery (Lied *et al.*, 2000). Herein lies a major technological shortcut, which has not been sufficiently explored or fully exploited yet.

REFERENCES

Esbensen, K. H. and Geladi, P. (1989) Strategy of multivariate image analysis (MIA). *Chemometrics and Intelligent Laboratory Systems*, 7, 67–86.

Esbensen, K. H., Geladi, P. and Grahn, H. (1992) Strategies for multivariate image regression (MIR). *Chemometrics and Intelligent Laboratory Systems*, 14, 67–86.

Esbensen, K. H., Edwards, G. and Eldridge, N. R. (1993) Multivariate image analysis in forestry applications involving high resolution airborne imagery. In: *8th Scandinavian Conference on Image Analysis*. NOBIM, Tromsø, pp. 953–963.

Geladi, P. and Esbensen, K. (1991) Regression on multivariate images: principal components regression for modelling, prediction and visual diagnostic tools. *Journal of Chemometrics*, 5, 97–111.

Geladi, P. and Grahn, H. (1996) *Multivariate Image Analysis*. John Wiley & Sons, Ltd, Chichester.

Kramer, H. J. (1996) Observation of the Earth and its environment. In: *Survey of Missions and Sensors*, 3rd Edn. Springer-Verlag, Berlin.

Lied, T. (1999) MIA software. http://wwwpors.hit.no/tf/forskning/kjemomet/kjemomet. html

Lied, T., Geladi, P. and Esbensen, K. (2000) Multivariate image regression, implementation of image PLS – first forays. *Journal of Chemometrics*, 14, 585–598.

Lowell, K. and Esbensen, K. H. (1993) Is image segmentation really a valid technique for information extraction. A re-examination of conventional reasons for classifying remotely sensed images. In: *8th Scandinavian Conference on Image Analysis*. NOBIM, Tromsø, pp. 973–979.

McColl, W. D., Till, S. M. and Neville, R. A. (1984) MEIS II operational sensor for multidisciplinary studies. In: *9th Canadian Symposium on Remote Sensing*, pp. 497–501.

Mortensen, P. P. and Esbensen K. H. (2005) Optimisation of an image analytical system for bulk materials using the Theory of Sampling. In: Holmes, R. (Ed.), *Proceedings of the 2nd World Conference on Sampling and Blending*. AUSIMM, Brisbane, pp. 45–54.

3

Clustering and Classification in Multispectral Imaging for Quality Inspection of Postharvest Products

Jacco C. Noordam and Willie H. A. M. van den Broek

3.1 INTRODUCTION TO MULTISPECTRAL IMAGING IN AGRICULTURE

3.1.1 Measuring Quality

Consumers use a combination of sensory inputs like appearance, aroma and hand-feel of the entire product to obtain a final judgement of the acceptability of that fruit or vegetable. A lot of instruments have been developed in the past (Abbot, 1999) to imitate human testing methods. Where aroma and hand-feel are more recent research topics, appearance in relation to fruit and vegetable quality has been a research topic for many decades. Appearance is not only related to the expected or required color of the fruit or vegetable but also to the presence of abnormalities, deformations or even visible defects on the fruit or vegetable. There have been many reviews of technologies in the literature (Brosnan and Sun, 2002; Graves and Batchelor, 2003) concerning the appearance quality attribute. A specific number of

Techniques and Applications of Hyperspectral Image Analysis   Edited by H. F. Grahn and P. Geladi
© 2007 John Wiley & Sons, Ltd

(a)   (b)   (c)

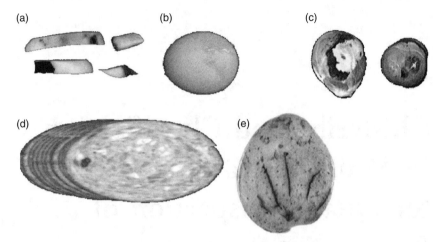

(d)   (e)

Figure 3.1   Typical appearance based quality defects on postharvest products: (a) defects on raw French fries; (b) faeces on eggs; (c) rot and alternaria on Brussels sprouts; (d) blood spots in cold meats; (e) cracks in consumer potatoes

technologies focus on replacing or imitating the visually human testing methods and are therefore based on imaging technologies as the spatial distribution of abnormalities on the products under consideration require the use of imaging systems. Figure 3.1 shows some typical appearance based quality defects on food products.

When a fruit or vegetable is exposed to light, about 4 % of the incident light is reflected at the outer surface, causing specular reflectance or gloss, and the remaining 96 % of incident energy is transmitted into the product (Birth, 1976). Most light energy penetrates only a very short distance and exits near the point of entry, which is the basis for color. As color is directly related to consumer perception of appearance, it explains why color imaging is preferred over gray value imaging. Therefore, color imaging has been the basis for sorting and grading all kind of products in different grades. Inspection of food products with color imaging has been reported for various applications like the detection of external defects on whole potato tubers (Noordam *et al.*, 2000), the detection of defects on apples (Tao *et al.*, 1995) and inspection of beef tenderness (Li *et al.*, 1999). Extended reviews about color inspection are reported in the literature (Jiminez *et al.*, 1999; Chen *et al.*, 2002).

3.1.2   Spectral Imaging in Agriculture

When light penetrates deeper into the material of the product it is altered by differences in path length and by different compound absorbencies, resulting in a signal containing valuable chemical information. Chemical bonds absorb light energy at specific wavelengths, so some compositional information can be determined from spectra measured by a spectrometer (Williams and Norris, 1987). However, these single spot spectrometers lack spatial resolution as only single spectral point measurements are used and thus no indication of disease or defect area can be obtained. This makes spectroscopy less suitable for appearance based quality inspection, as opposed to the more recently described spectral cameras. These cameras permit acquisition of images at multiple wavelengths, which is called spectral imaging, thereby providing both spectral and spatial information which makes them more suitable for appearance based quality control.

There are various methods to create spectral images, in the past spectral filters or liquid crystal tunable filters (LCTFs) were placed in front of the camera (Gat, 2000), nowadays an imaging spectrograph is used instead. More details and references about these techniques are summarized in the literature (Noordam, 2005).

Applications on food products with the spectral filter approach are described for the detection of wholesome birds in a poultry slaughter plant (Park et al., 2002), the detection of bruises on peaches and apricots (Zwingelaar et al., 1996), the status monitoring in vegetables such as cabbage (Chen and Li, 2001), detection of parasites in cod fillets (Wold et al., 2001) and for the spatial distribution of soluble solids in melons (Sugiyama, 1999).

Applications with LCTFs are the classification of blemishes on apples (Miller et al., 1998), the mapping of light penetration in apples and tomatoes (Thai et al., 1997b), the discrimination between different food products (Novales et al., 1996) and the detection of herbicide stress in crops (Thai et al., 1997a).

Applications of multispectral imaging systems with an imaging spectrograph has been reported for the quality inspection of poultry (Lawrence et al., 2003) and chickens (Chao et al., 2002), meat (Wold et al., 1999), the measurement of ripeness of tomatoes (Polder et al., 2002), the inspection of cherries (Guyer and Yang, 2000) and apples (Mehl et al., 2002), the spatial distribution of soluble solids in kiwi fruit (Martinsen and Schaare, 1998), for the quality evaluation in nuts and grains (Upchurch and Thai, 2000), detection of scab in wheat (Delwiche and Kim, 2000) and detection of aflatoxin in corn (Casasent and Che, 2003) and classification of defects on french fries (Noordam et al., 2005a).

3.2   UNSUPERVISED CLASSIFICATION OF MULTISPECTRAL IMAGES

In the majority of aforementioned multispectral imaging applications for appearance based quality control, the goal was to detect small defects or abnormalities in the object under inspection. To quantify the amount of defects or abnormalities in the inspected object classification of the spectra is required. In cases where only a single multivariate image or a few multivariate images with no a priori spectral information is available, no training set can be extracted from these images to train a supervised classifier. In those cases, unsupervised classification techniques can be applied as a segmentation technique for those images.

A widely used unsupervised technique is Fuzzy C-Means (FCM) (Bezdek, 1981). The main reason for its popularity as an image segmentation technique is its transparency and unsupervised nature (Rezaee, 1998) and its good performance in high dimensional spaces when the number of objects are limited (Malek *et al.*, 2002). This is an important feature as in pattern recognition literature, it is widely known that a finite number of training samples can lead to practical difficulties in designing a classifier for high dimensional spaces (Lee and Landgrebe, 1993).

3.2.1   Unsupervised Classification with FCM

When FCM is applied as a clustering technique to segment multivariate images for appearance based quality evaluation of food products, care should be taken when there is great difference in object area between objects of different classes. This is not uncommon for multivariate images acquired in post harvest quality inspection where small sized defects can occur on products that contain defects or diseases. The difference in object size between different classes affects the partitioning of the data space and thus the final image segmentation will be influenced by this cluster size sensitivity. A direct result of incorrect segmentation is an over- or underestimation of the classes on the inspected product. An example is shown in Figure 3.2 where a multivariate potato image is segmented in three classes with FCM.

The image contains the classes: potato skin, potato greening [large area in the top left of Figure 3.2(b)]; and the fungus rhizoctonia (black

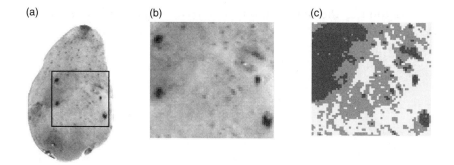

Figure 3.2 Segmenting a potato image with three classes: good skin; greening; and black spots. (a) Original potato image; (b) subimage; (c) results of clustering with FCM. Reprinted from Chemometrics and Intelligent Laboratory Systems, Vol 64, Noordam *et al.*, Multivariate image segmentation..., pp. 115–126 © 2005, with permission from Elsevier

spots). The FCM segmented image shows that FCM erroneously merges the classes greening and black spots and oversegments the larger potato skin class in two different classes in order to equalize the number of pixels per cluster. As a result, the black spot class remains undetected due to the small number of pixels in this class.

Therefore a new cluster size insensitive FCM (csiFCM) has been developed which overcomes the cluster size sensitivity of FCM and requires no parameters (Noordam *et al.*, 2002). csiFCM is based on a conditional FCM (cFCM), a modified version of FCM. First, FCM is briefly, explained.

3.2.2   FCM Clustering

Given a set of n data patterns, $X = x_1, \ldots, x_n$, the FCM algorithm minimizes the weighted within group sum of squared error objective function $J(U, V)$ (Bezdek, 1981):

$$J(U, V) = \sum_{k=1}^{n} \sum_{i=1}^{c} u_{ik}^{m} d^2(\mathbf{x}_k, \mathbf{v}_i) \qquad (3.1)$$

where $\mathbf{x}_k$ is the kth p-dimensional data vector, $\mathbf{v}_i$ is the prototype of the centre of cluster i, u_{ik} is the degree of membership of $\mathbf{x}_k$ in the ith cluster, m is a weighting exponent on each fuzzy membership, $d(\mathbf{x}_k, \mathbf{v}_i)$ is a distance measure between object x_k and cluster centre $\mathbf{v}_i$, n is the number of objects and c is the number of clusters.

A solution of the objective function $J(U,V)$ can be obtained via an iterative process where the degrees of membership u_{ik} and the cluster centres $\mathbf{v}_i$ are updated via:

$$u_{ik} = \frac{1}{\sum_{j=1}^{c} \left(\dfrac{d_{ik}}{d_{jk}} \right)^{\frac{2}{m-1}}} \tag{3.2}$$

with d_{ik} the distance between object k and cluster i, d_{jk} the distance between object k and cluster j,

$$\mathbf{v}_i = \frac{\sum_{k=1}^{n} u_{ik}^{m} \mathbf{x}_k}{\sum_{k=1}^{n} u_{ik}^{m}} \tag{3.3}$$

with the constraints:

$$u_{ik} \in [0, 1] \qquad \sum_{i=1}^{c} u_{ik} = 1 \quad \forall k \qquad 0 < \sum_{k=1}^{n} u_{ik} < N \quad \forall i \tag{3.4}$$

According to the update formula for the cluster prototypes [Equation (3.2)], objects with low membership values for that particular cluster have a small contribution to the final position of that particular cluster prototype. This is the general principle on which csiFCM is based: by weakening membership values of objects that belong to the larger cluster, the contribution of those weakened objects to the cluster centres of the smaller clusters will be small. cFCM (Pedrycz, 1996) is used to weaken the membership values of objects. The principle of cFCM is discussed in the next section.

3.2.3   cFCM Clustering

cFCM is a FCM based clustering technique where the clustering is influenced by an auxiliary variable to guide the outcome of the clustering process. A fixed value for this auxiliary variable is given beforehand or the value is continuously updated during clustering (Noordam and van den Broek, 2002).

The principle of cFCM is as follows: For each labelled pattern $\mathbf{x}_k$, there exists an auxiliary condition variable f_k, further referred to as condition, where f_k ranges from $[0,1]$. The update procedure for the partition matrix $\mathbf{U}$ is now changed into:

$$u_{ik} = \frac{f_k}{\sum_{j=1}^{c} \left(\dfrac{d_{ik}}{d_{jk}} \right)^{\frac{2}{m-1}}} \qquad (3.5)$$

with the modified constraint

$$\sum_{i=1}^{c} u_{ik} = f_k \qquad (3.6)$$

For condition values equal to 1, the object is unconditioned and the partition update procedure is similar to the partition update procedure of FCM. A small value of the condition results in a low membership value for all clusters, which minimizes the contribution of that particular object to all cluster centres.

3.2.4 csiFCM

The general principle of csiFCM is to weaken the contribution of objects from larger clusters to the prototypes of smaller clusters to prevent smaller clusters from drifting towards larger adjacent clusters. This is achieved by assigning low conditional values to objects of larger clusters and giving high conditional values to objects of smaller clusters. As a result, objects with low membership values u_k for cluster i contribute less to the final position of cluster prototype i, according to Equation (3.3). Objects of the smallest cluster are unconditioned, their contribution to the prototype of the smallest cluster is maximal when the condition f_k for these objects is set to 1. The objects of the remaining clusters receive a condition which is less then 1 ($f_k < 1$) to minimize their contribution to the prototype of the smaller clusters. The condition value f_k for cluster i is now determined after each iteration of the FCM algorithm via:

$$f_{k,i} = \frac{1}{(1 - P_{i_{\min}})} * (1 - P_i); \qquad i = 1, 2, \ldots, c \qquad (3.7)$$

where P_i is the a priori probability of selecting the ith cluster, $P_{i_{\min}}$ is the smallest prior probability. More detailed information is available in the literature (Noordam et al., 2003).

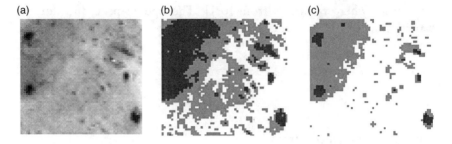

Figure 3.3 Segmenting a potato image with three classes: good skin; greening; and black spots. (a) Original potato image; (b) results of clustering with FCM; (c) results of clustering with csiFCM. Reprinted from Chemometrics and Intelligent Laboratory Systems, Vol 64, Noordam *et al.*, Multivariate image segmentation..., pp. 115–126 © 2005, with permission from Elsevier

Figure 3.3 shows the segmented potato image after clustering with csiFCM. The three different classes are correctly segmented, the black spot class and the greening class are no longer merged and the pixels of the potato skin class are in a single class.

In the second example, a multispectral image of minced meat is segmented with both FCM and csiFCM. The 12 band multispectral image (320×286 pixels) contains a Petri dish filled with a piece of minced meat with the classes: Petri dish; dark meat; light meat; and fat [Figure 3.4(a)]. The difference between dark meat and light meat is caused by the amount of blood in the meat. The dark gray pixels represent the dark meat class and the white spots represent the fat class. The light meat class surrounds the fat class and gradually turns into the dark meat class. The classes dark meat, light meat and fat contain overlap and are therefore hard to separate. The entire image is segmented in the experiment but only the subimages are shown to improve the visualization of the differences.

The results of FCM and csiFCM are listed in Table 3.1. The difference between the FCM segmented image and the csiFCM segmented image is most distinctive in the fat class. The FCM classifies 11.7 % of the pixels as fat [Figure 3.4(c)], whereas the csiFCM classifies merely 6.4 % of the pixels as fat [Figure 3.4(d)].

The percentages of the pixels per class in Table 3.1 show that not only the fat class is affected by the conditioning, also the number of pixels in the light meat and dark meat class is different compared with the FCM segmented image. Because the true class of the pixels is unknown, a product expert evaluated the segmented images. The expert considers the fat class as overestimated in the FCM segmented image, whereas

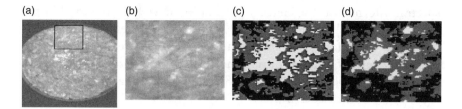

(a) (b) (c) (d)

Figure 3.4 (a) Original meat image; (b) subimage of minced meat; (c) results of clustering with FCM; (d) results of clustering with csiFCM. Reprinted from Chemometrics and Intelligent Laboratory Systems, Vol 64, Noordam *et al.*, Multivariate image segmentation..., pp. 115–126, © 2005, with permission from Elsevier

Table 3.1 Average percentage of the pixels per class after clustering the entire multivariate minced meat image with FCM and csiFCM. Reprinted from Chemometrics and Intelligent Laboratory Systems , Vol 64, Noordam *et al.*, Multivariate image segmentation..., pp. 115–126, © 2005, with permission from Elsevier

	Petri dish	Dark meat	Light meat	Fat
FCM	7.8	40	40.5	11.7
csiFCM	7.9	49.7	36	6.4

the csiFCM segmented image is considered to resemble the original image more.

3.2.5 Combining Spectral and Spatial Information

Although multivariate imaging offers possibilities to differentiate between both objects of similar spectra and different spatial correlations, existing unsupervised clustering techniques cannot utilize this property. The disadvantage of applying those techniques as a multivariate image segmentation technique is that the information from the spatial domain is ignored. Only the spectral information determines the partitioning of the measurement space. It is clear in the case of overlapping clusters in the spectral domain, only the extra spatial information can discriminate between the different objects. Even if no a priori information about the spectra is available, information about the shape of the objects in the image is sometimes known. However, the majority of supervised and unsupervised classification techniques including traditional FCM cannot use this a priori spatial information during the clustering process without modification. Papers about the combined use of spatial and spectral information for image segmentation have been reported in the literature before [see Noordam (2005) for detailed references], except for FCM. Therefore, two modifications of FCM are developed that utilize spatial information during the clustering (Noordam and van de Broek, 2002; Noordam *et al.*, 2003). Only the spatially guided FCM (sgFCM) will be discussed.

3.2.6   sgFCM Clustering

The rationale of sgFCM is to improve clustering by including spatial information, described by a geometrical shape description during the construction of the cluster prototypes. This means that both the spectral and spatial neighborhood of a pixel determine the contribution of a pixel to a cluster prototype. In theory, any arbitrary shape can be detected during clustering provided that there is a geometrical shape description (geometrical model) available. A geometrical model can vary from a simple local neighborhood model to describe outliers or small blobs, to a more complex model to detect any predefined shape. Figure 3.5 shows the principle of sgFCM.

During each FCM iteration, pixels or groups of pixels in the segmented image are compared with the predefined geometrical model. The difference between the geometrical model and the pixels in the segmented image described by the geometrical model result in a value for an additional mismatch variable. A high value for this variable indicates a mismatch between the geometrical model and pixels described by the model in segmented image, a low value indicates a match between

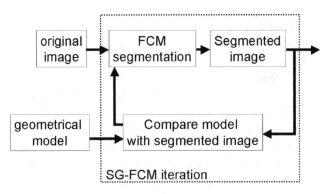

Figure 3.5   The principle of sgFCM. Reprinted from Unsupervised segmentation of predefined shapes in multivariate images, *Journal of Chemometrics* 17 (4), 216–224, © (2003) John Wiley & Sons Ltd

the model and pixels in the segmented image. This mismatch variable has been used by a modified FCM where the construction of cluster prototypes is influenced by the value of this variable. This results in a segmented image in which the construction of the cluster prototypes is influenced by spatial information, reflected by the geometrical model.

A modified version of FCM is used to add a priori information to influence the clustering via a modified update procedure for the membership u_{ik}:

$$u_{ik} = \frac{1}{1+\alpha} \left[\frac{1+\alpha(1-\sum_{l=1}^{c} f_{lk})}{\sum_{l=1}^{c} \frac{d_{ik}^2}{d_{lk}^2}} + \alpha f_{ik} \right] \tag{3.8}$$

High values of f_{ik} will enhance the membership of object k for cluster i and low values of f_{ik} will weaken the membership for cluster i. The value of f_{ik} is determined by the difference between a geometrical shape model and the segmented image. The geometrical model contains the a priori shape information which is used to influence the clustering via the variable f_{ik}. When no a priori information is present, f_{ik} is set to zero and the standard membership update formula of Equation (3.2) is used. However, if a priori information about some objects is available, the modified membership update formula of Equation (3.8) is used instead. Specific details can be found in the literature (Noordam et al., 2003).

In the following example, a multivariate RGB image is segmented with three different techniques: FCM; FCM and spatial majority filter (=FCM-majority); and sgFCM. The original image and the results are shown in Figure 3.6. The image consists of peppers of different shape, color and size. Several specular reflections are visible in the image owing to the shiny surface of the peppers. This results in spurious specular pixels and regions which are enveloped by spurious edges. The original image is shown in Figure 3.6(a) and contains the classes: dark green pepper; light green pepper; red pepper; shadow (black); and specular reflections (white). A 3 × 3 window is selected during the sgFCM clustering which serves as a simple geometrical model that describes the small edges and spurious pixels.

The image shown in Figure 3.6(b) is segmented by the traditional FCM. It shows that some of the segmented regions are indeed enveloped by spurious edges, especially those regions which correspond to specular reflections. Figure 3.6(c) shows the result after clustering with sgFCM with the 3 × 3 window. Both the spurious pixels and spurious envelopes around

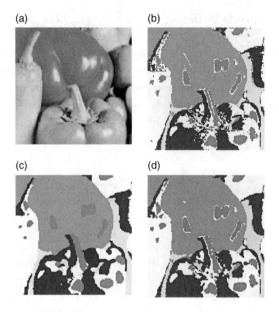

Figure 3.6 (a) Original pepper image; (b) FCM segmented image; (c) sgFCM segmented image; (d) FCM-majority segmented image. Reprinted from Unsupervised segmentation of predefined shapes in multivariate images, *Journal of Chemometrics* **17** (4), 216–224, © (2003) John Wiley & Sons Ltd

the specular regions are removed. The extra spatial information sees to it that spurious edge pixels are merged with the class of the surrounding pixels. The image depicted in Figure 3.6(d) is the result of the FCM-majority segmentation with a 3 × 3 window. It shows that the greater part of the spurious pixels is removed but almost all of the spurious envelopes are left intact. The images segmented by sgFCM and FCM-majority show the difference between both approaches. In sgFCM, pixels can gradually move to another class after each iteration, whereas the FCM-majority filter combination tries to correct all segmentation errors in a single iteration, which is not always possible. In this example a simple 3 × 3 window is used but is has been shown that even more complex spatial models, such as circles, can be evaluated during clustering (Noordam *et al.*, 2003).

3.3 SUPERVISED CLASSIFICATION OF MULTISPECTRAL IMAGES

Unsupervised clustering techniques when used as a segmentation technique are able to partition the image in regions based on a user selected number of classes as long as the number of spectral bands is limited. The huge amount of data in multispectral images requires computational demands which are not feasible in most applications. This is especially true for the multispectral images as discussed in the applications above where an imaging spectrograph (Imspector) is used for the image acquisition. For example, in certain applications, the number of bands ranges from 100 to 300 and a single 16 bit multispectral image contains about 84 Mbyte of data (Noordam *et al.*, 2005b). A widely used approach is to apply an unsupervised classifier to a small sample of the image data for the estimation of the classes and then classify the entire multivariate image using a supervised classifier trained with the estimated model (Banfield and Raftery, 1993). It has been shown that this strategy leads to under- or overestimated classes in the segmented image (Wehrens *et al.*, 2003). Furthermore, different samplings may result in different segmentation results due to the random nature of the sampling procedure. Therefore, in the case of high dimensional data, supervised classification techniques are often used to classify the multispectral image.

Supervised techniques require a labelled training set from which the class boundaries are determined. Usually, a representative training must be obtained from the multispectral image set which can be a challenging task as there are multiple image bands (>100) to explore.

3.3.1 Multivariate Image Analysis for Training Set Selection

A procedure for selecting a representative training set and test set from multivariate images is based on multivariate image analysis (MIA), as described thoroughly in the literature (Geladi and Grahn, 1996). MIA uses principal component analysis (PCA) for dimension reduction and with three selected score images a pseudo color image is constructed which is used for visualization and region selection. By selecting regions in the pseudo color image, corresponding scores in the score plots can be visualized. The opposite is also possible, by selecting regions or clusters in score space, the corresponding pixel position in the pseudo image can be visualized. Although it was originally introduced as an interactive tool for the explorative analysis of multivariate images, the selected regions can be used as a training or test set. The selected objects or scores can be extracted, backtransformed to original space, labeled and used as a labeled dataset. The procedure must be repeated for each class in the multivariate image and the classes are combined to a training or test set. This procedure has been followed in several applications (Wold *et al.*, 2001; van den Broek, 1997; Witjes, 2001) for the selection of training and test sets.

However, with respect to the explorative analysis and classification of huge multispectral images of agricultural and food products several difficulties might occur. To select a training set from the multivariate image with the MIA procedure, an exact selection of regions is required which can be a challenging task. Due to spectral and spatial mixing, overlapping classes occur which makes the selection of a single class difficult. Also, it is not possible to verify that all relevant classes have been selected, due to dense classes and class overlap small defect classes may be overlooked and therefore will not be included in a training set. To prevent this, a thoroughly coarse to fine exploration of the multivariate image is required which can be time consuming as the amount of data shown in the score plots makes it difficult to recognize structure or grouped data. Although one can state that the MIA procedure is an attractive method for global exploration, it is time consuming for training set selection and requires user experience.

Still, the MIA procedure inherently contains a very important evaluation characteristic as the user can immediately evaluate the quality of a selected region. The continuous switching between the spatial and spectral domain gives the user a visual feedback about the quality of the selected region and the user can adjust the selection if necessary. To overcome the aforementioned problems, a new Feedback Multivariate Model Selection (FEMOS) procedure for the estimation of classes in multivariate images has been developed (Noordam *et al.*, 2005b) in which this so-called visual feedback is automated. The principle of FEMOS is depicted in Figure 3.7.

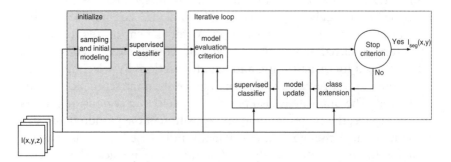

Figure 3.7 The principle of FEMOS. Reprinted from Chemometrics and Intelligent Laboratory Systems, Vol 64, Noordam *et al.*, Multivariate image segmentation..., pp. 115–126, © 2005, with permission from Elsevier

3.3.2   FEMOS

The initialization step consists of a sample and initial modeling stage and a supervised classifier. In the sample and initial modeling stage, a subset of pixels is taken from the multivariate image via random selection and an initial two-class model is estimated by an unsupervised classifier. The output of this sample and initial modeling stage is called a general model and consists of the centroid and covariance matrix for each class. With this general model, a supervised classifier is trained in the next stage and the entire image is classified with the supervised classifier. The result of the initialization stage is a segmented image with two classes and a corresponding general model consisting of a centroid and a covariance matrix for both classes.

With this general model and the corresponding segmented image, the model evaluation criterion is evaluated. The model evaluation criterion evaluates all classes of the segmented image with a predefined criterion. Usually, a similarity criterion is selected which assumes that the image consists of homogeneous classes. In FEMOS, variance is considered as a criterion for homogeneous regions. The variance (VAR) within class i can be calculated via:

$$\mathrm{VAR}_i = \frac{1}{N_i} \sum_{j=1}^{N_i} \{[\mathbf{I}(x, y, \lambda) \,\&\&\, I_i(x, y)] - \mathbf{v}_i\}^2 \qquad (3.9)$$

where i is the current class, N_i is the number of pixels in the segmented image with label i, $\mathbf{I}(x, y, \lambda)$ is the original 3-D multivariate image composed of a number of images $I(x, y)$ recorded at different wavelengths (λ), $I_i(x, y)$ is the image containing only pixels of class i, and $\mathbf{v}_i$ is the centroid of class i from the general model.

The class which gives the worst criterion performance is labeled as the candidate class and will be further explored in a subsequent class extension stage if the stop criterion is not fulfilled.

In the class extension stage, a two-class unsupervised classification is performed on a subset of pixels from the segmented image which are labeled with the candidate class label. The subset is obtained again via random sampling. The underlying idea is that the bad performance of the candidate class is due to the fact that this class contains two or more classes. The outcome of the class extension stage is an estimated two class model consisting of a centroid and a covariance matrix for the two classes. The csiFCM is used here as an unsupervised classifier in order to overcome the sensitivity of FCM for clusters which have an unequal number of points (Noordam *et al.*, 2002).

In the model update stage, the original centroid and covariance matrix of the candidate class in the general model are replaced by the two new centroids and covariance matrices of the class extension stage. Now, the general model has been refined as the number of classes has increased by one.

In the last stage of the feedback loop a supervised classifier is trained with the updated general model and the entire multivariate image is classified. The linear discriminant analysis (LDA) is used here as a supervised classifier. In the next run, a new iteration is carried out in which the updated general model and the segmented image are evaluated by the model evaluation criterion again.

As the FEMOS procedure is a general procedure for the modeling of classes it is not dependent on a particular classification technique or image type which makes it widely applicable. In the following examples, the previously described csiFCM will be used as the unsupervised classifier and a LDA will be used as the supervised classifier.

3.3.3   Experiment with a Multispectral Image of Pine and Spruce Wood

In this experiment, a 6 band multispectral image (680, 740, 800, 840, 1010, 1110 nm) is explored with the FEMOS procedure. The image is recorded with a PbS camera and a filter wheel (detailed information of the measurement set-up can be found in the literature; Geladi *et al.*, 1994). The image consists of two wooden disks and a pseudo color image is shown Figure 3.8(a).

The upper disk in the image is cut from a spruce tree (with a crack), the lower disk in the image is cut from a pine tree and both disks contain several rings of wood and bark. The multispectral image is previously explored with the MIA procedure and the corresponding score plot is shown in Figure 3.8(b). The goal of the original work was to investigate if three specific spatial regions in both disks contain the same chemical information. The spatial regions are the two types of wood in the center region (dead xylem and a thin outer layer of living xylem), two types of bark (dead floem and a thin layer of living floem) and the thin layer under the bark (living xylem).

The (nonselected) upper left cluster in the score plot corresponds to the background class and is ignored in the analysis. The three clusters

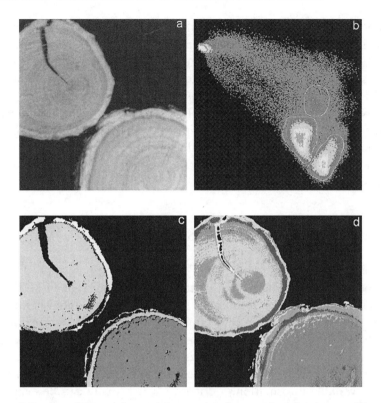

Figure 3.8 A multispectral wood image: (a) the pseudo-color image; (b) score plot; (c) MIA combined score image; and (d) the FEMOS 9-class segmented image. Reprinted from Chemometrics and Intelligent Laboratory Systems, Vol 64, Noordam *et al.*, Multivariate image segmentation..., pp. 115–126, © 2005, with permission from Elsevier (see Plate 14)

visible in the lower right corner of the score plot are selected according to the MIA method and are backprojected in a single combined score image of Figure 3.8(c). This MIA score image shows corresponding regions in both disks, the center heart of the upper disk (compression wood) is shown as a ring in the lower disk (living xylem) and also the bark wood of both disks shows spectral resemblance.

The result of the FEMOS procedure is shown in Figure 3.8(d). By using the Variance criterion the similarity based segmented image reveals more detailed information about the outer and inner regions of the wood compared with the MIA analysis. Both disks reveal additional rings below the bark wood which are visible in the other disk, either as additional bark wood rings or as regions in the internal heart. Furthermore, the FEMOS segmented image shows that the physical–chemical composition of the center part of the upper disk is not homogeneous, parts in the center of the upper pine disk are spectrally similar to some of the bark wood rings of the spruce disk. This is not visible in the MIA segmented images, due to mixed pixels and the small number of pixels in these regions are not visible as individual clusters in the score plot with any combination of the PCs which makes the reproduction of the FEMOS segmented image with MIA analysis a challenging task. This example shows the superior value of the FEMOS procedure, where just a single run reveals detailed information of present classes in the multispectral wood image.

3.3.4   Clustering with FEMOS Procedure

In the following example, the FEMOS procedure is applied as an unsupervised clustering technique to show the ability of FEMOS as a general class estimation technique. The FEMOS procedure is compared with csiFCM, as described in Section 3.2.4. For csiFCM, the examined number of classes ranges from 5 to 14 and for each number of classes the experiment is repeated five times to reduce initialization influences.

The focus of the example is on the robust and accurate representation of all available information, a criterion which easily can be examined by a visual comparison between the original and the segmented image. Therefore, a three-variable spectral color image (RGB) of a known object is used. The used RGB image is an African Violet plant (*Saintpaulia ionantha* Wendl) and the original RGB color image is shown in Figure 3.9(a).

The image contains five classes: background white; black soil; purple flower; yellow heart; and green leaves. Although the yellow hearts in the purple flower color have a bright color, the hearts are difficult to segment in a single class owing to the small number of yellow pixels.

In Figure 3.9 the results of csiFCM and FEMOS are shown. Only the graphs with the covariance and centroids are shown, the ellipse represent 1σ confidence intervals of the pixels of that class (68 %). The colors in the segmented images and graphs are selected according to a method called class coloring; this method will be explained in detail in Section 3.4.

The increase in classes from 5 to 14 has only resulted in an over-segmented green leaves class and an oversegmented blue flower class. This is also visible in the 13-class graph of Figure 3.9(c) where an abundance of blue and green covariance matrices is located in the lower left

(a)

(c)

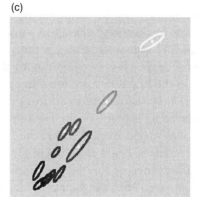

(b)

(e)

(d)

Figure 3.9 (a) The original RGB image of the African violet with the small yellow hearts in the blue flowers; (b) the 13-class csiFCM segmented image; (c) the 13-class model of csiFCM; (d) the 8-class FEMOS segmented image; (e) the 8-class model of the FEMOS procedure. Reprinted from Chemometrics and Intelligent Laboratory Systems, Vol 64, Noordam *et al.*, Multivariate image segmentation . . . , pp. 115–126, © 2005, with permission from Elsevier (see Plate 15)

corner, representing a number of dark colored classes green leaves and purple flowers. In the upper right corner of the 13 class graph, a single white background class and a gray background class is visible. In all 45 csiFCM segmented images the yellow flower heart class is completely missed; the class is merged with the white or gray background class.

In the FEMOS segmented image [Figure 3.9(d)] the yellow heart class is clearly visible and in all five runs, the class is correctly segmented and labeled as a single class. The number of clusters in the obtained general model varied from 8 to 11, the 8-class graph is shown in Figure 3.9(e). The yellow heart class is clearly visible as an extra yellow class in the right corner of the 8-class graph, next to the white background class. Compared with the 8-class model, the 11-class FEMOS segmented image (not shown) has two additional background classes and one extra green leaves class.

The difference between csiFCM and FEMOS is clearly visible if the graphs in Figure 3.9(c) and (e) are compared. The FEMOS procedure is consequently successful in estimating the yellow heart class. The csiFCM clustering technique is not capable of properly modeling all the classes of this African violet image.

The example shows that when the FEMOS procedure is applied as an unsupervised classifier the results are accurate and precise compared with a traditional unsupervised classifier like csiFCM. Although the number of classes in the general model may vary, the represented information does not. For all five runs, the FEMOS procedure extracts the relevant information from the image.

3.4   VISUALIZATION AND COLORING OF SEGMENTED IMAGES AND GRAPHS: CLASS COLORING

In order to visualize the multispectral image for the user, a pseudo color image has to be created. In the MIA procedure, three score images are selected to form a pseudo color image (Geladi and Grahn, 1996). However, if no PCA is performed, score images are not available and three variables from the original image must be selected to represent these as red, green and blue. A drawback of this method is that only three variables at a time can be shown at once. Finding out which components to choose and how to combine them may lead to testing many combinations. A faster and very simple procedure to create a pseudo color image from a raw N-dimensional multivariate image is to distribute the N variables over the three color bands red, green and blue of the color image. The distribution of the N variables over the three color bands can be done equally or according to a specific function, like the function that describes the color sensitivity of the human eye. For each color band, the assigned variables are averaged and scaled to the dynamic range of the color band, usually between 0 and 255. The advantage of this procedure is that no variable selection is required and that all the variables are included in the pseudo color image.

A common method to visualize the segmented image is to represent the segmented pixels with a color that corresponds to a predefined label. This way of labeling can give good contrast but the colors for the labels must be assigned beforehand. Another method which shows more visual resemblance to the original (pseudo color) image is to use the centroid of the class as color for the label instead. This method is called class coloring. In the case of a multivariate image with N variables, the N-dimensional centroid must be reduced to a 3-D pseudo centroid. This dimension reduction can be determined in a similar way as described above.

The result is that both the multivariate image and the segmented image are represented as pseudo color images, which share the same color mapping, which makes them easy to compare. The same coloring procedure can be used for the visualization of data, centroids and covariance matrices in graphs. During a clustering process or after training a supervised classifier, the original image, the segmented image and the corresponding class information can all be visualized in similar colorings.

The major advantage of class coloring is that it is independent of the number of classes in the segmented image. Traditionally, if an extra class is required to describe the data, a new label is created with a corresponding assigned color for the segmented image. With the presented class coloring method, a label color is automatically created as it is just the (pseudo color) centroid of the new class. In Figure 3.9(b), a csiFCM segmented African violet image is shown and in Figure 3.9(c) the cluster plot with 1-σ ellipses is shown according to this color labeling. The segmented violet image clearly shows that the segmentation of the original image [Figure 3.9(a)] is not correct. The yellow flower class has been missed by the csiFCM algorithm and is misclassified as background. The yellow heart class that has been detected with FEMOS is not only visible in the segmented image [Figure 3.9(d)] but also in the cluster plot [Figure 3.9(e)] where the yellow ellipse represents this class.

3.5   CONCLUSIONS

We have demonstrated the applicability of multispectral images for quality inspection of agricultural products. Three typical problems that are encountered in the clustering and classification process of multispectral images are discussed. To illustrate the context of these problems a straightforward, transparent and broadly accepted clustering technique called Fuzzy C-Means (FCM) is chosen. The selected problems are: (1) small sized cluster detection; (2) insufficient spectral or spatial discrimination between image features of interest; and (3) the estimation of the absolute minimum number of classes to describe the features of interest.

For each typical problem a chemometric technique is presented to show directly or indirectly that clustering or classification improvement can be obtained when both spectral and spatial information are incorporated into the clustering and/or classification algorithm.

The first technique overcomes the sensitivity of FCM for unequal cluster sizes and as a result the final segmentation will be influenced by this cluster size sensitivity. Incorrect segmentation gives incorrect information about the defect percentages of the product under inspection. The presented csiFCM overcomes this cluster size sensitivity and shows good performance compared with traditional FCM.

The second technique presented in this chapter is called spatially guided FCM (sgFCM) and combines both spectral and spatial information to improve discrimination between objects of similar spectra. The performance of sgFCM is compared with both FCM and the combination of FCM and a majority filter in an experiment with a multivariate pepper image where spectral information is not sufficient to discriminate between objects. The sgFCM segmented images show more homogeneous regions and less spurious pixels compared with FCM and FCM-majority results.

The third technique described in this chapter is a procedure for the estimation of classes in multivariate images. The FEMOS procedure combines unsupervised and supervised classifiers with a model evaluation criterion to extract classes from the multivariate image in an iterative manner. The procedure uses a subset of the multivariate image to estimate a general model and evaluates this model with the original multivariate image via a model evaluation criterion. The FEMOS procedure can be applied for the unsupervised segmentation of multivariate images or for training and test set estimation from multivariate images. The performance of FEMOS procedure is shown in two examples with real world multivariate images and the results are compared with csiFCM, a clustering algorithm. The results show that the FEMOS procedure outperforms traditional routines in terms of robustness and accuracy when applied as either unsupervised segmentation technique or class modeling technique.

A visualization technique for segmented images or graphs is also presented. This technique is called class coloring and automatically generates color labels for segmented images or graphs. The color labels correspond to the colors in the original image which improves visual comparison between the segmented image and the original image.

The ongoing expansion of the number of bands in multispectral images and the spatial resolution of the images emphasizes the need for supporting exploration and analysis techniques. Further improvements are expected for visualization and classification to allow effective image exploration and representative dataset extraction.

REFERENCES

Abbott, J. (1999) Quality measurement of fruits and vegetables,*Postharvest Biology and technology*, 15, 207–225.

Banfield, J. and Raftery, A. (1993) Model-based gaussian and non-gaussian clustering, *Biometrics*, 49, 803–821.

Bezdek, J. (1981) *Pattern Recognition with Fuzzy Objective Functions*, Plenum Press, New York.

Birth, G. (1976) How light interacts with food, in: *Quality Detection in Foods* (J. Gaffney Jr, ed.), American Society for Agricultural Engineering, St Joseph, MI, pp. 6–11.

van den Broek, W. (1997) Chemometrics in Spectroscopic Near Infrared Imaging for Plastic Material Recognition, PhD thesis, Catholic University of Nijmegen.

Brosnan, T. and Sun, D. (2002) Inspection and grading of agricultural and food products by computer vision systems – a review, *Computers and Electronics in Agriculture*, 36, 193–213.

Casasent, D. and Che, X. (2003) Aflatoxin detection in whole corn kernels using hyper-spectral methods, in: *Food Safety and Agricultural Monitoring* (Y. R. Chen and G. E. Meyer, eds), Vol. 5271, SPIE, Providence, RI.

Chao, K., Mehl, P. and Chang, Y. (2002) Use of hyper- and multi-spectral imaging for detection of chicken skin tumors, *Applied Engineering in Agriculture*, 18, 113–119.

Chen, S. and Li, M. (2001) Multispectral imaging of chlorophyll content for vegetable status monitoring, in: *Fruit, Nut and Vegetable Production Engineering* (M. Zude, B. Herold and M. Geyer, eds), Institut fur Agrartechnik Bornim eV, Potsdam, pp. 603–608.

Chen, Y., Chao, K. and Kim, M. (2002) Machine vision technology for agricultural applications, *Computers and Electronics Agriculture*, 36, 173–191.

Delwiche, S. and Kim, Y. (2000) Hyperspectral imaging for detection of scab in wheat, *Proceedings of SPIE*, 4203, 13–20.

Gat, N. (2000) Imaging spectroscopy using tunable filters: a review, *Wavelet Applications VII*, **4056**, 50–64.

Geladi, P. and Grahn, H. (1996) *Multivariate Image Analysis*, JohnWiley & Sons, Ltd, Chichester.

Geladi, P., Swerts, J. and Lindgren, F. (1994) Multiwavelength microscopic image analysis of a piece of painted chinaware, *Chemometrics and Intelligent Laboratory Systems*, **24**, 145–167.

Graves, M. and Batchelor, B. (2003) *Machine Vision for the Inspection of Natural Products*, Springer, London.

Guyer, D. and Yang, X. (2000) Use of genetic artificial neural networks and spectral imaging for defect detection on cheries, *Computers and Electronics in Agriculture*, **29**, 179–194.

Jimenez, A., Jain, A., Ceres, R. and Pons, J. (1999) Automatic fruit recognition: a survey and new results using range/attenuation images, *Pattern Recognition*, **32**, 1719–1736.

Lawrence, K., Windham, W., Park, P. and Smith, D. (2003) Comparison between visible/NIR spectroscopy and hyperspectral imaging for detecting surface contaminants, in: *Food Safety and Agricultural Monitoring* (Y. R. Chen and G. E. Meyer, eds), Vol. 5271, SPIE, Providence, RI.

Lee, C. and Landgrebe, D. (1993) Analyzing high dimensional multispectral data, *IEEE Transactions on Geoscience and Remote Sensing*, **31**, 792–800.

Li, J., Tan, J., Martz, F. and Heyman, H. (1999) Image texture features as indicators of beef tenderness, *Meat Science*, **53**, 17–22.

Malek, J., Alim, A. and Tourki, R. (2002) Problems in pattern classification in high dimensional spaces: behavior of a class of combined neuro-fuzzy classifiers, *Fuzzy Sets and Systems*, **128**, 15–33.

Martinsen, P. and Schaare, P. (1998) Measuring soluble solids distribution in kiwifruit using near-infrared imaging spectroscopy, *Postharvest Biology and Technology*, **14**, 271–281.

Mehl, P., Chao, K., Kim, M. and Chen, Y. (2002) Detection of defects on selected apple cultivars using hyperspectral and multispectral image analysis,*Applied Engineering in Agriculture*, **18**, 219–226.

Miller, W., Throop, J. and Upchurch, B. (1998) Pattern recognition models for spectral reflectance evaluation of apple blemishes, *Postharvest Biology and Technology*, **14**, 11–20.

Noordam, J. (2005) Chemometrics in Multispectral Imaging for Quality Inspection of Postharvest Products, PhD thesis, Radboud University, Nijmegen; http://www.wur.nl/GreenVision

Noordam, J. and van den Broek, W. (2002) Multivariate image segmentation based on geometrically guided fuzzy c-means clustering, *Journal of Chemometrics*, **16**, 1–11.

Noordam, J., Olten, G., Timmermans, A. and van Zwol, B. (2000) High-speed potato grading and quality inspection based on a color vision system, *Proceedings of SPIE*, **3966**, 206–220.

Noordam, J., van den Broek, W. and Buydens, L. (2002) Multivariate image segmentation with cluster size insensitive Fuzzy C-means, *Chemometrics and Intelligent Laboratory Systems*, **64**, 65–78.

Noordam, J., van den Broek, W. and Buydens, L. (2003) Unsupervised segmentation of predefined shapes in multivariate images, *Journal of Chemometrics*, **17**, 216–224.

Noordam, J., van den Broek, W. and Buydens, L. (2005a), Detection and classification of latent defects and diseases on french fries with multispectral imaging, *Journal of the Science of Food and Agriculture*, 85, 2249–2259.

Noordam, J., van den Broek, W., Geladi, P. and Buydens, L. (2005b) A new strategy for the modelling and representation of classes in multivariate images, *Chemometrics and Intelligent Laboratory Systems*, 75, 115–126.

Novales, B., Bertrand, D., Devaux, M., Robert, P. and Sire, A. (1996) Multispectral fluorescence imaging for the identification of food products, *Journal of the Science of Food and Agriculture*, 71, 376–382.

Park, B., Lawrence, K. C., Windham, W. R., Chen, Y.-R. and Chao, K. (2002) Discriminant analysis of dual-wavelength spectral images for classifying poultry carcasses, *Computers and Electronics in Agriculture*, 33, 219–231.

Pedrycz, W. (1996) Conditional Fuzzy C-Means, *Pattern Recognition Letters*, 17, 625–631.

Polder, G., van der Heijden, G. and Young, I. (2002) Spectral image analysis for measuring ripeness of tomatoes, *Transactions of the ASAE*, 45, 1155–1161.

Rezaee, M. (1998) Application of Fuzzy Techniques in Image Segmentation, PhD thesis, University of Leiden.

Sugiyama, J. (1999) Visualisation of sugar content in the flesh of a melon by near-infrared imaging, *Journal of Agricultural and Food Chemistry*, 47, 2715–2718.

Tao, Y., Heinemann, P., Varghese, Z., Morrow, C. and Sommer, H. (1995) Machine vision for color inspection of potatoes and apples, *Transactions of the ASAE*, 38, 1551–1561.

Thai, C., Evans, M. and Grant, J. (1997a), Herbicide stress detection using liquid crystal tunable filter, *ASAE Annual International Meeting*, Minneapolis, MN.

Thai, C., Kays, S. and Grant, J. (1997b), Mapping light penetration into fruits using a liquid crystal tunable filter, *ASAE Annual International Meeting*, Minneapolis, MN.

Upchurch, B. & Thai, C (2000), Spectral characterization of Pecan weevil larvae and Pecan nutmeat using multispectral imaging, *ASAE Annual International Meeting*, Milwaukee, WI.

Wehrens, R., Buydens, L., Fraley, C. and Raftery, A. (2003) Model-based clustering for image segmentation and large datasets via sampling, *Technical Report 424*, Department of Statistics, University of Washington.

Williams, P. and Norris, K. (1987) *Near-Infrared Technology in the Agricultural and Food Industries*, American Association of Cereal Chemists, St Paul, MN.

Witjes, W. (2001) Explorative Analysis of Chemometrics in Magnetic Resonance (Spectroscopic) Imaging of Human Brain Tumors PhD thesis, Catholic University of Nijmegen.

Wold, J., Kvaal, J. and Egelandsdal, B. (1999) Quantification of intramuscular fat content in beef by combining autofluorescence spectra and autofluorescence, *Applied Spectroscopy*, 53, 448–456.

Wold, J., Westad, F. and Heia, K. (2001) Detection of parasites in cod fillets by using SIMCA classification in multispectral images in the visible and NIR region, *Applied Spectroscopy*, 55, 1025–1034.

Zwingelaar, R., Yang, S., Garcia-Pardon, E. and Bull, C. (1996) Use of spectral information and machine vision for bruise detection on peaches and apricots, *Journal of Agricultural Engineering Research*, 63, 323–332.

4
Self-modeling Image Analysis with SIMPLISMA

Willem Windig, Sharon Markel and Patrick M. Thompson

4.1 INTRODUCTION

Multivariate image analysis can be performed with techniques such as principal component analysis (PCA) (Geladi and Grahn, 1996). The major advantage is that the many images (one image per wavelength) can be reduced to a few, while preserving all the significant information in the dataset. Furthermore, noise reduction is obtained. The data reduction obtained with PCA, however, is an orthogonal system explaining the maximum variance. Therefore the set of reduced images is difficult to interpret.

Another approach is to use self-modeling mixture analysis. In this case, the dataset will be reduced to a limited number of images based on mathematical criteria that result in chemically meaningful results. For example, a chemical image of a cross-section of three different polymers will result in three images, each describing one polymer. Each of the images will have an associated spectrum describing the pure polymer. Examples of self-modeling image analysis (SIA) have been described by Sasaki *et al.* (Sasaki *et al.*, 1989), where entropy minimization was used to resolve 13 images (400–700 nm at 25 nm intervals) of biological samples. Andrew and Hancewicz (Andrew and Hancewicz, 1998)

Techniques and Applications of Hyperspectral Image Analysis Edited by H. F. Grahn and P. Geladi
© 2007 John Wiley & Sons, Ltd

analyzed Raman image data by selecting the most dissimilar spectra in a dataset. Batonneau *et al.* (Batonneau *et al.*, 2001) selected pure variables from Raman spectroscopic images of industrial dust particles using simple-to-use interactive self-modeling mixture analysis (SIMPLISMA). The latter technique will also be used for this study. SIMPLISMA is a technique that selects pure variables from the spectral dataset. Pure variable values are proportional to concentrations and can therefore be used to resolve data. When no pure variables are present, 2nd derivative data can be used. SIMPLISMA has been used in image related projects: to resolve Fourier transform infrared (FTIR) microscopy data of a polymer laminate (Windig and Markel, 1993; Guilment *et al.*, 1994) and to resolve FTIR microscopy data of a KBr tablet with a mixture of three powders (Guilment *et al.*, 1994).

A newly developed SIMPLISMA addition was used in this study (Windig *et al.*, 2002). The new addition allows the use of a combination of conventional and 2nd derivative data, which enables the user to extract baseline related problems. Examples will be given of:

- FTIR microscopy transmission data of a cross-section of a polymer laminate.
- FTIR microscopy reflectance data of a mixture of aspirin and sugar.
- Secondary ion mass spectrometry (SIMS) of a two-component mixture of palmitic acid and stearic acid.

In the latter case the differences will be shown between selecting pure spectra and pure variables (masses) as a starting point for resolving and selecting. Furthermore, it will be shown how to obtain a pure variable solution with the pure spectra as a starting point.

4.2   MATERIALS AND METHODS

4.2.1   FTIR Microscopy

Transmission experiment

A laminate sample, consisting of four different polymers, including a 3 μm inner layer, was analyzed in cross-section using a Spectral Dimensions, Inc. (Olney, MD) IR imaging microscope. The imaging system at the Molecular Microspectroscopy Laboratory at Miami University consists of a step-scan spectrometer interfaced with an IR microscope equipped with a 64 × 64 focal plane array detector. The spectral range of the system is 5000–900 cm^{-1}. The system magnification using a 15x IR objective is approximately 6 μm per pixel. Spatial resolution for the imaging system is wavelength limited. The cross-section was analyzed in a compression cell to gently flatten the sample for viewing and imaging. IR spectra were collected in transmission mode using 4, 8 and 16 cm^{-1} resolution to determine the optimum spectral resolution for IR imaging experiments.

The individual layers of the laminate had been previously identified using conventional IR mapping techniques. The laminate consists of a 10 μm outer layer identified as ethylene vinyl acetate with talc filler, a 50 μm layer of colorless polyethylene (PE), a 3 μm inner layer identified as an isophthalic polyester and a 160 μm layer of black polyethylene terephthalate (PET).

Reflectance experiment

Elemental silver membranes may be used as a reflective substrate for *in situ* microreflectance IR analysis. In this example, a small amount of sugar and baby aspirin were ground together in a mortar and added to approximately 50 ml of iso-octane solvent. The solid particulate was filtered using a 5.0 μm silver membrane and analytical filtration techniques. The fine particulate was analyzed on the surface of the silver membrane using the Spectral Dimensions, Inc. IR imaging microscope described above. Spectra were collected in reflectance mode at 8 cm^{-1} resolution. A gold mirror was used to collect a background spectrum for ratio purposes.

4.2.2   SIMS Imaging of a Mixture of Palmitic and Stearic Acids on Aluminum foil

Simulated data

A simulated image of 8 × 8 pixels was generated as follows (Figure 4.1):

(a) A background labeled B was generated with a contribution range of 0.4 to 0.2. The contributions gradually change over the image in a diagonal manner. The top left pixel has the highest contribution; the bottom left pixel has the lowest contribution.

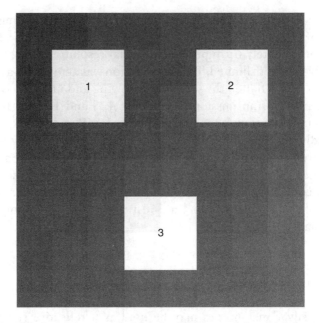

Figure 4.1 The 8 × 8 image consists of a background B, which has a diagonal contribution pattern. The 2 × 2 area labeled with a 1 contains component A and the background B. The 2 × 2 area labeled with a 2 contains component C and the background B. The 2 × 2 area labeled with a 3 contains a mixture of components A and C and the background B

- (b) A component labeled A has a contribution of 0.8 in the 2 × 2 pixel square labeled with a 1 and a contribution of 0.4 in the 2 × 2 pixel square labeled with a 3.
- (c) A component labeled C has a contribution of 0.8 in the 2 × 2 pixel square labeled with a 2 and a contribution of 0.4 in the 2 × 2 pixel square labeled with a 3.
- (d) The spectral dataset was generated by multiplying the contribution profiles with Gaussian profiles of 100 points with a standard deviation of 6 and a maximum value of 1. For the background B, the mean value of the Gaussian profile is 50, for component A the mean value is 25, and for component C the mean value is 75.
- (e) Random uniformly distributed noise with values between −0.01 and +0.01 was added to the dataset.

Actual data

The 99+% purity organic acids were purchased from Aldrich Chemical Company. Aldrich product numbers for the stearic and palmitic acid used were 26,838-0 and 25,872-5, respectively. Each of the pure solutions of the stearic and palmitic acids were produced by dissolving approximately 10 mg of the respective acid in a vial containing 3 ml of USP ethanol. The mixed organic acid solution was similarly made by dissolving approximately 10 mg of each acid in a single vial of 3 ml USP ethanol. A 10 μl syringe was used to place a separate drop of each solution on an approximately 1 cm² piece of Reynolds Wrap aluminum foil. The volume of each dispensed drop was not kept constant in order to prevent cross-contamination due to the different wetting characteristics of each solution while still maintaining a reasonably short distance between each drop to facilitate sample imaging. A clean watch glass was placed over the sample while the ethanol was allowed to evaporate at room temperature.

Once the ethanol had fully evaporated the sample was mounted in the standard backside mounting sample holder supplied with the ION TOF IV time of flight secondary ion mass spectrometer manufactured by ION-TOF GmbH. The design of this energy-focusing spectrometer has been reported elsewhere (Terhorst *et al.*, 1997). A 12 mm × 12 mm area of the sample was imaged using the ION TOF IV large area imaging capability. The final image data file obtained contains the information required to regenerate complete high-resolution spectra at each pixel element of the 128 × 128 pixel image. In order to keep the size of the total dataset within the restrictions imposed by the 1 Mbyte memory of the personal computer, only the data for the 1–600 m/z range were used; these high-resolution data were further compressed into 1 m/z bins centered on each nominal mass. Although not required for this experiment, static-SIMS conditions were maintained during the image acquisition (Benninghoven *et al.*, 1987).

4.2.3 Data Analysis

The programs for this project were written in MATLAB Version 5.3.1 (The MathWorks, Inc., Natick, MA, USA). The computer configuration was a PENTIUM III, 500 MHz, 1 GB of RAM.

4.3 THEORY

The details of the data analysis have been described in Windig (Windig *et al.*, 2002). A summary will be given here. The principle behind the pure variable approach is that the intensity at a pure variable provides an estimate of the concentration of its associated component. As a consequence, the pure variable intensities can be used as a concentration estimate in the following relation:

$$D = CP^T + E \qquad (4.1)$$

D represents the original data matrix with the spectra in rows, and C represents the 'concentration' matrix, obtained by using the columns of D that represent pure variables. Since the values of the pure variables are not concentrations, just proportional to concentrations, they will be referred to as contributions. P represents the matrix with the spectra of the pure components in its columns, P^T represents the transpose of P. E represents the residual error, which is minimized in the least squares equations shown below. When the matrices D and C are known (the pure variables are known through the SIMPLISMA algorithm) the estimate of the pure spectra $\hat{P}$ can be calculated by standard matrix algebra.

$$\hat{P} = D^T C (C^T C)^{-1} \qquad (4.2)$$

In a next step, the contributions are now calculated from $\hat{P}$, which are basically a projection of the original pure variable intensities in the original dataset. This step reduces the noise in the contributions. The equation is:

$$C^* = D\hat{P}(\hat{P}^T\hat{P})^{-1} \qquad (4.3)$$

Where C^* stands for the projected C.

When no pure variables are present in the data, the inverted 2nd derivative data can be used to determine the pure variables. The inverted 2nd derivative data are calculated as follows:

(a) Calculate 2nd derivative spectra.
(b) Change the sign.
(c) Ignore the negative parts.

For details about SIMPLISMA with Matlab code and demonstration files see Windig (Windig, 1997).

In the inverted 2nd derivative data baseline problems are minimized and peaks are narrowed. Examples will be shown later. In the case of the 2nd derivative approach in SIMPLISMA, the conventional (in contrast to 2nd derivative) data are used in **D**. However, columns of **C** consist of intensities from 'inverted 2nd derivative data'.

In the latest SIMPLISMA approach, the pure variable intensities from conventional data as well as the inverted 2nd derivative data are used. For example, for a baseline type component, the conventional data intensities are used, for sharp peaks, mixed with a baseline, the inverted 2nd derivative data intensities are used.

In this study the use of pure spectra will also be described. In this case the pure spectra will be used to resolve the contributions [according to Equation (4.3)] and the projected pure spectra will be calculated from the resulting contributions [according to Equation (4.2)].

4.4   RESULTS AND DISCUSSION

4.4.1   FTIR Microscopy Transmission Data of a Polymer Laminate

In Figure 4.2, a microscopic image is shown of the cross-sectioned polymer laminate. In Figure 4.3, an image of one dataset at $1465\,cm^{-1}$ is shown ($8\,cm^{-1}$ resolution). The image clearly shows four discrete layers in the laminate, previously identified as PET, PE, IPE and ethylene

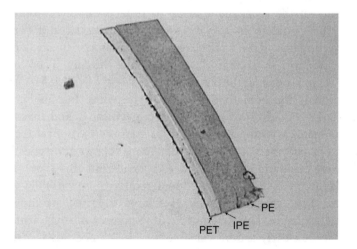

Figure 4.2   A microscopic image of the polymer laminate. The three layers are indicated: polyethylene terephthalate (PET), isophthalic polyester (IPE) and polyethylene (PE). The ethylene vinyl acetate with talk layer is at the right edge.

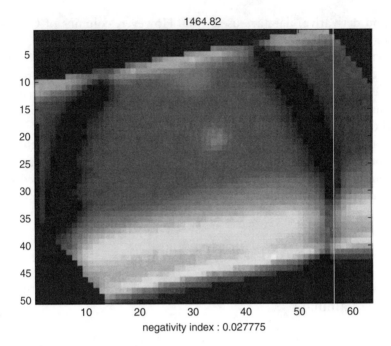

Figure 4.3 The 64 × 64 pixels image at 1465 cm⁻¹

vinyl acetate with talc. The dark ring visible in the image is believed to be related to an inappropriately sized cold shield installed inside the housing of the focal plane array to reduce emissive radiation and improve the signal to noise ratio of data (Bhargava and Fernandez, 2000). Preliminary data analysis indicated that the image within the ring showed more noise than the area outside the ring. Therefore, only a small region outside the ring was used for SIMPLISMA analysis. That region is the area to the right of the white line in Figure 4.3. In Figure 4.4, an overlay plot of the conventional and inverted 2nd derivative spectra within the small area is shown. Spectral features of the different polymers are visible, but there is also a varying baseline of relatively high intensity. In order to deal with this type of data a combination of conventional and 2nd derivative pure variables was used to analyze these spectra. The pure variables selected are indicated in Figure 4.4 by lines. The first pure variable spectrum has a maximum at 1267 cm⁻¹, clearly a pure variable for PET. Since this is a relatively narrow peak on top of a baseline, the 2nd derivative intensities will be used for this component. The second pure variable is at 1465 cm⁻¹, which is representative of PE. Again, the 2nd derivative intensities will

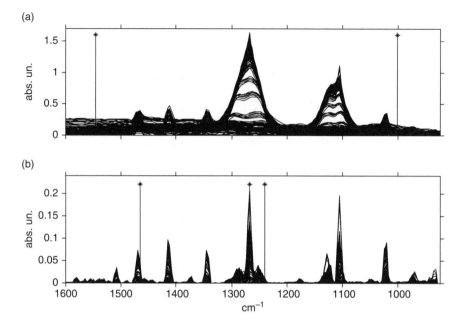

Figure 4.4 (a) Conventional data and (b) the inverted 2nd derivative (smoothing window 5) spectra. The selected pure variables are indicated by lines

be used for this pure variable. The next two pure variables at $1546\,\text{cm}^{-1}$ and $1001\,\text{cm}^{-1}$ clearly represent the baseline, for which the intensities of the conventional data will be used. Next a pure variable (2nd derivative intensities) for isophthalic polyester is chosen at $1236\,\text{cm}^{-1}$. With the five pure variables chosen, the original dataset can be reproduced with only 4 % difference from the original dataset, which is a clear indication that five pure variables are needed to describe the dataset properly. Furthermore, selecting another pure variable results in an additional spectrum with negative intensities, which is another indication that five pure variables are needed to properly resolve this dataset.

The result of this pure variable selection in terms of resolved spectra is shown in Figure 4.5. As explained above, only a small part of the whole image was used to derive the pure variables. In order to obtain resolved images for the whole dataset, the resolved spectra from the reduced dataset were used to calculate the images for the whole dataset (Figure 4.6).

A study of the resolved images and associated spectra indicate that Figure 4.6(a) represents PET, Figure 4.6(b) represents PE and

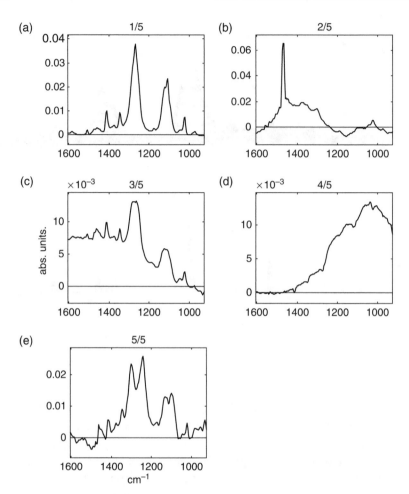

Figure 4.5 Resolved spectra of the dataset: (a) PET; (b) PE; (c) and (d) are dominated by a scattering pattern; and (e) IPE

Figure 4.6(e) represents the isophthalic polyester. The spectra in Figure 4.6(c) and (d) are dominated by scattering pattern and the images show high contributions in areas where scattering is likely to occur, at sample edges and interfaces. It was not possible to extract a spectrum of ethylene vinyl with talc. This is probably due to the fact that this component is at the edges and is therefore dominated by scatter. As a consequence, the ethylene vinyl with talc spectrum my well be hidden in one of the resolved background spectra. It has to be noted that this component is not visible in the original spectra in Figure 4.4.

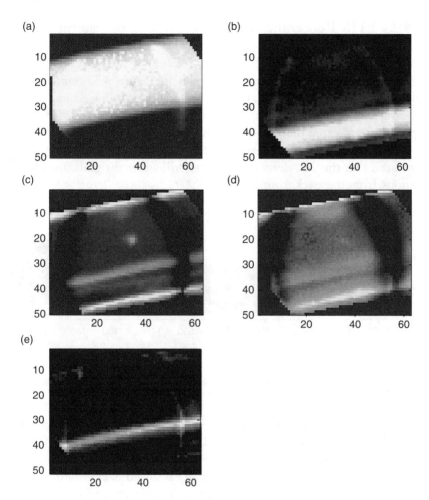

Figure 4.6 Resolved images of the dataset: (a) PET; (b) PE; (c) and (d) are dominated by a scattering pattern and present at the interfaces between different layers; and (e) IPE

The data collected at $16\,cm^{-1}$ resolution were poorly resolved (not shown). Due to the low resolution, it was not possible to obtain a proper spectrum of the isophthalic polyester. A closer examination showed that this spectrum of isophthalic polyester has a contribution from the adjacent PET. This is an indication that the spectral resolution of $16\,cm^{-1}$ is too low to resolve spectra with highly overlapping patterns.

The images collected with $4\,cm^{-1}$ resolution appeared too noisy to extract a spectrum of the isophthalic polyester.

4.4.2 FTIR Reflectance Data of a Mixture of Aspirin and Sugar

The original reflectance data are shown in Figure 4.7. Some of the spectra at the edges of the image have been deleted, followed by deleting spectra with negative intensities. After trying the different SIMPLISMA options, it appeared that the use of 2nd derivative data did not improve the results. Since the background is of a very complex structure, it appeared that the 2nd derivative spectra still showed significant contributions from the background.

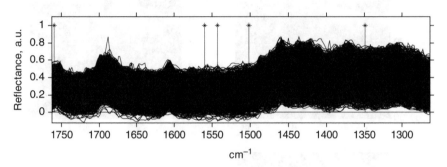

Figure 4.7 Conventional reflectance spectra of the sugar and aspirin mixture. The selected pure variables are indicated by lines

The offset used for this dataset was 3. The following pure variables were chosen: 1502, 1759, 1543, 1349 and $1560\,cm^{-1}$. The difference between the original and reconstructed dataset was 0.066459.

The reflectance spectra resulted in many resolved background spectra that exhibited interference like patterns. This is probably due to the granular sample. The resulting spectra from this dataset are shown in Figure 4.8. The complexity of the background spectra is amazing, but from comparison of the reference spectra in Figure 4.9(a) and (b) with the resolved spectra in Figure 4.8(d) and (b) it is clear that good quality spectra were extracted.

Because of the nature of the original samples, the extracted images do not show clear patterns and therefore are not shown.

4.4.3 SIMS Imaging of a Mixture of Palmitic and Stearic Acids on Aluminum Foil

The pure variable approach for the actual dataset appeared problematic. Possible reasons for this are:

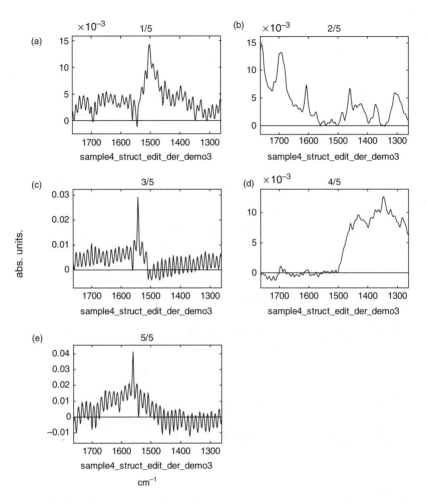

Figure 4.8 Resolved spectra of the reflectance data: (a) is dominated by an interference like pattern, which is also the case for (c) and (e); (b) aspirin [cf. Figure 4.9(b)]; and (d) sugar [cf. Figure 4.9(a)]

(a) The quality of pure variables is more difficult to judge in discrete mass spectra than in continuous spectra, such as FTIR spectra.

(b) The programs that bin the original mass spectral data with several channels per mass into unit resolution data cause artifacts in the multivariate data structure.

(c) The 'contributions', secondary ion intensities in this case, are seldom linear with concentration for each species present, but they are usually related to their respective concentration through a smooth but unknown function.

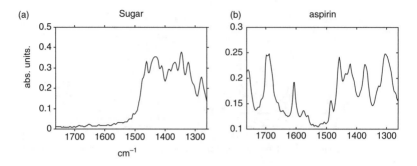

Figure 4.9 Reference spectra of (a) sugar and (b) aspirin

A similar problem was described before with time resolved mass spectral data (Phalp *et al.*, 1995). The solution found was to resolve the data in the following steps:

(a) Determine pure spectra and calculate the resolved contribution profiles.
(b) Determine, for each contribution profile, the mass with the highest correlation. This mass is a pure mass.
(c) Resolve the data using these pure variables.

The differences of a pure spectrum and a pure variable solution will be explained using the simulated dataset described in Section 4.2, after which the actual dataset will be discussed.

Simulated dataset

Figure 4.10(a), (b) and (c) show the pure spectrum solution. The spectra that were determined to be pure as determined by SIMPLISMA, with an offset of 3, are indicated by stars. The projected pure spectra are shown in Figure 4.10(d), (e) and (f).

The projected purest spectrum of component A is not really pure, as Figure 4.10(d) clearly shows; there is a contribution from the background spectrum B. Similarly, there is no appropriate pure spectrum for component C [Figure 4.10(f)]. Spectrum B of the background is really pure, as Figure 4.10(e) shows. The lack of appropriate pure spectra is reflected in the resolved images. The resolved image of the background [Figure 4.10(b)] shows 'holes' where the other components are present. This is exactly what could be expected: since the spectra in the squares with components

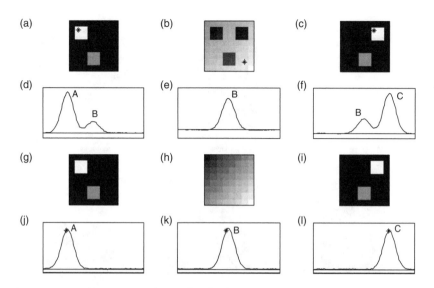

Figure 4.10 The mixture solution for the dataset visualized in Figure 4.1. The pure spectrum solution results in the resolved images (a)–(c). The selected pure spectra are indicated by plus signs. The projected pure spectra are shown in (d)–(f). The pure variable solution results in the resolved images (g)–(i). The resolved spectra are shown in (j)–(l). The selected pure variables are indicated by stars.

A and C are considered pure, these squares must by definition be absent in the background image. Summarizing, since the purest spectra did not represent appropriate pure spectra, the data can not be resolved properly.

The pure variable solution is shown in Figure 4.10(g), (h) and (i). Since there are appropriate pure variables for each component, the data can be resolved properly.

Actual dataset

In order to select pure spectra instead of pure variables, the dataset was simply transposed prior to analysis. The offset was set at 20, which is relatively high. This was necessary because of the relatively high amount of noise in the data. The successive series of purity images are displayed in Figure 4.11(a), (d), (g), (j) and (m). The images resulting from the pure pixel solution are shown in Figure 4.11(b), (e), (h) and (k). The resolved images resulting from the pure variable solution are shown in Figure 4.11(c), (f), (i) and (l).

Although this actual dataset seems on first sight very similar to the simulated dataset, it appeared to be much more complex. The

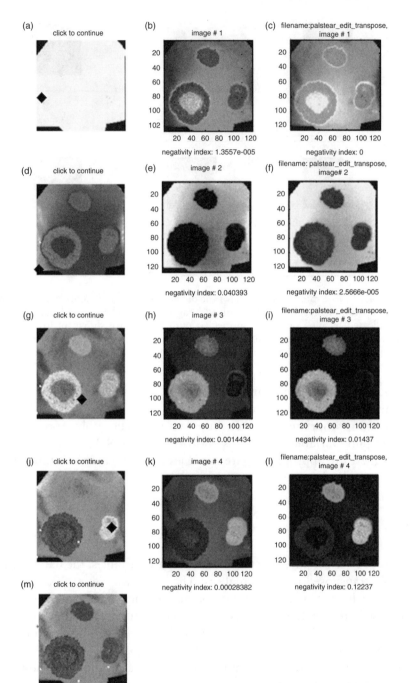

Figure 4.11 The successive series of purity images are shown in (a),(d),(g),(j) and (m). The selected pure pixels are indicated by diamonds. The resolved images based on the pure pixels solution are shown in (b), (e), (h) and (k). The resolved images based on the pure variable solution are shown in (c), (f), (i) and (l)

interpretation of the successive selection of pure pixels/spectra and the resolved images is as follows.

The first pure pixel/spectrum in Figure 4.11(a) represents a coating on the aluminum foil that interacts with the chemical components in such a way that a ring with a higher concentration of this coating around the original droplets originated. The drop of palmitic acid became donut shaped, which resulted in an area of higher concentration of the coating in the center. The second pure pixel/spectrum in Figure 4.11(d) represents a second component of the aluminum foil. The third pure pixel/spectrum in Figure 4.11(g) represents palmitic acid. The fourth pure pixel/spectrum in Figure 4.11(j) represents stearic acid. After selecting four pure pixels/spectra, it seemed the image was dominated by noise. Furthermore, the selection of a fifth pure variable results in a fifth resolved spectrum with an unacceptable amount of negative peaks.

The resolved spectra, based on the pure pixel solution of the components of interest, palmitic acid and stearic acid, are shown in Figure 4.12, as are the spectra based on the pure variable solution. As mentioned above, the pure variables are the variables with the highest correlation with the pure spectrum-based resolved images. The differences between the pure pixel/spectrum solution [Figure 4.12(a) and (c)] and the pure variable solution [Figure 4.12(b) and (d)] are not as clear and easy to interpret as with the simulated dataset. However, for both components, it is clear that their contributions, as expressed by their M+H and M+H– H_2O ion intensities, where M is the molecular weight of the parent acid, does increase significantly. The intensities in the lower mass area are shared with other components, and their lower contribution in the pure variable solution is to be expected.

4.5   CONCLUSIONS

The application of SIMPLISMA to image data enabled the resolution of complex chemical images. It was possible to extract separate spectra for background spectra, which resulted in good quality spectra for the other components. The differences between a pure spectrum/pixel solution and a pure variable solution were shown with a simulated dataset and an actual dataset. The demonstrated combination of pure spectrum/pixel approach and pure variable approach has the following advantages:

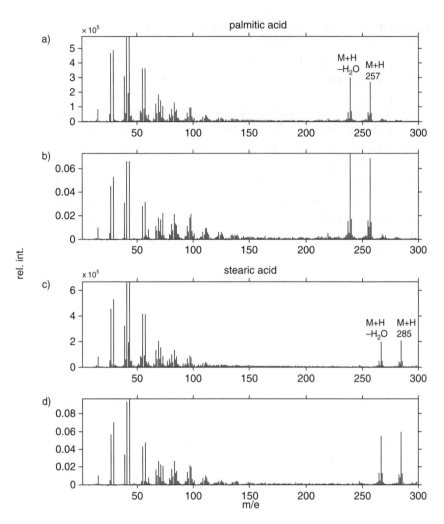

Figure 4.12 Spectra resulting from the pure pixel solution (a,c) and the derived pure variable solution (b,d)

(a) The images are easier to relate to than the complex and noisy mass spectra, which favors the selection of pure spectra/pixels.
(b) The pure variable assumption is realistic for mass spectral data and for a proper resolution a pure variable solution is required.

The approach demonstrated here combines both techniques in an optimal manner.

REFERENCES

Andrew, J. J and. Hancewicz, T. M (1998) *Rapid analysis of Raman image data using two-way multivariate curve resolution*, Appl. Spectrosc., **52**, 797–807.

Batonneau, Y., Laureyns, J., Merlin, J. C. and Bremard, C. (2001) *Self-modeling mixture analysis of Raman microspectrometric investigations of dust emitted by lead and zinc smelters*, Anal. Chim. Acta, **446**, 23–37.

Benninghoven, A., Rudenauer, F. G. and Werner, H. W. (1987) *Secondary Ion Mass Spectrometry, Basic Concepts, Instrumental Aspects, Applications and Trends*, ed. by P. J. Fleming *et al.*, John Wiley & Sons, Ltd, New York, p. 679.

Bhargava, R. and Fernandez, D. (2000) *Effect of focal plane array cold shield aperture size on Fourier transform infrared micro-imaging*, Appl. Spectrosc., **54**, 1743–1750.

Geladi, P. and Grahn, H. (1996) *Multivariate Image Analysis*, John Wiley & Sons, Ltd, Chichester.

Guilment, J., Markel, S. and Windig, W. (1994) *Infrared chemical micro-imaging assisted by interactive self-modeling multivariate analysis*, Appl. Spectrosc., **48**, 320–326.

Phalp, J. M., Payne, A. and Windig, W. (1995) *The resolution of mixtures using data from automated probe mass spectrometry*, Anal. Chim. Acta, **318**, 43–53.

Sasaki, K., Kawata, S. and Minami, S. (1989) *Component analysis of spatial and spectral patterns in multispectral images. II. Entropy minimization*, J. Opt. Soc. Am. A, **6**, 73–79.

Terhorst, M., Cramer, H.-G. and Niehuis, E. (1997) *Secondary Ion Mass Spectrometry, SIMS X*, ed. by A. Benninghoven *et al.*, John Wiley & Sons, Ltd, New York, p. 427.

Windig, W. (1997) *Spectral data files for self-modeling curve resolution with examples using the SIMPLISMA approach*, Chemom. Intell. Lab. Syst., **36**, 3–16.

Windig, W. and Markel, S. (1993) *Simple-to-use interactive self-modeling mixture analysis of FTIR microscopy data*, J. Mol. Struct., **292**, 161–170.

Windig, W., Antalek, B., Lippert, J. L., Batonneau, Y. and Bremard, C. (2002) *Combined use of conventional and second-derivative data in the SIMPLISMA self-modeling mixture analysis approach*, Anal. Chem., **74**, 1371–1379.

5

Multivariate Analysis of Spectral Images Composed of Count Data

Michael R. Keenan

5.1 INTRODUCTION

Spectral images are datasets for which a complete spectrum is available at each point in a spatial array. The goal of spectral image analysis, then, is to extract the chemical information from these typically large, high-dimensional datasets into a limited number of components that describe the spectral and spatial characteristics of the imaged sample. Frequently, the spectra are acquired by counting particles: photons, electrons, ions, etc. In these cases, a typical 1000-channel spectrum might contain only tens to hundreds of total counts. Thus, the individual spectra have low signal to noise ratios (SNRs) and the noise is highly nonuniform. Properly accounting for these spectral characteristics is essential for a successful, comprehensive analysis of the image. In this chapter, the shortcomings of standard factor analysis based methods as applied to count data will be illustrated, and several approaches that consider the statistical properties of this type of data during spectral image analysis will be described, compared and contrasted.

Several full-spectrum imaging techniques have been introduced in recent years that promise to provide rapid and comprehensive chemical

Techniques and Applications of Hyperspectral Image Analysis Edited by H. F. Grahn and P. Geladi
© 2007 John Wiley & Sons, Ltd

characterization of complex, heterogeneous samples. These new spectroscopic imaging systems enable the collection of a complete spectrum at each point in a 1-, 2- or 3-D spatial array. A typical spectral image is acquired by rastering a focused probe across the sample and observing the probe/sample interaction in a manner that is both spatially and spectrally resolved. Many imaging techniques familiar to the surface analysis and microanalysis communities form their respective spectra by counting particles such as photons, electrons or ions. Table 5.1 lists several common surface and near-surface analytical techniques that rely

Table 5.1 Some common spectroscopic techniques that compose their spectra by counting particles

Technique	Acronym	Probe	Detect
Energy dispersive X-ray spectroscopy	EDS	electrons	photons
Time of flight-secondary ion mass spectrometry	TOF-SIMS	ions	ions
Matrix-assisted laser desorption ionization SIMS	MALDI-SIMS	photons	ions
X-ray fluorescence	XRF	photons	photons
X-ray photoelectron spectroscopy	XPS	photons	electrons
Auger electron spectroscopy	AES	electrons	electrons
Proton(particle)-induced X-ray emission	PIXE	protons	photons

on particle counting, together with their specific probes and types of particles detected. Dealing with the special problems posed for spectral image analysis by the statistical properties of count data is the subject of this chapter.

It is not uncommon for spectral image datasets to contain tens of thousands of individual spectra, or more, with each spectrum being comprised of hundreds to thousands of spectral channels. Spectral images containing more than 10^9 individual data elements can be readily acquired with commercial instruments. One of the major remaining obstacles to widespread adoption of spectral imaging techniques for routine materials characterization is the task of reducing the vast quantities of raw spectral data to meaningful chemical information. In general, the goal of spectral image analysis is to provide an unbiased and comprehensive picture of a given sample. That is, we seek to discover all sources of chemical variation, from major phases to single pixel impurities, with limited foreknowledge of the sample's composition. It is also desirable that the data analysis be computationally tractable, requiring a reasonable amount of time on commonly available laboratory computers, and that the methods provide easily interpretable representations of the data. In the end, we are interested in solving particular chemical problems, not in discovering some abstract 'truth.' The multivariate image analysis methods discussed should be evaluated with that ultimate goal in mind, namely, by their ability to provide reliable chemical insights.

Factor analysis based techniques are essential tools for spectral image analysis. Methods such as principal component analysis (PCA) have proven effective for extracting the essential chemical information from high dimensional spectral image datasets into a limited number of components that describe the spectral characteristics and spatial distributions of the chemical species comprising the sample. Ideally, the number of components found by factor analysis will equal the number of independent sources of chemical variation in the sample. At root, factor analysis attempts to discriminate chemically induced signals from experimental noise. Standard factor analysis methods assume that the noise is uniform, that is, that the uncertainty in a given measurement is constant, independent of the magnitude of the signal. This assumption is violated in the case of count data where experimental variability is governed by Poisson statistics. Given that the true spectral intensity in a particular data element is μ, the probability p of observing d counts in that element is described by the Poisson probability distribution function:

$$p(d|\mu) = \frac{\mu^d}{d!}e^{-\mu} \tag{5.1}$$

It is noteworthy that μ is necessarily non-negative. Several relevant properties of the Poisson distribution have been summarized in Thompson (Thompson, 2001). Central to the present discussion is the fact that the mean and variance of the Poisson probability function are both equal to the single parameter μ. In other words, the estimated variance of a given measurement will be equal to the magnitude of the measurement itself. Additionally, we will make use of the fact that a sum of Poisson variables is also a Poisson variable, with mean and variance equal to the sum of the individual means. Differences of Poisson variables, on the other hand, are not Poisson since differences can be negative. This implies that care must be taken when performing data preprocessing steps, such as background subtraction, that will destroy the Poisson nature of the data (Thompson, 1999).

In the remainder of this chapter, the errors and difficulties that can arise from standard analysis when Poisson noise is a predominant feature of the data will be illustrated, and two basic approaches to accounting for nonuniform Poisson noise will be discussed. One approach explicitly considers the Poisson probability distribution while estimating the parameters of the data model using the maximum likelihood method. In the other approach, the data are transformed in such a way that the variance of the transformed data is approximately constant prior to standard factor analysis. Finally, the usefulness of factor based techniques as filtering operations to exclude noise prior to subsequent analysis will be explored.

5.2 EXAMPLE DATASETS AND SIMULATIONS

Two datasets that are representative of spectral imaging applications will be used to illustrate the multivariate analysis methods for count data. The first sample is described in Figure 5.1. The sample consists of series of six types of wires embedded in an epoxy matrix, with the wire alloys themselves being composed from a pallet of six different elements. The sample was imaged in a scanning electron microscope (SEM) with an attached energy dispersive X-ray spectrometer (EDS). Typical data acquisition conditions for this type of spectral image are described in Kotula *et al.* (Kotula *et al.*, 2003). Figure 5.1(a) shows a standard SEM image of the sample together with the composition key. The image dimensions are 128×128 pixels, and a complete 1024-channel spectrum was collected at each pixel. The mean spectrum, obtained by averaging over all of the pixels in the image, is shown in Figure 5.1(b). While the mean spectrum has good signal to noise, this is not the case for any individual spectrum. Figure 5.1(c) shows a typical single-pixel spectrum for the Cu/Mn/Ni wire. The discrete nature of the data is clearly evident, and the SNR is sufficiently low that the presence of Ni cannot be detected.

The EDS dataset is typical of spectroscopies for which a single spectral feature, an X-ray emission peak in this case, is comprised of several spectral channels. EDS spectra also exhibit chemically nonspecific backgrounds due to Bremsstrahlung radiation. The second example spectral image, a TOF-SIMS mass spectral dataset having unit mass resolution, is typical of low-background spectroscopies where a single spectral feature maps to a single spectral channel. This specific dataset is distributed by PHI (Physical Electronics USA, Chanhassen, MN, USA) as part of

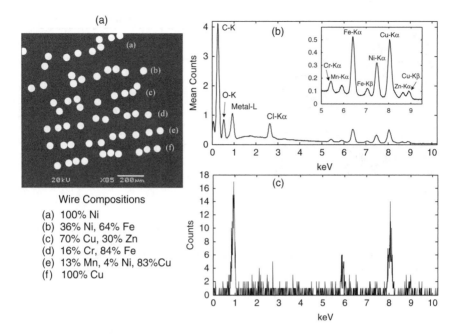

Wire Compositions
(a) 100% Ni
(b) 36% Ni, 64% Fe
(c) 70% Cu, 30% Zn
(d) 16% Cr, 84% Fe
(e) 13% Mn, 4% Ni, 83%Cu
(f) 100% Cu

Figure 5.1 (a) An SEM image of the wires sample consisting of metal wires embedded in an epoxy matrix, together with the composition key. (b) The mean EDS spectrum computed from the dataset. A 1024-channel spectrum was acquired each pixel in the 128-pixel × 128-pixel image. (c) A single-pixel spectrum from the Cu/Mn/Ni wire

the WinCadence software for their TRIFT TOF-SIMS instrument. The sample is a copper grid that has been plated on aluminum, and whose central portion has been sputtered with a gallium ion beam. The TOF-SIMS spectral image, which is 256 × 256 pixels in size, was compiled by binning individually detected ions in the mass range from 7 to 149 amu into mass channels having unit mass resolution. Each mass spectrum contains about 33 counts, on average. The total-ion image and mean mass spectrum corresponding to this spectral image are displayed in Figure 5.2.

The wires sample and the copper grid sample are sufficiently simple that the results of multivariate analysis can be anticipated. One would expect that the wires sample will require seven components to describe, together, the background and six wires (either by alloy or element). In fact, the sample had been sputtered over part of its area prior to the EDS analysis, and two components are required to describe the background, giving a total of eight components for the entire dataset. The copper grid sample might be expected to consist of four components: copper and aluminum, native surface and sputtered. The reality is somewhat

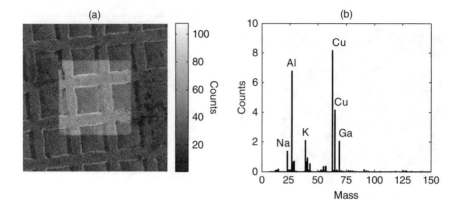

Figure 5.2 (a) The 256-pixel × 256-pixel total ion image from the copper-grid TOF-SIMS dataset. The image covers an area of 100 μm × 100 μm. (b) The mean mass spectrum having unit mass resolution (Keenan and Kotula, 2004a). Accounting for Poisson noise in the multivariate analysis for ToF-SIMS spectrum images, Michael R. Keenan & Paul G Kotula, Surface and Interface Analysis, 2004. © John Wiley & Sons Ltd. Reproduced with permission

more complex. Na, K and Ca associated with the plating process are found in the sputtered region, and two components are required to describe their spatial distributions. In addition, a six-pixel inclusion can be identified. Thus, seven components are required to fully describe the chemical information in this dataset.

In order to evaluate, in a quantitative manner, the ability of a data analysis method to discriminate real chemical information from noise, it is essential that the true composition of the sample be known. Since such knowledge is difficult to achieve with real data, simulated spectral images based on the two real sample datasets were constructed. For the wires sample, the real spectral image was segmented into wire and background pixels using standard image processing techniques. The true pure component spectra were then obtained by averaging the real spectra over the background and the respective wire pixels yielding seven total components. The relative abundances at each pixel were taken to be unity, that is, no pixels have a mixed composition. The true pure-component spectra and abundances for the copper-grid simulation were derived from the real dataset using the non-negative matrix factorization algorithm described later in this chapter. Pixels can have mixed composition in this case. For simplicity of exposition, the six-pixel inclusion was omitted from the simulation, resulting in a six-component model. Detecting the six-pixel inclusion out of the 65536 total pixels in the image, however, is a truly impressive demonstration of the power of multivariate statistical methods to solve the 'needle-in-a-haystack' problem. The interested reader is referred to Keenan and Kotula (Keenan and Kotula, 2004a) for a more comprehensive analysis of this sample. Finally, noise was added to the simulated spectral images using a Poisson random number generator such that the number of counts closely approximates that of the original data.

Matlab Version 7 (The Mathworks, Natick, MA, USA) was used to perform all of the calculations presented here. Several custom mex-files were employed to maximize memory efficiency. These were developed with the Intel C++ Compiler Version 9 (Intel, Santa Clara, CA, USA), and made use the Intel Math Kernel and Integrated Performance Primitives Libraries. A dual 2.2 GHz Xeon processor workstation with 2 GB of memory and running Windows 2000 was used for all of the timing comparisons.

5.3 COMPONENT ANALYSIS

Component analysis is based on the premise that the observed spectral image data are a realization of an underlying physical process. For the types of spectral data considered here, the data generation process is assumed to follow a linear additive model. That is, the expected spectral intensity for a mixture of spectrally active chemical species is simply the abundance-weighted sum of the spectral intensities characteristic of those same components in the pure state. Mathematically, linear additivity implies that an m-pixel × n-channel matrix of spectra data $\mathbf{D}$ can be approximated by the matrix product:

$$\mathbf{D} \cong \mathbf{A}\mathbf{S}^{\mathrm{T}} \tag{5.2}$$

where $\mathbf{A}$ is an m-pixel × p-component matrix describing the abundances of the pure components at each spatial location, and $\mathbf{S}$ is an n-channel × p-component matrix representation of the pure component spectra. Here, it will be assumed, without loss of generality, that $m \geq n$. The goal of component analysis is three-fold: to determine the minimum number of components p needed to fully describe the chemistry of the sample, and to estimate the unknown pure component abundance and spectral matrices, $\mathbf{A}$ and $\mathbf{S}$, respectively. Typically, $p \ll n$, and the factorization in Equation (5.2) accomplishes a large reduction in the dimensionality of the dataset. In the ideal case, p components will capture all of the chemically relevant sources of variation in the data, while the remaining n − p dimensions represent experimental noise, and can be eliminated from further consideration.

It is well known methods based on matrix factorization suffer from a 'rotational ambiguity.' Given any invertible p × p transformation matrix **R**, **D** can be equally well expressed as:

$$\mathbf{D} \cong (\mathbf{AR})\,(\mathbf{R}^{-1}\mathbf{S}^{T}) = \tilde{\mathbf{A}}\tilde{\mathbf{S}}^{T} \tag{5.3}$$

That is, an infinite number of factor pairs $\tilde{\mathbf{A}}$ and $\tilde{\mathbf{S}}$ will provide equally good fits to the data. The key to deriving relatively unique components is to select those solutions that satisfy additional optimization criteria. That is, the different methods attempt to exploit different characteristics of the data. In general, one expects that the extent to which these criteria or constraints reflect the physical reality of a given sample and spectroscopic technique will largely determine the ease with which the derived components can be interpreted.

Two basic approaches to component analysis will be discussed here in the context of count-based spectral images. One approach includes the basic techniques of traditional factor analysis, which often involve direct orthogonal factorization of a matrix. The second set of techniques that will be discussed derives from a maximum likelihood estimation approach.

5.4 ORTHOGONAL MATRIX FACTORIZATION

Factor analysis (FA) techniques have enjoyed a long and rich history in the analysis of multivariate data, and can be applied to the estimation problem (5.2). General references to FA in the context of chemical problems, or the natural sciences, in general, include Malinowski (Malinowski, 2002), and Reyment and Jöreskog (Reyment and Jöreskog, 1996). These methods routinely employ orthogonal matrix factorization techniques familiar to linear algebra, either alone or as a data preprocessing step. Many numerical algorithms for matrix factorization are presented in Golub and Van Loan (Golub and Van Loan, 1996).

Datasets are frequently pretreated in a variety of ways prior to analysis (Kramer, 1998). Common procedures include mean centering, and various types of scaling. In the present work, the data is not mean-centered, and, unless explicitly noted, the data are not scaled. Bro and Smilde (Bro and Smilde, 2003) have shown that component analysis of mean-centered data provides a fit that is equivalent to fitting the original data to the model (5.2) plus a common offset. For the types of spectra considered here, zero counts represents a true zero, so there is no common offset and equivalent fits are obtained. A detailed discussion about the effects of data pretreatment on the analysis of the copper grid sample can be found in Keenan and Kotula (Keenan and Kotula, 2004b).

Obviously, determining the appropriate number of components p to retain in a factor model is an essential part of the analysis. We will defer discussion of this important issue until later. The next several sections will assume that p is known.

5.4.1 PCA and Related Methods

Probably the most ubiquitous and familiar tool of factor analysis is principal component analysis (PCA), which has been thoroughly reviewed by Jollife (Jollife, 2002), and by Geladi and Grahn (Geladi and Grahn, 1996) for its application to multivariate image analysis. PCA describes the data in terms of the matrix factorization:

$$D \cong TP^T \tag{5.4}$$

As conventionally performed, T is an $m \times q$ matrix whose columns are mutually orthogonal and P is an $n \times q$ matrix having orthonormal columns. Furthermore, the components are constructed and ordered such that they serially maximize the variance in the data that each accounts for. In this formulation, the spectrum at any pixel can be represented by a linear combination of the spectral basis vectors in P. If $q = n$, PCA provides a mathematically exact representation of the data matrix D. More commonly, the component matrices are truncated to have p columns, which are assumed to contain the chemically significant signal, and the remaining $n - p$ columns are discarded as noise. Neither the orthogonality constraint nor the enforced variance-maximization property has any basis in physical reality, so the factors obtained via PCA are typically abstract and not easily interpreted. A p-component PCA model does have the useful property, however, that it is the best possible rank p approximation to the dataset, in a least squares sense.

There are close connections between PCA, the singular value decomposition (SVD) of D and eigenanalysis of the spectral cross-product matrix $\mathbf{D}^T\mathbf{D}$. SVD decomposes the data matrix $\mathbf{D}$ into the product of three matrices:

$$\mathbf{D} = \mathbf{U}\mathbf{\Sigma}\mathbf{V}^T \qquad (5.5)$$

$\mathbf{U}$ and $\mathbf{V}$ are $m \times m$ and $n \times n$ orthogonal matrices containing the left and right singular vectors, respectively, and $\mathbf{\Sigma}$ is an $m \times n$ diagonal matrix with the singular values placed along the diagonal. The singular values and corresponding singular vectors are ordered by significance. Comparison with Equation (5.4) shows that the p-component PCA factorization can be computed simply as:

$$\mathbf{P} = \mathbf{V}_p \quad \text{and} \quad \mathbf{T} = (\mathbf{U}\mathbf{\Sigma})_p \qquad (5.6)$$

where the subscript p indicates that only the first p columns of the respective matrices are being retained. Returning to Equation (5.5), the spectral cross-product matrix can be expressed as:

$$\mathbf{D}^T\mathbf{D} = \left(\mathbf{U}\mathbf{\Sigma}\mathbf{V}^T\right)^T \left(\mathbf{U}\mathbf{\Sigma}\mathbf{V}^T\right) = \mathbf{V}\mathbf{\Sigma}^2\mathbf{V}^T \qquad (5.7)$$

which is the eigen decomposition of $\mathbf{D}^T\mathbf{D}$. Thus, the right singular vectors of $\mathbf{D}$ can be obtained as the eigenvectors of $\mathbf{D}^T\mathbf{D}$, and the eigenvalues of $\mathbf{D}^T\mathbf{D}$ are the squares of the corresponding singular values. In this formulation, the p-component PCA decomposition in Equation (5.6) can be written as:

$$\mathbf{P} = \mathbf{V}_p \quad \text{and} \quad \mathbf{T} = (\mathbf{U}\mathbf{\Sigma})_p = \mathbf{D}\mathbf{V}_p = \mathbf{D}\mathbf{P} \qquad (5.8)$$

The eigenanalysis approach to computing PCA has substantial appeal in spectral imaging applications where, frequently, the number of pixels is much greater than the number of spectral channels. In this case the cross-product matrix is much smaller than the original data matrix and the computations proceed more quickly. This approach is akin to the kernel PCA method discussed by Wu et al. (Wu et al., 1997), which contains a much fuller discussion of these points.

Given the PCA factorization in Equation (5.4), a reconstruction of the dataset $\hat{\mathbf{D}}$ can be computed as:

$$\hat{\mathbf{D}} = \mathbf{T}\mathbf{P}^T \qquad (5.9)$$

This estimate minimizes the reconstruction error $\left\|\mathbf{D} - \hat{\mathbf{D}}\right\|_F^2$, where $\|\cdot\|_F^2$ is the squared Frobenius norm, which is the sum of the squared differences over the entire dataset. While PCA tries to account for as much variance as possible in the fewest number of factors, it cannot, in the general case, distinguish variance due to the underlying spectral process from that due to experimental noise. This is amply demonstrated for the case of Poisson noise using the wires example. Figure 5.3 compares the mean background (epoxy) spectrum in the region of the carbon emission

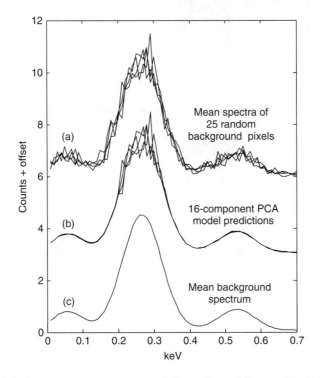

Figure 5.3 (a) Mean spectra computed from four different 25-pixel random samples taken from the epoxy background of the wires sample. (b) Mean spectra for the same random samples as reconstructed from a 16-component PCA model. (c) Mean spectrum over all of the background pixels

peak, with the means, and mean 16-component PCA reconstructions, for four different 25-pixel random samples. It is obvious that while PCA has the effect of denoising low count channels, it quantitatively accounts for the noise in the most intense spectral channels. This PCA model, on the other hand, has difficulty fully describing the chemistry

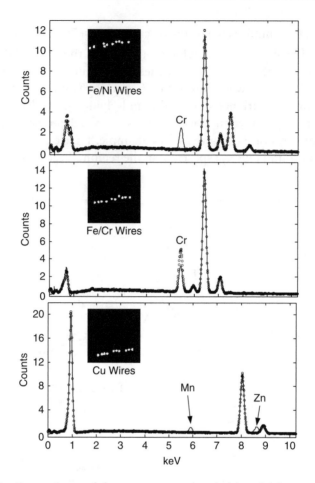

Figure 5.4 Comparisons of the mean spectra (symbols) and 16-component PCA model predictions (solid line) for three of the wires in the wires dataset

of the sample. Figure 5.4 compares the mean spectra for three different wires with the PCA model predictions. PCA finds substantial Cr in wires containing only Fe and Ni, while underestimating the amount of Cr in the Fe/Cr wires. PCA also finds Mn and Zn in wires that are pure Cu. The conclusion to be drawn from this example is that by not considering the nonuniform nature of Poisson, PCA finds it more profitable, in a least squares sense, to fit noise associated with high intensity signals than to extract small, yet chemically meaningful, features.

Further evidence for the relative lack of sensitivity of PCA to minor spectral features in the presence of nonuniform noise is illustrated in

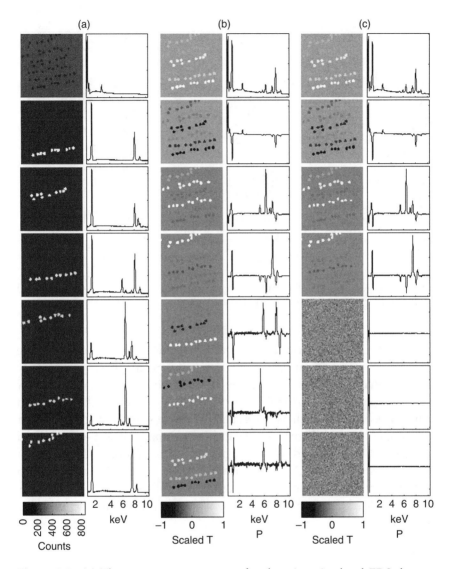

Figure 5.5 (a) The true pure components for the wires simulated EDS dataset. (b) PCs of the noise-free wires simulation, ordered by significance from top to bottom. (c) The first seven PCs computed by standard PCA of the wires simulation

Figure 5.5. Figure 5.5(a) shows the true pure components for the wires simulation, and Figure 5.5(b) displays the seven principal components (PCs) computed from the noise-free dataset. In the presence of Poisson noise, the first four PCs obtained by standard PCA are essentially identical, as shown in Figure 5.5(c), to the true ones. These components

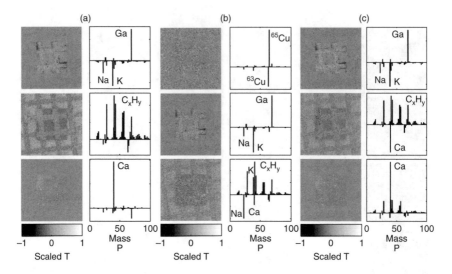

Figure 5.6 (a) True PCs 4–6 for the copper grid simulation. (b) PCs 4–6 estimated by standard PCA. (c) PCs 4–6 estimated by WPCA

describe the major chemical constituents of the sample. The next three PCs, however, represent only noise. While it is relatively easy to identify noise-related PCs here because the spectral noise components do not resemble physically reasonable spectra, that may not be the case with other types of data, such as mass spectra, where spectral features consist of a single channel. Figure 5.6(a) and (b) compares PCs 4–6 computed from the noise-free and noisy copper grid simulation, respectively. In this case, standard PCA finds variation in the copper isotope ratio to be the fourth most important source of spectral variation when noise is present. Clearly, care must be taken when trying to interpret such PC models.

5.4.2 PCA of Arbitrary Factor Models

This chapter describes and compares several different approaches to factoring the spectral data matrix **D**. Quantitative comparison of these various factor models is facilitated by putting them on a common basis. Since the PCA factorization is unique (within the signs of the components), it provides such a convenient basis. Given that **D** is represented by an arbitrary factorization such as Equation (5.2), the PCA model can be computed very efficiently from the component matrices without having to reconstruct the data matrix. Equations (5.2), (5.7) and (5.8) can be combined to yield:

$$\mathbf{D}^T\mathbf{D} = \mathbf{S}\left(\mathbf{A}^T\mathbf{A}\right)\mathbf{S}^T = \mathbf{P}\mathbf{\Sigma}^2\mathbf{P}^T \qquad (5.10)$$

For a p-component model, $\mathbf{\Sigma}^2$ will contain p non-zero eigenvalues. The computations can be greatly facilitated by first computing the Cholesky factorization of $\mathbf{A}^T\mathbf{A}$:

$$\mathbf{A}^T\mathbf{A} = \mathbf{G}^T\mathbf{G} \qquad (5.11)$$

where the Cholesky factor $\mathbf{G}$ is a $p \times p$ upper triangular matrix. Since, in general, the non-zero eigenvalues of a matrix $\mathbf{XX}^T$ equal the eigenvalues of $\mathbf{X}^T\mathbf{X}$, this formulation allows the eigenvalues of $\mathbf{D}^T\mathbf{D}$ to be computed from the much smaller eigen-problem:

$$\mathbf{G}\left(\mathbf{S}^T\mathbf{S}\right)\mathbf{G}^T = \tilde{\mathbf{P}}\tilde{\mathbf{\Sigma}}\tilde{\mathbf{P}}^T \qquad (5.12)$$

Here, $\tilde{\mathbf{\Sigma}}^2$ is a $p \times p$ diagonal matrix containing the p non-zero eigenvalues of $\mathbf{\Sigma}^2$ along the diagonal. $\tilde{\mathbf{P}}$ contains the right singular vectors of $\mathbf{SG}^T$. The left singular vectors can be recovered as $\mathbf{S}\left(\mathbf{G}^T\tilde{\mathbf{P}}\tilde{\mathbf{\Sigma}}^{-1}\right)$. Of course, the left singular vectors of $\mathbf{SG}^T$ are the right singular vectors of $\mathbf{GS}^T$, which, after substituting Equation (5.11) into Equation (5.10), are seen to be precisely the eigenvectors $\mathbf{P}$ of the data cross-product matrix. The matrix $\mathbf{T}$ can then be recovered in a manner analogous to Equation (5.8). To summarize:

$$\mathbf{P} = \mathbf{S}\left(\mathbf{G}^T\tilde{\mathbf{P}}\tilde{\mathbf{\Sigma}}^{-1}\right)$$
$$\mathbf{T} = \mathbf{A}\left(\mathbf{S}^T\mathbf{P}\right) \qquad (5.13)$$

A simple Matlab function that implements this factor PCA algorithm (fPCA) is provided in Figure 5.7. The computational performance in terms of both time and memory use are greatly improved by this algorithm. For the wires simulation, for instance, PCA of a 7-component factored representation of the data matrix required 0.012 s. Reconstructing the dataset from the factors and using either SVD (150 s) or the eigenanalysis approach (15 s), required orders of magnitude more computation time.

```
function [T,P,Evals]=fPCA(A,S)
% PCA of a factored data set D=AS'

G=chol(A'*A);
[U,Evals]=eig(G*(S'*S)*G');
[Evals,idx]=sort(diag(Evals),1,'descend');
U=U(:,idx);
P=S*(G'*U*diag(1./sqrt(Evals)));
T=A*(S'*P);
```

Figure 5.7 Matlab function to compute the PCs of a matrix that is expressed in terms of an arbitrary factor model: $D = AS^T$. T has mutually orthogonal columns and the columns of P are orthonormal. Evals contains the non-zero eigenvalues of D^TD

5.4.3 Maximum Likelihood PCA (MLPCA)

Wentzell *et al.* (Wentzell *et al.*, 1997) have described a method for PCA that attempts to account, in a general way, for nonuniform measurement errors. For the case of uncorrelated errors, this method, which they have termed 'maximum likelihood principal component analysis', minimizes a weighted sum of squared residuals:

$$\min_{\hat{D}|p} \left\| W * \left(D - \hat{D} \right) \right\|_F^2 \qquad (5.14)$$

where the * operator represents Hadamard, or element-by-element, multiplication. The elements of the weighting matrix W are estimates of inverse standard deviations of the corresponding data elements:

$$W_{ij} = \frac{1}{\sigma_{ij}} \qquad (5.15)$$

Each data element is allowed to have its own independent uncertainty, and a major challenge faced by practitioners of MLPCA is estimating W from a priori information. Given W, MLPCA performs a matrix factorization similar to SVD:

$$\hat{D} = \hat{U}\hat{S}\hat{V}^T \qquad (5.16)$$

where $\hat{U}$ and $\hat{V}$ are orthogonal matrices, $\hat{S}$ is a diagonal matrix, and the carets represent maximum likelihood estimates. As with standard

SVD, a dimensional reduction is realized by truncating these matrices to contain only the most significant p components. Wentzell *et al.* (Wentzell *et al.*, 1997) showed that the maximum likelihood estimate for the jth column of D could be expressed as:

$$\hat{D}_j = \hat{U}\left(\hat{U}^T\Psi_j^{-1}\hat{U}\right)^{-1}\hat{U}^T\Psi_j^{-1}D_j \tag{5.17}$$

Again assuming uncorrelated error, Ψ_j is a diagonal matrix that is proportional to the error covariance matrix corresponding to the jth column of D. In terms of W, $\Psi_j^{-1} = \text{diag}\left(W_j * W_j\right)$. Similarly, the maximum likelihood estimate for the ith row of D, or, equivalently, the ith column of D^T, can be written:

$$\left(\hat{D}^T\right)_i = \hat{V}\left(\hat{V}^T\Phi_i^{-1}\hat{V}\right)^{-1}\hat{V}^T\Phi_i^{-1}\left(D^T\right)_i \tag{5.18}$$

where $\Phi_i^{-1} = \text{diag}\left(\left(W^T\right)_i * \left(W^T\right)_i\right)$ is proportional to the error covariance matrix for the ith row of D. After making an initial estimate of $\hat{D}$ by SVD of D, MLPCA proceeds via an alternating regression algorithm. The maximum likelihood estimate of the data is computed, alternately, by Equations (5.17) and (5.18), and the process is repeated until the two estimates of $\hat{D}$ coincide to a given level of precision. After each estimate is made, $\hat{U}$, $\hat{V}$ and $\hat{S}$ are updated through the SVD of Equation (5.16).

In principle, the task of estimating W is relatively simple in the case of count data since the estimated variance of a data element is equal to the data element itself. Keenan (Keenan, 2005) showed, however, that in the case of low-count-rate, sparse datasets, this naïve approach to estimating the errors caused MLPCA to perform poorly. The best performance was achieved when a reduced-rank model for W was used. In particular, the rank of the error estimate should be $\leq$ p. MLPCA is also computationally expensive, requiring two SVDs per iteration. For the copper grid example, as discussed by Keenan (Keenan, 2005), the benefits provided by MLPCA did not outweigh its computational costs.

5.4.4 Weighted PCA (WPCA)

While the full MLPCA method is difficult to use effectively, one special case is of great importance. When the weighting matrix W in Equation (5.14) is rank-1, the Hadamard product can be rewritten as a combination of row scaling and column scaling with diagonal matrices.

Any rank-1 matrix can be expressed as the outer product of two vectors. Letting $\mathbf{W} = \mathbf{gh}^T$, $\mathbf{G} = \text{diag}(\mathbf{g})$ and $\mathbf{H} = \text{diag}(\mathbf{h})$, the minimization problem (5.14) becomes:

$$\min_{\hat{D}|p} \left\| \mathbf{G}\left(\mathbf{D} - \hat{\mathbf{D}}\right)\mathbf{H} \right\|_F^2 = \min_{\hat{D}|p} \left\| \mathbf{\underset{\sim}{D}} - \hat{\mathbf{\underset{\sim}{D}}} \right\|_F^2 \tag{5.19}$$

Over all factor models having p components, Equation (5.19) is minimized by the ordinary PCA factorization of $\mathbf{\underset{\sim}{D}}$.

$$\hat{\mathbf{\underset{\sim}{D}}} = \mathbf{\underset{\sim}{T}}\mathbf{\underset{\sim}{P}}^T \tag{5.20}$$

The PCs of the unscaled data matrix can then be recovered by inverse transformation:

$$\hat{\mathbf{D}} = \mathbf{G}^{-1}\hat{\mathbf{\underset{\sim}{D}}}\mathbf{H}^{-1} = \left(\mathbf{G}^{-1}\mathbf{\underset{\sim}{T}}\right)\left(\mathbf{H}^{-1}\mathbf{\underset{\sim}{P}}\right)^T = \mathbf{TP}^T \tag{5.21}$$

The components are no longer orthogonal after inverse scaling. They can, however, be made to satisfy the PCA orthogonality and variance maximization constraints by applying the fPCA algorithm of Section 5.4.2.

There are a number of ways to construct the scaling matrices $\mathbf{G}$ and $\mathbf{H}$. One obvious choice is to base the scaling on the error covariance matrices for the rows and columns of $\mathbf{D}$. In the development of the MLPCA method, each row and each column of $\mathbf{D}$ was allowed to have a corresponding covariance matrix. The assumption that $\mathbf{W}$ is rank-1 is equivalent to assuming that a single covariance matrix Φ can describe the variance structure for all columns of $\mathbf{D}$, and a single covariance matrix Ψ can describe the variance for all rows of $\mathbf{D}$. With this choice, the scaling matrices become:

$$\mathbf{G} = \Phi^{-\frac{1}{2}} \quad \text{and} \quad \mathbf{H} = \Psi^{-\frac{1}{2}} \tag{5.22}$$

This is a particularly convenient choice for count data since the diagonal error covariance matrix can be easily estimated from the data. Recalling that the sum of Poisson variables is also Poisson with mean and variance equal to the sum, the expected total variance of a column of $\mathbf{D}$ is simply the column sum. Thus, Ψ can be estimated as:

$$\Psi = \frac{1}{m}\text{diag}\left(\mathbf{1}_m^T\mathbf{D}\right) \tag{5.23}$$

Here, 1_m is an m-vector of ones, and Ψ is simply a diagonal matrix with the mean spectrum placed along the diagonal. Likewise, for each row, the total variance is estimated by the row sum and Φ is computed as:

$$\Phi = \frac{1}{n}\text{diag}\,(D1_n) \qquad (5.24)$$

This is a diagonal matrix with the properly unfolded mean image placed along the diagonal. The foregoing development parallels that of Cochran and Horne (Cochran and Horne, 1977), and an alternate derivation of Equations (5.22)–(5.24) was provided by Keenan and Kotula (Keenan and Kotula, 2004a). In the latter reference, the covariance matrix estimates are shown to be maximum likelihood estimates given the rank-1 constraint on W.

When applied to the simulated wires dataset, WPCA yields spatial and spectral components that are the same, within noise, as the true components shown in Figure 5.5. Some differences between the true components and the WPCA-estimated components are observed for the simulated copper grid dataset. These are illustrated in Figure 5.6(c). In particular, components 5 and 6 are not completely unmixed. These differences are not great enough to seriously hamper the interpretation of this dataset, however.

5.4.5 Principal Factor Analysis (PFA)

Another approach to accounting for Poisson noise is inspired by the idea underlying the PFA method briefly outlined in Reyment and Jöreskog (Reyment and Jöreskog, 1996). By explicitly incorporating the error term in the factor model, Equation (5.2) can be rewritten:

$$D = AS^T + E \qquad (5.25)$$

Assuming the error is uncorrelated and has zero mean, the expected value of the data cross-product matrix is:

$$E\{D^TD\} = S(A^TA)S^T + E\{E^TE\} = S(A^TA)S^T + m\Psi \qquad (5.26)$$

where $E\{\cdot\}$ is the expectation operator and Ψ is the diagonal error covariance matrix given by Equation (5.23). For noisy data, $m\Psi$ may not be small with respect to the diagonal of D^TD, and a better representation of the model can be obtained by analyzing a diagonally modified data cross-product matrix:

$$\mathbf{D^T D} - m\boldsymbol{\Psi} = \mathbf{S}\left(\mathbf{A^T A}\right)\mathbf{S^T} \tag{5.27}$$

Analogous to the discussion surrounding Equation (5.10), the eigenvectors computed from the modified cross-product matrix provide an estimate of the orthogonal spectral basis $\mathbf{P}$ for the original linear model. The spatial components can then be estimated by standard weighted least squares using the original data:

$$\mathbf{T^T} = \left(\mathbf{P^T \Psi^{-1} P}\right)^{-1}\mathbf{P^T \Psi^{-1} D^T} \tag{5.28}$$

At this point, the spatial components in $\mathbf{T}$ are not, in general, orthogonal; the fPCA method of Section 5.4.2 can be used, however, to transform $\mathbf{T}$ and $\mathbf{P}$ to a representation that satisfies the constraints of PCA. Applying the PFA approach to the wires simulated dataset yielded components that are visually indistinguishable, within the noise, from the true PC representation shown in Figure 5.6. PFA performed similarly to WPCA on the copper grid simulation, although details of the mixing of components 5 and 6 differed.

An alternative estimation method can be derived by pre- and post-multiplying each side of Equation (5.27) by $(m\boldsymbol{\Psi})^{-1/2}$ and performing an eigenanalysis of $m^{-1}\boldsymbol{\Psi}^{-1/2}\mathbf{D^T D}\boldsymbol{\Psi}^{-1/2} - \mathbf{I}$. The spectral components are then recovered from the eigenvectors by applying an inverse column-scaling with $(m\boldsymbol{\Psi})^{1/2}$. The spatial factors can then be estimated by weighted least squares, as before. In this case, neither the spectral or spatial components will be orthogonal but, again, PCA of this factor model can be computed using the method of Section 5.4.2. This latter estimation method possesses scale-invariance properties not shared by the method employing the unscaled, modified cross-product matrix. These are not particularly germane for count data, however. Both methods achieve similar results, and experience has shown that while one method might outperform the other in any particular case, neither method can be deemed superior in all cases.

5.4.6   Selecting the Number of Components

Up to this point, we have examined several approaches to estimating reduced p-dimensional component, or factor, models for high dimensional spectral image data. The remaining question to be answered is how to choose the number of components that should be retained in these models. Nor have we addressed the fundamental reasons that standard PCA performs erratically or poorly on spectral data arising from counting experiments. Considerable insight into both of these questions can be gained by inspecting the eigenvalues (or singular values) obtained as part of a PCA.

The eigenvalues of the data cross-product matrix describe the distribution of spectral variance among the PCs. The largest eigenvalue is the amount of variance accounted for by the first PC, the second largest eigenvalue, that for the second PC, and so on. For noncentered data, the sum of the eigenvalues includes all of the spectral variance arising from the underlying physical process together with that from random measurement error. For the case of count data, the total random error variance can be estimated simply as the sum over the entire dataset. This suggests a metric for the overall signal to noise ratio (SNR) of Poisson spectral images:

$$\text{SNR} = \frac{\text{sum of eigenvalues of } \mathbf{D}^\text{T}\mathbf{D}}{\text{sum of all data elements in } \mathbf{D}} \quad (5.29)$$

By this metric, the wires and copper grid datasets have SNR = 3.4 and 8.5, respectively. A greater level of understanding can be achieved, however, by studying the distribution of the eigenvalues.

When the spectral data are contaminated with uniform gaussian noise, Malinowski (Malinowski, 1987, 1988) has shown that the eigenvalues of $\mathbf{D}^\text{T}\mathbf{D}$ associated with noise are statistically equal after accounting for the degrees of freedom inherent in their calculation. For components representing systematic spectral variation, the total variance is the sum of the systematic variation plus colinear noise. This implies that a plot of Malinowski's reduced eigenvalues should exhibit a break at the point where the eigenvalues begin accounting for real systematic variation in addition to noise. Furthermore, all of the chemical information should be embedded in the first p eigenvectors. Unfortunately, these arguments do not hold for nonuniform noise, as was amply demonstrated in Figures 5.3 and 5.4 for a PCA model of the wires dataset.

The basic premise that the least significant eigenvalues account for noise, while the more significant ones account for noise plus the spectral signals of interest is still valid, however. Recalling Equation (5.26), the data cross-product matrix sums the cross-product of the linear additive model together with $m\mathbf{\Psi}$, a multiple of the error covariance matrix. In the following discussion, λ_i will be the ith largest eigenvalue of $\mathbf{D}^T\mathbf{D}$, λ_j^0 will be the jth largest eigenvalue of $\mathbf{S}(\mathbf{A}^T\mathbf{A})\mathbf{S}^T$ with $j \leq p$, and λ_k^N will be the kth largest eigenvalue of $m\Psi$. For eigenvectors describing real systematic variation with high SNR, noise is a small perturbation and it would be expected that $\lambda_i \approx \lambda_i^0 >> \lambda_i^N$. At the other extreme, eigenvectors describe only noise and $\lambda_i \approx \lambda_i^N$. When $\lambda_i^0 \lesssim \lambda_i^N$ eigenvectors describing the spectral model become dispersed among, and commingled with those describing noise. This effect is illustrated in Figure 5.8 for the simulated wires dataset. The four most significant eigenvalues describe major chemical components and standard PCA does an excellent job of estimating the true spectral components as shown in Figure 5.5. The variances associated with the fifth and higher components, however, fall below the apparent noise level. Since standard PCA is only concerned with accounting for variance in the most parsimonious way irrespective of its origin, eigenvectors associated with noise begin to dominate

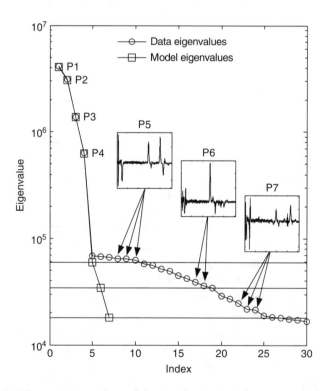

Figure 5.8 The true eigenvalues of the simulated wires data cross-product matrix are compared with the eigenvalues obtained from PCA of the noisy data. The dispersal of the minor chemically significant PCs among those describing noise is also indicated

the results. It is not until $\lambda_j^0 \approx \lambda_k^N$ for some $k > j$ that the jth model component becomes evident in the eigenvectors. In the wires simulations, three clusters of eigenvectors are found that describe the spectral characteristics of the three least significant chemical components. These clusters occur in order of decreasing variance, and each eigenvector in a given cluster describes the same systematic variation but different aspects of the noise. A comparison with Figure 5.5(b) shows, in fact, that these clusters of eigenvectors are providing good representations of the true components. Unfortunately, at least 12 factors are required to fully explain the chemical information in a dataset known to contain 7 components a priori.

The foregoing discussion pointed out that when noise is nonuniform, the eigenvalues associated with noise provide a floor below which the systematic spectral information is teased out only with difficulty. Interestingly, for count data the eigenvalues associated with noise can be easily estimated since the estimated error covariance matrix can be obtained, according to Equation (5.23), in terms of the mean spectrum. Because $m\Psi$ is a diagonal matrix, the diagonal elements themselves are the eigenvalues, and the sorted eigenvalues of the noise can be estimated simply by sorting the sum of all of the spectra in the dataset. The first 50 elements of the sorted spectral sum are compared with the corresponding eigenvalues of D^TD for the wires dataset, in Figure 5.9. Unlike the simulation, the first five PCs are estimated correctly using standard PCA. Eigenvectors representing the remaining three chemically relevant components are distributed throughout several less significant PCs as shown, and recognizable spectral features can be found out as far as PC 42.

Two strategies come to mind to make eigenanalysis more sensitive to minor, yet real, spectral features. Either noise can be excluded from the analysis, thus lowering the noise floor, or the data can be transformed so the noise appears more uniform. In the latter case, the assumptions of Malinowski's analysis are more closely approximated, and the systematic variation will tend to be concentrated in the most significant eigenvectors. Excluding noise was the basis for the PFA factorization method described in Section 5.4.5. The eigenvalues of $D^TD - m\Psi$ are compared, in Figure 5.9, with the eigenvalues of D^TD for the wires dataset. The noise level is substantially reduced, and the 8 eigenvalues associated with real spectral features are distinctly present. The approach taken by the WPCA algorithm of Section 5.4.4 is to make the variance appear more uniform by weighting the data. The eigenvalues in the weighted space, when displayed on a semilog plot, also show a sharp break at

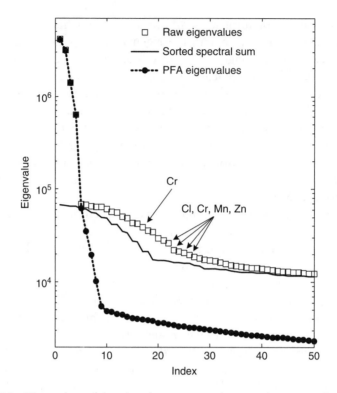

Figure 5.9 Eigenvalues of the wires data cross-product matrix computed with PCA
are compared with the corresponding eigenvalues computed with the PFA algorithm,
and eigenvalues of the error covariance matrix estimated as the sorted spectral sum

the ninth component, thus, the presence of eight real components is
easily deduced. As noted before, PFA and WPCA perform comparably
on the simulation of this dataset, and derive spectral and spatial components that are indistinguishable from the true components within the
noise. A similar eigenvalue plot for the copper grid example is shown
in Figure 5.10. Only three eigenvalues of $D^T D$ clearly rise above the
noise floor, which is consistent with standard PCA finding noise in the
copper isotope ratio to be the fourth most important source of spectral
variation. The PFA method is, again, much more sensitive, allowing
seven components to be unambiguously identified. The WPCA method
does not perform as well in this case. The eigenvalue plot suggests that
only five non-noise components are present, although an examination
of the eigenvectors hints that the sixth component does contain real
information about the 6-pixel inclusion.

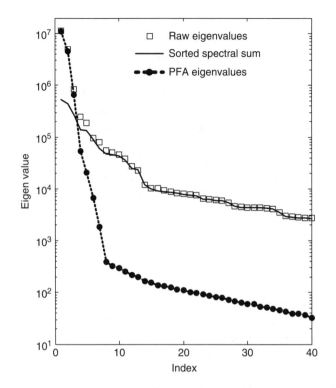

Figure 5.10 Eigenvalues of the copper grid data cross-product matrix computed with PCA are compared with the corresponding eigenvalues computed with the PFA algorithm, and eigenvalues of the error covariance matrix estimated as the sorted spectral sum

5.5 MAXIMUM LIKELIHOOD BASED APPROACHES

The methods surveyed in the previous section are largely concerned with minimal subspace approximation. That is, a low dimensional representation of the data is sought that extracts all of the chemically relevant information into as small a number of components as possible. Typically, the components, or factors, are forced to be orthogonal and, as a result, they are neither physically plausible nor easily interpreted. Standard matrix factorization techniques, such as SVD, are the tools of choice for accomplishing the dimensional reduction. Maximum likelihood estimation, on the other hand, explicitly assumes that the measured spectral data are samples from an underlying probability distribution, and seeks to find that set of model parameters that most likely gave rise to the observed data in a probabilistic sense. Models are also made to comply with the restrictions imposed by particular probability distributions. Count data and the Poisson parameter μ from Equation (5.1) are fundamentally non-negative. Thus, non-negativity of the components is imposed as a condition for model acceptability. Of course, as before, eigenanalysis will play a key role in estimating the proper number of components, and the remaining discussion will assume that the number of components p has already been determined. More information about maximum likelihood estimation can be found, for example, in Pawitan (Pawitan, 2001), and maximum likelihood estimation of Poisson models is discussed in Cameron and Trivedi (Cameron and Trivedi, 1998).

5.5.1 Poisson Non-negative Matrix Factorization (PNNMF)

The following development will parallel that of Sajda *et al.* (Sajda *et al.*, 2003), which extends the non-negative matrix factorization (NNMF) algorithm of Lee and Seung (Lee and Seung, 1999) to Poisson data. Referring back to the Poisson probability model of Equation (5.1) and the linear additive model in Equation (5.2), given component matrices $\mathbf{A}$ and $\mathbf{S}$, the probability of observing the measured number of counts in data element D_{ij} is:

$$p\left(D_{ij}|\mathbf{A},\mathbf{S}\right) = \frac{\left(\mathbf{AS}^{T}\right)_{ij}^{D_{ij}}}{D_{ij}!} e^{-\left(\mathbf{AS}^{T}\right)_{ij}} \qquad (5.30)$$

Assuming that all of the data elements are independent, the probability of the observed dataset is simply the product of all of the individual probabilities. Generally, it is more convenient to work with the log of the probability, which is maximized for the same set of parameters as the probability itself. For the Poisson model of Equation (5.30), this so-called log-likelihood function is given by:

$$\log p\left(\mathbf{D}|\mathbf{A},\mathbf{S}\right) = \sum_{i=1}^{m}\sum_{j=1}^{n}\left[D_{ij}\log\left(\mathbf{AS}^{T}\right)_{ij} - \left(\mathbf{AS}^{T}\right)_{ij} - \log\left(D_{ij}!\right)\right] \qquad (5.31)$$

The goal of maximum likelihood estimation, then, is to find that set of parameters $\mathbf{A}$ and $\mathbf{S}$ that maximize the log-likelihood. There are a great many ways to attempt solving this problem. The approach of Sajda *et al.*

(Sajda *et al.*, 2003) is essentially a gradient descent method. By making a judicious choice of step size, the components can be estimated in an iterative fashion using multiplicative update rules:

$$A \leftarrow A * \left[\left(\frac{D}{AS^T} \right) S \, diag \left(1_n^T S \right)^{-1} \right] \quad (5.32)$$

$$S \leftarrow S * \left[\left(\frac{D}{AS^T} \right)^T A \, diag \left(1_m^T A \right)^{-1} \right] \quad (5.33)$$

Here, the quotients are computed element-wise. Given initial estimates of the component matrices A and S, Equations (5.32) and (5.33) are used to alternately update these parameters until an acceptable level of convergence is achieved. By using multiplicative updates, A and S are guaranteed to remain non-negative as long as the initial estimates are non-negative, as well. Figure 5.11 shows the wire-related components of the wires simulation obtained when the PNNMF algorithm was initialized with the

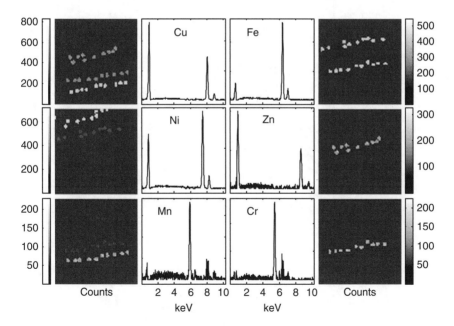

Figure 5.11 Wire-related abundance and spectral components computed for the wires simulation using the PNNMF algorithm. The true components of Figure 5.5(a) were used to initialize the algorithm

true pure components of Figure 5.5(a). Interestingly, the method changed the representation of the data from one in which the wires are separated by alloy to one where the components are broken out by element.

Proof that the NNMF updating scheme converges has been given by Lee and Seung (Lee and Seung, 2001). Unfortunately, only convergence to a local maximum can be expected, thus, the component estimates tend to be very sensitive to the initial conditions. Figure 5.12 illustrates the convergence behavior of this algorithm for the copper grid simulation. A and S were initialized with either the known true pure components or with uniformly distributed random numbers. Convergence to multiple different end points is clear, and the derived components varied substantially in quality. Only the start from the known pure components and a single start from random numbers gave accurate representations of all six components in the simulation. In practice, good results can be

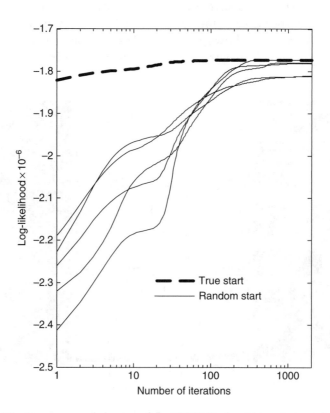

Figure 5.12 Convergence behavior of the PNNMF algorithm when applied to the simulated copper grid dataset. Algorithm initialization with the true components, as well as, several random starts, are included

achieved by using initial component estimates derived from an orthogonal factorization such as PCA.

5.5.2 Iteratively Weighted Least Squares (IWLS)

As noted by Cameron and Trivedi (Cameron and Trivedi, 1998), the maximum likelihood estimate for a Poisson model can be computed using an IWLS procedure. The goal is to find abundance and spectral components $\mathbf{A}$ and $\mathbf{S}$ that minimize:

$$\left\| \frac{1}{\left(\mathbf{AS}^T\right)^{1/2}} * \left(\mathbf{D} - \mathbf{AS}^T\right) \right\|_F^2 \quad \text{with } \mathbf{A}, \mathbf{S} \geq 0, \ \mathbf{AS}^T > 0 \qquad (5.34)$$

Equation (5.34) is very similar to the cost function (5.14) employed by Wentzell's MLPCA method except that the Poisson-specific error model replaces the general weight matrix $\mathbf{W}$. In addition, the estimated components are constrained to be non-negative in keeping with the restrictions imposed by the Poisson probability model. Starting with initial values of $\mathbf{A}$ and $\mathbf{S}$, the maximum likelihood estimates are obtained by alternately solving sequences of constrained least squares problems

$$\mathbf{A}_i^T \leftarrow \text{solve} \left(\min_{\mathbf{A}_i^T | \mathbf{S}} \left\| \mathbf{\Psi}_i^{-1/2} \left(\mathbf{D}_i^T - \mathbf{SA}_i^T \right) \right\|_F^2 \right) \text{ with } \mathbf{A}_i^T \geq 0; \ i = 1, \ldots, m,$$

$$\mathbf{\Psi}_i = \text{diag} \left(\mathbf{SA}_i^T \right) \qquad (5.35)$$

$$\mathbf{S}_j^T \leftarrow \text{solve} \left(\min_{\mathbf{S}_j^T | \mathbf{A}} \left\| \mathbf{\Phi}_j^{-1/2} \left(\mathbf{D}_j - \mathbf{AS}_j^T \right) \right\|_F^2 \right) \text{ with } \mathbf{S}_j^T \geq 0; \ j = 1, \ldots, n,$$

$$\mathbf{\Phi}_j = \text{diag} \left(\mathbf{AS}_j^T \right) \qquad (5.36)$$

where i indexes the rows of $\mathbf{A}$ and rows of $\mathbf{D}$, and j indexes the rows of $\mathbf{S}$ and columns of $\mathbf{D}$. As with MLPCA, each row and each column of $\mathbf{D}$ is assigned its own covariance matrix, $\mathbf{\Psi}_i$ and $\mathbf{\Phi}_j$, respectively. Unlike MLPCA, however, estimates of the covariance matrices are derived from the factor model itself, and are continually updated as the iteration proceeds. The individual non-negativity-constrained least squares problems can be solved using algorithms such as the one presented by Bro and De Jong (Bro and De Jong, 1997). The IWLS technique will not be considered further here other than to provide motivation for the approximate maximum likelihood methods that follow.

5.5.3 NNMF: Gaussian Case (Approximate Noise)

The IWLS approach to component analysis in the Poisson case, much like the corresponding MLPCA method, provides a mathematically rigorous accounting of nonuniform error. The cost of this rigor is high; computations tend to be lengthy, for instance. In addition, the component models lack robustness to misspecification of the error. Poor estimates of the error can lead to models that are decidedly worse than what might have been obtained without considering the nature of the error, at all. During the development of the WPCA method in Section 5.4.4, it was shown that these difficulties could be largely avoided by approximating the structure of the errors. In particular, it was assumed that a single covariance matrix Φ adequately describes the error covariance of all columns of the data matrix and, likewise, that a single error covariance matrix Ψ is sufficient for all rows. Furthermore, for Poisson noise, Ψ and Φ can be easily estimated from the data matrix D itself according to Equations (5.23) and (5.24), respectively. Within these approximations, maximum likelihood estimates of the component matrices can be obtained by analyzing a scaled version of the data: $\underset{\sim}{D} = \Phi^{-\frac{1}{2}} D \Psi^{-\frac{1}{2}}$ and the components can be recovered in physical space by inverse scaling. The general procedure is:

$$\text{Scale the data: } \underset{\sim}{D} = \Phi^{-\frac{1}{2}} D \Psi^{-\frac{1}{2}}$$

$$\text{Factor } \underset{\sim}{D}: \underset{\sim}{D} = \underset{\sim}{A}\underset{\sim}{S}^{T} \tag{5.37}$$

$$\text{Inverse-scale } \underset{\sim}{A} \text{ and } \underset{\sim}{S}: A = \Phi^{\frac{1}{2}}\underset{\sim}{A}, \ S = \Psi^{\frac{1}{2}}\underset{\sim}{S}$$

The goal of scaling, or weighting, the data is to make the noise appear more uniform in the scaled space. Thus, it is appropriate to apply methods that assume uniform noise to the task of factoring $\underset{\sim}{D}$ in algorithm (5.37). This was exactly the approach taken when $\underset{\sim}{D}$ was factored by standard PCA in the WPCA algorithm. The analogous approach for NNMF is to use the original algorithm of Lee and Seung (Lee and Seung,

1999). This method employs multiplicative update rules akin to Equations (5.32) and (5.33) for the Poisson case. They take on a particularly simple and compact form, however. Assuming we have scaled data, the updates become:

$$\underset{\sim}{A} \leftarrow \underset{\sim}{A} * \frac{\underset{\sim}{D}\underset{\sim}{S}}{\underset{\sim}{A}\left(\underset{\sim}{S}^T\underset{\sim}{S}\right)}$$

$$\underset{\sim}{S} \leftarrow \underset{\sim}{S} * \frac{\underset{\sim}{D}^T\underset{\sim}{A}}{\underset{\sim}{S}\left(\underset{\sim}{A}^T\underset{\sim}{A}\right)}$$

(5.38)

As before, given initial estimates of $\underset{\sim}{A}$ and $\underset{\sim}{S}$, the pair of Equations (5.38) are alternately updated until an acceptable level of convergence is achieved.

5.5.4 Factored NNMF: Gaussian Case (Approximate Data)

During the discussion of subspace approximation by orthogonal factorization in Section 5.4, two strategies were outlined to account for Poisson noise in spectral images arising from count data. The first approach, the one taken by the WPCA algorithm, is to preprocess the data to make the noise look more uniform. The other strategy is to exclude noise prior to factorization, and that was the motivation for the PFA approach. These same basic ideas can be combined and used to advantage here. The primary objective of subspace approximation is to exclude noise. That is, we attempt to concentrate all of the chemically relevant, systematic spectral variation into a limited number of components p, and discard the remaining n − p factors as noise. This suggests that in addition to approximating the noise structure, it would be reasonable to approximate the data, as well. Using WPCA as an example, the subspace approximation algorithm can be incorporated into algorithm (5.37) to give:

$$\text{Scale the data: } \underset{\sim}{D} = \Phi^{-1/2}D\Psi^{-1/2}$$

$$\text{Factor } \underset{\sim}{D}: \underset{\sim}{D} = \underset{\sim}{T}\underset{\sim}{P}^T$$

$$\text{Estimate } \underset{\sim}{A} \text{ and } \underset{\sim}{S} \text{ from: } \underset{\sim}{T}\underset{\sim}{P}^T = \underset{\sim}{A}\underset{\sim}{S}^T$$

(5.39)

$$\text{Inverse-scale } \underset{\sim}{A} \text{ and } \underset{\sim}{S}: A = \Phi^{1/2}\underset{\sim}{A}, \ S = \Psi^{1/2}\underset{\sim}{S}$$

NNMF can be used to estimate the component matrices from the scaled principal components. For the scaled data in factored form, the update rules (5.38) become:

$$\underset{\sim}{A} \leftarrow \underset{\sim}{A} * \frac{\underset{\sim}{T}\left(\underset{\sim}{P}^T \underset{\sim}{S}\right)}{\underset{\sim}{A}\left(\underset{\sim}{S}^T \underset{\sim}{S}\right)}$$

$$\underset{\sim}{S} \leftarrow \underset{\sim}{S} * \frac{\underset{\sim}{P}\left(\underset{\sim}{T}^T \underset{\sim}{A}\right)}{\underset{\sim}{S}\left(\underset{\sim}{A}^T \underset{\sim}{A}\right)}$$

(5.40)

Note that these updates only involve operations on small matrices, so they can be computed very quickly and efficiently. One caution, here, is that data reproductions from PCA are not necessarily positive. The non-negativity of the updated component matrices can be enforced, however, by simply setting any negative values that result to zero.

5.5.5   Alternating Least Squares (ALS)

PNNMF and IWLS, which explicitly incorporate the Poisson probability distribution, are two computational approaches for obtaining equivalent maximum likelihood estimates of the component matrices when spectral images are contaminated by Poisson noise. The two previous sections have considered making approximations to the noise structure and/or data in the context of the NNMF algorithm. Analogous approximations can be made when estimating the component models by least squares. In the latter case, the resulting algorithms are broadly termed ALS. Performing non-negativity constrained ALS on scaled data was the approach taken by Kotula *et al.* (Kotula *et al.*, 2003) in their analysis of EDS spectral images.

Two algorithms, which are direct analogues of (5.37) and (5.39), differ from them only in the manner whereby the scaled components $\underset{\sim}{A}$ and $\underset{\sim}{S}$ are estimated from the scaled data. Rather then using multiplicative update rules, the components are estimated iteratively by non-negativity constrained least squares. Given initial estimates of $\underset{\sim}{A}$ and $\underset{\sim}{S}$, and the scaled data $\underset{\sim}{D}$, the iteration is:

$$\underset{\sim}{A} \leftarrow \text{solve} \left(\min_{\underset{\sim}{A}|\underset{\sim}{S}} \left\| \underset{\sim}{D} - \underset{\sim}{A}\underset{\sim}{S}^T \right\|_F^2 \right) \quad \text{with} : \underset{\sim}{A} \geq 0$$

$$\underset{\sim}{S} \leftarrow \text{solve} \left(\min_{\underset{\sim}{S}|\underset{\sim}{A}} \left\| \underset{\sim}{D} - \underset{\sim}{A}\underset{\sim}{S}^T \right\|_F^2 \right) \quad \text{with} : \underset{\sim}{S} \geq 0$$

(5.41)

Convergence of this sequence of conditional least squares problems is guaranteed by the properties of least squares and the fact that a least squares solution is obtained at each step. Thus, the iteration (5.41) is continued until $\underset{\sim}{A}$ and $\underset{\sim}{S}$ have been estimated to the desired level of precision. A and S are then recovered by unweighting as in Equation (5.37). Finally, a reduced-dimension model of the data can be used with the ALS approach by simply replacing $\underset{\sim}{D}$ in Equation (5.41) with, for instance, $\underset{\sim}{T}\underset{\sim}{P}^T$. This algorithm will be termed factored ALS (fALS). A very efficient algorithm for solving the large-scale non-negativity constrained least squares problems posed by algorithm (5.41) has been given by Van Benthem and Keenan (Van Benthem and Keenan, 2004).

5.5.6 Performance Comparisons

This section has presented several methods for component analysis that ultimately derive from maximum likelihood considerations. All of the methods are capable of generating accurate component representations, but the computational effort varies widely. In contrast to the orthogonal matrix factorization approaches, all of these estimation algorithms are iterative in nature. Thus, sensitivity to the initial conditions is also a key criterion for assessing performance.

Quantitative evaluations of component accuracy can be made for the simulated spectral images since the true components are known. The metric to be used is the sum of the squared errors (SSE):

$$\text{SSE} = \left\| D_{\text{True}} - AS^T \right\|_F^2$$

(5.42)

where D_{True} is a data matrix constructed from the noise-free true components. SSEs attained by the various iterative algorithms are plotted in Figure 5.13 for the wires simulation and in Figure 5.14 for the copper

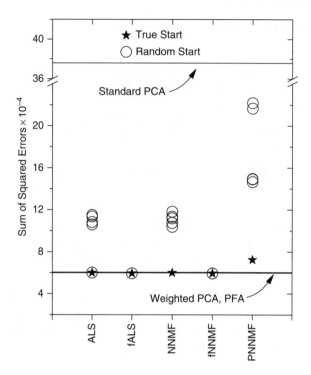

Figure 5.13 Assessments of accuracy and sensitivity to algorithm initialization for various factorization algorithms applied to the wires simulation. Five random-number starts are shown for each of the iterative algorithms

grid simulation. The reconstruction errors of the corresponding PCA, WPCA and PFA models are also shown on these plots for comparison. For the iterative algorithms, the component matrices were initialized with either the true pure components, or with uniformly distributed random numbers. When initialized with the true pure components, all algorithms generate acceptable representations of all components, and the model accuracies are similar. The accuracies are also compa-rable to those achieved by the orthogonal factorization methods that account for the Poisson noise. The iterative methods differ markedly in their sensitivities to the starting point, however. Specifically, the maximum likelihood based methods that employ the entire dataset to estimate the components (PNNMF, NNMF and ALS) are prone to finding local maxima that incorrectly estimate one or more compo-nents. Their performances in this regard are summarized in Table 5.2 for five different random starts each. For the wires simulation, to take one example, the PNNMF algorithm never did better than estimating five of the seven components correctly, and, in two cases, only four components were estimated correctly. The latter results, after putting them on a PC basis using the fPCA algorithm, are essentially the same as the standard PCA estimates shown in Figure 5.5(c). By way of contrast, the two methods that operate on a factored representation of the data (fNNMF and fALS), arrived at the correct component descriptions from all of the different random starting points. Presumably, the improved performance results because these techniques do not need to search the full n-dimensional data space, but rather, a greatly reduced subspace.

The final performance comparison concerns computational performance. Table 5.2 lists the computational effort required to achieve convergence, for the two simulations, in terms of both time and number of iterations. The time needed to factor the scaled data matrix is not included in the fALS and fNNMF times, but is a small fraction of the total in both cases. While it cannot be assured that all of the algorithms were coded to be equally optimized, the trends are clear. The ALS-based

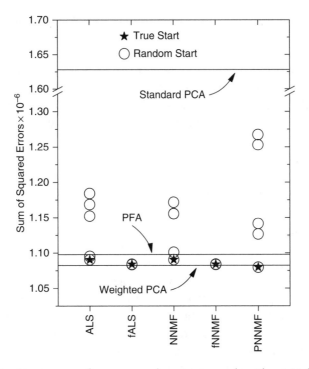

Figure 5.14 Assessments of accuracy and sensitivity to algorithm initialization for various factorization algorithms applied to the copper grid simulation. Five random-number starts are shown for each of the iterative algorithms

Table 5.2 Performance comparisons for the iterative maximum likelihood based algorithms when initialized with uniformly distributed random numbers. The copper grid simulation contains $p = 6$ chemical components, while the wires simulation contains $p = 7$ components. Five different random starts were made for each algorithm

Simulation	Method	Iterations	Time (s)	No. of components correct from random start			
				p of p	p−1 of p	p−2 of p	p−3 of p
	fALS	75	36	5	0	0	0
	ALS	75	416	1	2	2	0
Cu grid	fNNMF	2000	219	5	0	0	0
	NNMF	2000	1660	1	2	2	0
	PNNMF	2000	5480	1	2	1	1
	fALS	300	46	5	0	0	0
	ALS	300	217	1	4	0	0
Wires	fNNMF	5000	235	5	0	0	0
	NNMF	5000	5280	0	5	0	0
	PNNMF	5000	7350	0	0	3	2

methods require more work per iteration but need many fewer iterations to converge, and working with the data in a factored form yields a large reduction in computation time. Depending on the choice of method, model estimation for the same dataset can take from seconds to hours. That the most robust methods, those employing factored representations of the data, are also the quickest, is a fortunate circumstance, and enables component analysis to be applied routinely to spectral images.

5.6 CONCLUSIONS

Spectral imaging techniques that count particles to generate their respective spectra are commonly encountered in surface characterization and microanalysis laboratories. Often, particularly when spectra are acquired at low count rates, the nonuniform Poisson noise characteristic of counting experiments becomes a predominant feature of the spectral image. Factor based data analysis methods, such as PCA, which simply try to account for spectral variance in a parsimonious way, have limited sensitivity in the presence of nonuniform noise. The fundamental reason for this is that variance arising from small, yet chemically important spectral components, can become insignificant with respect to large magnitude noise associated with major spectral features. Thus, a fully comprehensive analysis of a spectral image requires proper accounting for the nonuniform noise characteristics of the data.

This chapter has surveyed several methods that account for Poisson noise in the analysis of spectral images composed of count data. These fall into two broad categories depending on the goals of the analysis. Orthogonal matrix factorizations are useful for computing approximations to the subspace spanned by the data. Maximum likelihood based techniques, on the other hand, can incorporate constraints that enable physically realistic components to be estimated, thus facilitating the interpretation of the spectral image. In both cases, excellent results can be achieved by scaling the data with matrices that are easily computed from the data itself. Finally, the two approaches can be combined to yield robust and computationally efficient algorithms for spectral image analysis. Through these developments, data analysis time has become commensurate with data acquisition time, and component analysis is poised to assume a prominent place in the analyst's toolbox.

ACKNOWLEDGEMENTS

The author would like to thank Paul Kotula for providing the wires EDS dataset. This work was completed at Sandia National Laboratories, a multiprogram laboratory operated by Sandia Corporation, a Lockheed Martin Company, for the United States Department of Energy's National Nuclear Security Administration under contract DE-AC04-94AL85000.

REFERENCES

Bro, R. and De Jong, S. (1997) A fast non-negativity-constrained least squares algorithm, *Journal of Chemometrics*, **11**, 393–401.

Bro, R. and Smilde, A. (2003) Centering and scaling in component analysis, *Journal of Chemometrics*, **17**, 16–33.

Cameron, A. C. and Trivedi, P. K. (1998) *Regression Analysis of Count Data*, Cambridge University Press, Cambridge.

Cochran, R. N. and Horne, F. H. (1977) Statistically weighted principal component analysis of rapid scanning wavelength kinetics experiments, *Analytical Chemistry*, **49**, 846–853.

Geladi, P. and Grahn, H. (1996) *Multivariate Image Analysis*, John Wiley & Sons, Ltd, Chichester.

Golub, G. H. and Van Loan, C. F. (1996) *Matrix Computations*, 3rd Edn, The Johns Hopkins University Press, Baltimore.

Jollife, I. T. (2002) *Principal Component Analysis*, 2nd Edn, Springer-Verlag, New York.

Keenan, M. R. (2005) Maximum likelihood principal component analysis of time-of-flight secondary ion mass spectrometry spectral images, *Journal of Vacuum Science and Technology A (Vacuum, Surfaces, and Films)*, 23, 746–750.

Keenan, M. R. and Kotula, P. G. (2004a) Accounting for Poisson noise in the multivariate analysis of ToF-SIMS spectrum images, *Surface and Interface Analysis*, 36, 203–212.

Keenan, M. R. and Kotula, P. G. (2004b) Optimal scaling of TOF-SIMS spectrum-images prior to multivariate statistical analysis, *Applied Surface Science*, 231–2, 240–244.

Kotula, P. G., Keenan, M. R. and Michael, J. R. (2003) Automated analysis of SEM X-ray spectral images: a powerful new microanalysis tool, *Microscopy and Microanalysis*, 9, 1–17.

Kramer, R. (1998) *Chemometric Techniques for Quantitative Analysis*, Marcel Dekker, New York.

Lee, D. D. and Seung, H. S. (1999) Learning the parts of objects by non-negative matrix factorization, *Nature*, 401, 788–791.

Lee, D. D. and Seung, H. S. (2001) Algorithms for non-negative matrix factorization, *Proceedings 14th Annual Neural Information Processing Systems Conference (NIPS)*, 13, 556–562.

Malinowski, E. R. (1987) Theory of the distribution of error eigenvalues resulting from principal component analysis with application to spectroscopic data, *Journal of Chemometrics*, 1, 33–40.

Malinowski, E. R. (1988) Statistical f-tests for abstract factor analysis and target testing, *Journal of Chemometrics*, 3, 49–60.

Malinowski, E. R. (2002) *Factor Analysis in Chemistry*, 3rd Edn, John Wiley & Sons, Ltd, New York.

Pawitan, Y. (2001) *In all Likelihood: Statistical Modelling and Inference Using Likelihood*, Clarendon Press, Oxford.

Reyment, R. and Jöreskog, K. G. (1996) *Applied Factor Analysis in the Natural Sciences*, Cambridge University Press, Cambridge.

Sajda, P., Du, S. Y. and Parra, L. (2003) Recovery of constituent spectra using non-negative matrix factorization, *Proceedings of SPIE, Wavelets – Applications in Signal and Image Processing X*, 5207, 321–331.

Thompson, W. (1999) Don't subtract the background, *Computing in Science and Engineering*, 1, 84–88.

Thompson, W. (2001) Poisson distributions, *Computing in Science and Engineering*, 3, 78–82.

Van Benthem, M. H. and Keenan, M. R. (2004) Fast algorithm for the solution of large-scale non-negativity-constrained least squares problems, *Journal of Chemometrics*, 18, 441–450.

Wentzell, P., Andrews, D., Hamilton, D., Faber, K. and Kowalski, B. (1997) Maximum likelihood principal component analysis, *Journal of Chemometrics*, 11, 339–366.

Wu, W., Massart, D. L. and de Jong, S. (1997) The kernel PCA algorithms for wide data. I. Theory and algorithms, *Chemometrics and Intelligent Laboratory Systems*, 36, 165–172.

6

Hyperspectral Image Data Conditioning and Regression Analysis

James E. Burger and Paul L. M. Geladi

6.1 INTRODUCTION

The quality of an image or any other dataset may depend on the choice of any physical or numerical based processing applied to the data, as well as sample selection, i.e. which data to include or reject from the dataset. In the specific case of hyperspectral image datasets, data quality depends on using the proper transform functions to convert instrument signal counts into reflectance or absorbance units. Data quality can also be aided by incorporation of internal reference standards directly into the image field of view (FOV), thus enabling instrument standardization between images. Various spectral pre-processing treatments may be applied to correct for nonlinearity or baseline offsets. The data quality is improved by all of these transformation techniques. Data quality can also be improved by the application of spatial image masks to select image data subsets, and by cleaning or removing bad data values. The abundance of data in hyperspectral images enables a liberal approach to outlier detection and removal. Classical image thresholding type data screening can of course be applied to individual data values. In addition, the large

Techniques and Applications of Hyperspectral Image Analysis   Edited by H. F. Grahn and P. Geladi
© 2007 John Wiley & Sons, Ltd

numbers of data samples permit computation of accurate population statistic estimates for use in probabilistic based trimming algorithms.

The abundance of data also complements multivariate statistics based chemometrics tools. Data subset collections may be first combined and averaged to improve signal to noise ratios (SNRs) before being subjected to further calibration modeling techniques. Large sample populations also permit computation of confidence intervals and provide uniformity information not easily obtained with bulk analysis based instrumentation. Univariate visualization tools can be extended to sets of prediction or residual values. For example, image prediction results may be presented as either histogram distributions indicating uniformity profiles, or as image maps that reveal the actual spatial distribution within imaged samples.

The following sections describe the implementation of some of the aspects of hyperspectral image data conditioning, exploration and regression analysis. Examples are provided to demonstrate the benefits of different processing or data mining approaches.

6.2   TERMINOLOGY

This chapter deals extensively with the manipulations of numerical data. It is important to distinguish between the actions of transformation, standardization, and calibration. Data transformation involves the conversion of data from one set of units to another: e.g. raw instrument data may be transformed to reflectance; reflectance data may be transformed to absorbance. Standardization is a correction process used to move data towards a targeted value: e.g. internal standards may be used for instrument standardization; multiplicative scatter correction (MSC) moves data towards a targeted spectrum. Calibration is used in the context of chemometrics: calibration data and calibration models are used to allow the prediction of dependent variable values from related independent variables.

6.3   MULTIVARIATE IMAGE REGRESSION

Regression on spectra has been presented in many contexts. A very popular application is relating near infrared (NIR) spectra to chemical or physical information (Martens and Næs, 1989; Osborne et al., 1993; Beebe et al., 1998; Næs et al., 2002). The spectra can come from bulk sample measurements or be localized in hyperspectral images where the data values for an individual pixel location make up a single spectrum. The equations below are given for one concentration and one type of optical spectrum to keep the notation simple. Generalizations to other spectra (mass spectra) and to more concentrations and physical properties are easily done.

The most general equation useful for all types of spectra and all types of chemical or physical information is:

$$y = b_0 + b_1 x_1 + b_2 x_2 \ldots + b_K x_K + f \qquad (6.1)$$

where K is the number of independent variables (wavelengths) in a spectrum, y is the concentration in the sample or pixel, b_i are the regression coefficients $i = 0, \ldots, K$, x_i are the absorbances at the K wavelengths used, measured in a bulk sample or pixel and f is a residual.

Calibration regression models are computed based on sets of known spectra and concentrations y in order to find the coefficients b_i.

Equation (6.1) can also be written with vectors and matrices:

$$y = Xb + f \qquad (6.2)$$

where y is a vector of concentrations for I samples or pixels, size $I \times 1$, X is a matrix with the spectra as rows and a column of ones, size $I \times (K+1)$, b is a vector of regression coefficients, size $K + 1$, $b = [b_0\, b_1\, b_2 \ldots . b_K]^T$ and f is a vector of residuals, size $I \times 1$.

It has become customary to mean-center y and X to simplify Equation (6.2) into:

$$y = Xb + f \qquad (6.3)$$

where y is a vector of concentrations for I samples or pixels, X is a matrix with the spectra as rows, size $I \times K$, b is a vector of regression coefficients of size K and f is a vector of residuals.

With Equation (6.3), b_0 is not calculated explicitly. This can speed up calculations and increase precision. The means of y and X are kept as y_m and x_m. In Equation (6.3), y is decomposed to a part called *model* (Xb) and a part called *residual* (f).

The calculation and interpretation of b are not always straightforward. This is only easy if the data in X come from an orthogonal experimental design. In many other cases special methods are needed to find useful values of b. A popular method is Partial Least Squares (PLS) regression (Martens and Næs, 1989; Beebe *et al.*, 1998; Brown,

1993; Martens and Martens, 2001). For reading this chapter, knowing how PLS works is not important, it is enough to know that it can solve Equation (6.3).

The most important property of PLS is that it uses *components* also called *factors* or *latent variables* and that the number of components used is very important. Not using enough components leads to *underfitting* (bad models) and using too many components leads to *overfitting* or bad prediction [Equation (6.4)]. The concept of factors was explained in earlier chapters (3–5) for principal component (PC) factors. This topic is further discussed in Section 6.5.2, pseudorank determination.

It is important to construct experiments where y and X are well known (precise and accurate) and can lead to the desired coefficients b with a vector f of sufficiently small norm. This means that the range of values in y spans the expected range in future applications of Equations (6.1)–(6.3) and that X contains all spectral phenomena that can be encountered in future measurements.

The goal in making Equations (6.1)–(6.3) is to be able to calculate unknown concentrations from unknown spectra of new samples or pixels:

$$y_{hat} = X_{unk}b \qquad (6.4)$$

where y_{hat} are predicted concentrations, offset by y_m, X_{unk} is a matrix with spectra of unknowns as rows, with x_m subtracted and b are coefficients from Equation (6.3).

If all the spectra in X_{unk} belong to pixels in an image, then y_{hat} can be used to construct a new image of concentration predictions: a concentration map, using the pixel location of each spectrum. This additional spatial resolution makes regression on hyperspectral images a very powerful analytical technique.

6.3.1 Regression Diagnostics

Equation (6.3) must be tested because there are many ways of calculating b and not all of them are error-free or give satisfactory results. Carrying out these tests is called regression diagnostics.

The easiest diagnostic is based on the sum of squares of the parts of Equation (6.3):

$$SS_{tot} = SS_{mod} + SS_{res} \qquad (6.5)$$

where SS_{tot} is $y^T y$ for mean-centered y, SS_{mod} is $b^T X^T X b$ for Equation (6.3) and SS_{res} is $f^T f$.

If the sum of squares of y is the total sum of squares to be explained (or 100 %), the model sum of squares needs to be high and the residual sum of squares needs to be low. Seventy percent of the total sum of squares modeled is an absolute minimum and many models reach between 85 % and 99.9 %. A rule of thumb is that if five PLS components give less than 70 % of the total sum of squares in the model, then no usable model is possible.

A plot of Xb against y can also be used as a diagnostic. In the ideal case, all points in the plot are on the diagonal. Deviations from the diagonal can be interpreted as flaws of the model. Such deviations can be due to noise, nonlinearity, outliers, heteroscedasticity and bias.

The residual vector f can be used in many ways for making diagnostics. It can be shown as a histogram, or as a normal probability plot. Also a plot of f against Xb is sometimes used.

One important concept for calibration is that of validation, or testing the model. This is best done by using a test set:

$$y_{test} = X_{test} b + f_{test} \qquad (6.6)$$

where y_{test} are the concentration values for the test set, X_{test} are spectra for the test set and f_{test} is the residual from the test set.

b is from the calibration model. In traditional calibration it is sometimes difficult to find a meaningful test set, but for images where thousands of pixels are available there is no problem in finding test set pixels. The diagnostics described earlier can be used for test images (f_{test} instead of f). These diagnostics are more realistic because they test the calibration models on truly independent data.

A summary diagnostic that is often used is root mean square error of prediction (RMSEP):

$$RMSEP = [f_{test}{}^T f_{test}/J]^{1/2} \qquad (6.7)$$

Where J is the number of test set objects. RMSEP has the same units as the items in y and has the form of a standard deviation.

Other forms of Equation (6.7) are used:

$$RMSEC = [f^T f/I]^{1/2} \qquad (6.8)$$

RMSEC stands for root mean square error of calibration. RMSECV is a similar diagnostic but is based on internal or cross-validation.

More about regression diagnostics can be found in the literature (Belsley *et al.*, 1980; Cook and Weisberg, 1982; Fox, 1990; Meloun *et al.*, 1992; 2006).

6.3.2   Differences between Normal Calibration and Image Calibration

In the normal regression literature, examples are given where the number of samples I is approxemately equal to the number of variables K in Equation (6.3). In image regression, I >> K is valid. This provides a number of advantages, but also a number of limitations. In the classical multivariate calibration literature, I typically ranges between 20 and 250 while for imaging, a value of 20 000 is no problem and 200 000 is also easily achieved. This is advantageous in that large populations of calibration and test objects can be made. This enables more opportunities for statistical testing, for example by making histograms of residuals. Another advantage of images is that all I objects have spatial coordinates; this makes it possible to construct images from prediction or residual values enabling additional visual inspection and interpretation.

A disadvantage of multivariate image regression is that in principle the Y-values for all I calibration objects should be known accurately. These values are often determined by wet chemistry or by slow instrumental methods; the additional difficulty of sampling at each pixel in an image means that obtaining the complete set of dependent Y-values becomes an impossible task. But this limitation can sometimes be overcome by clever sampling techniques.

6.4   DATA CONDITIONING

Instrument signal transformation, standardization, and calibration are common topics in spectroscopy. These topics can be extended to hyperspectral image spectra as well. This section explores transformation and standardization issues that are unique to hyperspectral images. First, since a hyperspectral image contains thousands of spectra, how can the raw instrument signal be transformed to best ensure that a uniform response has been achieved for all pixel locations at all wavelengths within a single hyperspectral image? Second, can subtle image to image or day to day variations in imaging system response be detected in such an extremely large dataset, and if so, how can they be best corrected?

The answers to these questions are dependent on having valid standard reference materials (SRMs) to use for routine measurements. SRMs are needed to calibrate and correct raw spectra for variations in both wavelength and intensity axes. Ideally spectral band features in SRMs should be independent of spectral resolution. They should be stable, inert, and insensitive to environmental changes such as temperature and humidity. They should also be easy to handle, easily cleaned, and be affordable.

Spectroscopic reflectance standards are generally provided in one of three forms: powdered, thin films or coatings, or solid tiles. For imaging applications it is desirable to have materials that exhibit high Lambertian reflection, diffuse reflection that is uniform in all angles. The predominant material in use is Spectralon, a sintered PTFE material (Labsphere, USA). This material is also available with increasing amounts of black carbon added to produce gray reflectance standards. Spectralon tiles with 99, 75, 50, 25, and 2 % reflectance values were used in the research reported in this chapter. These gray body reflectors are appropriate for area integrating spectrometers; however, at some level of magnification this material no longer appears homogenous. This can cause severe problems for image calibration. This issue can be partially alleviated by randomly moving the Spectralon tiles during image acquisition, providing a physical averaging effect. Further development of acceptable SRMs for high magnification imaging needs to be addressed, but is beyond the scope of this chapter.

6.4.1 Reflectance Transformation and Standardization

The first aspect of data transformation addresses the issue of conversion of spectral intensities from raw instrument signal counts to percent reflectance or absorbance units. This conversion is routinely performed in spectroscopy utilizing a single point transform function based on a single measurement of a totally reflecting material such as Spectralon and a dark or background instrument measurement. For the example datasets in this chapter, a set of five Spectralon tiles was used to build transformation models. These models were based on acquisition of complete independent hyperspectral images of individual tiles of 99, 75, 50, 50, and 2 % reflectance Spectralon. The objective was to construct models for transformation of the raw instrument signal from any future sample images into reflectance. The average spectrum from each of the Spectralon hyperspectral images can be used to create transform models for each wavelength channel, thereby accounting for any general wavelength response dependencies in the camera system. However, to account for spatial variation in illumination and other system responses, independent reflectance transform models were computed for every pixel location at every wavelength channel.

Calibration reference spectra were available for each of the Spectralon SRMs, consequently regression models were computed where the independent variables x represent the five hyperspectral image spectral values, and the dependent y variables were computed from polynomial interpolation of standard reference spectra supplied by the SRM vendor. Since baseline offsets are determined automatically by regression, no dark current or background measurement is needed. Individual second order regression models were computed based on the measured values at each pixel, for each wavelength. These transformation models could then be applied to any subsequent sample image to convert instrument signal counts to percent reflectance.

The use of SRMs embedded within the image FOV has been previously investigated for use in instrument standardization (Burger and Geladi, 2005). The hyperspectral images described in this chapter each contained over 80 000 pixels or spectra. Some of these spectra could be specifically used for calibration purposes, while retaining large areas for sample imaging. Figure 6.1 shows a typical image indicating the inclusion of a gray scale containing pieces of Spectralon ranging from 99 to 2 % reflectance, positioned along the right hand edge of the image FOV. Since this gray scale was included in all images, instrument standardization based on signal intensity could be achieved. But the Spectralon SRMs purposely have a very flat spectral response and are therefore inappropriate for wavelength calibration. A rectangular piece of the prototype SRM-2035 glass was obtained from NIST which could be positioned directly on top of the Spectralon gray scale. This configuration provided a series of wavelength calibration spectra with decreasing signal intensity. This material can also be seen as the overlapping darker rectangle on the far right in Figure 6.1.

Corrections based on these internal standards have been discussed in previous publications (Burger and Geladi, 2005, 2006). In general it was found that the intensity corrections had very little effect on quantitative prediction results. However, the spectra were useful in monitoring the health of the camera system. For example, slight changes in the perfor-

Figure 6.1 A typical image (320 x 256 pixels) indicating the placement of internal standards for intensity calibration (gray scale on right) and wavelength calibration (small rectangle on top of gray scale). The powdered sample is contained in a circular holder with glass window

mance of the camera filters could be determined by examination of PCA plots of the median spectra from a region of the 99 % standard for a series of hyperspectral images acquired.

6.4.2   Spectral Transformations

Data errors may occur which effect many or all spectra in the sample population contributing to an overall shift or disturbance to the quality of the dataset. Sources of these errors could be instrumental noise, sample preparation issues, or simply sample contamination or introduction of a new chemical component. Various filters or transformations may be applied to spectra to reduce the effect of these variances. Transformations may be applied to reduce or counter the effect of noise, account for baseline offsets or slopes, or simply change nonlinear components of the signal. Collectively these data transformations are called spectral pre-processing treatments, and are often categorized based on their numerical or computational form as linearization, additive, or multiplicative transforms. While these treatments were initially developed for processing small sets of individual spectra from integrating spectrometers, they can also be applied to the thousands of spectra within hyperspectral images. For general discussions of spectral pre-processing treatments, see Katsumoto et al. (Katsumoto et al., 2001), Næs et al. (Næs et al., 2002) and Thennadil and Martin (Thennadil and Martin, 2005).

Reflectance values are not linearly proportional to chemical constituent concentrations, frequently the objective of spectral analysis. Two spectral pretreatments are often applied to linearize reflectance responses: conversion of spectral units to absorbance, or application of the Kubelka – Munk transform. For diffuse reflection spectroscopy, the absorbance A is simply estimated as the base 10 logarithm of the reflectance R. The Kubelka – Munk theory was developed to account for the effect of both absorbance and scattering in thin films of paint (Kubelka and Munk, 1931). This transform should be applied when scattering dominates over absorbance. Scattering can be increased by grinding samples to a very fine powder. As an alternative, samples can also be diluted with nonabsorbing materials such as potassium bromide.

Changes in instrumentation such as lamp intensity, temperature, or detector response, or changes in sample orientation, particle size distributions, or packing may result in a background signal that is added equally throughout the spectrum. These constant additive effects can be compensated for by applying first and second derivative transforms (Giesbrecht et al., 1981; Norris and Williams, 1984; Hopkins, 2001). When discrete spectral data points are evenly spaced, derivatives can be computed using standard Savitsky – Golay polynomial filters (Savitzky and Golay, 1964; Steiner et al., 1972; Madden, 1978). These filters have an additional smoothing effect resulting from fitting the multiple data points used within the filter window.

Martens et al. (Martens et al., 1983) and Geladi et al. (Geladi et al., 1985) proposed an alternative correction for scatter effects, namely multiplicative scatter correction (MSC). A target spectrum is first computed, typically the mean spectrum x_m from the sample dataset. For each sample spectrum x, it is assumed that the scatter effect is constant for all wavelengths. A plot of x versus x_m for all wavelengths should therefore result in a straight line. The offset and slope estimates from a linear regression of x versus x_m can be used to compute a corrected spectrum. This correction process computes a different offset and slope unique to each sample spectrum. But the scatter effect may not be exactly the same for all wavelength ranges. Isaksson and Kowalski (Isaksson and Kowalski, 1993) proposed correcting the spectral value at each wavelength with independent offset and slope correction terms. These terms are also computed by linear regression, using a small window of data centered about each wavelength. This technique is called piecewise multiplicative scatter correction (PMSC). Another variant, extended multiplicative signal correction (EMSC) was presented by Martens and Stark (Martens and Stark, 1991) and Martens *et al.* (Martens *et al.*, 2003).

The standard normal variate (SNV) transform was proposed by Barnes *et al.* (Barnes *et al.*, 1989) as an alternative correction for multiplicative scatter effects. Although this technique does not require the use of a target spectrum, it has been shown to be linked directly to the MSC transform (Dhanoa *et al.*, 1994). This transform normalizes each spectrum x by first subtracting its mean value x_m followed by dividing by the standard deviation x_s. This process effectively creates a common baseline offset (zero) and variance (one) for all spectra.

6.4.3   Data Clean-up

An outlier is an observation that lies at an unusually large distance from other values observed in a small random population sample. Outliers can distort simple population statistics such as the mean and standard deviation, and have high leverage effects on chemometric techniques (Martens and Næs, 1989; Beebe *et al.*, 1998; Næs *et al.*, 2002). This is especially important in linear regression where a single point can greatly perturb the slope of the regression line. It is important to identify dataset outliers and selectively remove them from both calibration model generation and prediction processes. Since hyperspectral images contain millions of data values, it is also important to utilize automatic outlier detection and removal schemes.

Hyperspectral images often contain random elements with extremely erroneous data values, known as dead or bad pixels. Often these points occur with either a constant zero or maximum signal value, but elements that are simply stuck at an intermediate value may occur as well. A simple thresholding approach can be used to identify any pixels with an unusually high or low value. This cleaning process removes the problematic data outliers with extreme signals, but it does not detect or remove the pixels with constant intermediate values within the acceptable data range. Diagnostics such as distances or correlations between sample points can be computed and trimmed of extreme values based on probabilistic tests. The large sample set sizes provided by hyperspectral image datasets enables accurate estimation of the population means and standard deviations used in setting limits for outlier detection.

6.4.4 Data Conditioning Summary

The results of any analytical technique depend on the quality of the data. It is imperative to optimize both the spatial and spectral quality of hyperspectral image datasets. Carefully chosen SRMs must be imaged to account for spatial variations in sample illumination and detector system response. Inclusion of these same materials as internal standards in all images enables detection and correction of instrument instabilities due to environmental or system aging processes.

In addition to providing space for internal standards, the immensity of the dataset within a single hyperspectral image provides two distinct advantages: First, since the spectral sample population size is quite large, probabilistic population statistic estimates are reliable. Second, this large sample size permits the liberal removal of outliers; if hundreds or even thousands of samples are thrown out, tens of thousands of samples still remain. This permits the selection of very clean calibration sets which lead to more reliable and robust calibration models.

Additional sources of data error can be corrected using conventional spectroscopic transformations to compensate for nonlinear, additive, or multiplicative effects. These corrections rely exclusively on the spectral content of hyperspectral images. Spatial transformations and corrections such as median filters, sharpening, blurring, or edge detection, wavelets, or fast Fourier transforms could be applied but are not addressed in this chapter.

6.5 PLS REGRESSION OPTIMIZATION

6.5.1 Data Subset Selection

Since PLS models the variance in sample spectra, it is critical that spectral datasets be carefully selected to ensure maximum correlation between variance in independent variables (spectra) and dependent variables (concentration). Consequently it is important to remove any extreme spectral outliers which enact a high leverage on the calibration model. Spatial masking utilizing object recognition, thresholding, or user defined regions of interest should also be used to restrict spectral data to local subsets highly correlated to the dependent variable of interest.

6.5.2 Pseudorank Determination

The number of latent variables or PLS components (A) to include in the PLS calibration model is called the *pseudorank*. If A is too small, termed underfitting, not enough signal is incorporated into the calibration model resulting in poor predictions. Overfitting occurs when A is too large and excessive noise is included in the calibration model. This results in a good fit of the spectra used for modeling, but poor predictions of additional spectra (Martens and Næs, 1989; Beebe *et al.*, 1998; Næs *et al.*, 2002).

To assist with determining the pseudorank various approaches have been suggested for splitting experimental data into a calibration or training set $\{y_{cal}X_{cal}\}$ and a test set $\{y_{test}X_{test}.\}$ As explained in Section 6.3.1 various regression diagnostics including RMSEC, RMSECV, and RMSEP can be computed to assess the goodness of fit of the calibration model. PRESS (Predicted residual error sum of squares) plots can be made, plotting these error diagnostics versus the number of latent variables included in the model. A minimum value or breaking point in the plot may be observed when a sufficient number of variables have been included.

Additional diagnostic information can be obtained based on the predicted values and residual errors of hyperspectral image spectra. When an image region of interest (ROI) contains a single chemical constituent with a known concentration, accurate population statistics can be computed from the large collection of concentration predictions obtained from the ROI spectra. As the number of latent variables is increased, the prediction bias should decrease because of better model fit. At some point, the standard deviation of ROI predictions should begin to increase because of model overfitting.

A new diagnostic D-metric was introduced by Burger and Geladi (Burger and Geladi, 2006) based on pooling the average biases and the standard deviations of predictions from N different image ROIs:

$$\mathrm{BIAS}_{pool} = [\mathbf{r}_{ave}{}^T\mathbf{r}_{ave}/N]^{1/2} \qquad (6.9)$$

$$S_{pool} \approx [\mathbf{s}_{ave}{}^T\mathbf{s}_{ave}/N]^{1/2} \qquad (6.10)$$

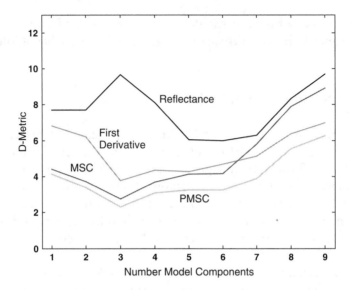

Figure 6.2 D-metric combining $BIAS_{pool}$ and S_{pool} prediction errors (as in Figures 6.3 and 6.4). Only three components are required for the pretreated data, as opposed to five for the original reflectance data. PMSC yields the best result

$$D\text{-metric} = [w_1 BIAS_{pool}{}^2 + w_2 S_{pool}{}^2]^{1/2} \qquad (6.11)$$

Here r_{ave} and s_{ave} represent the vectors containing the average bias and average standard deviation estimates from each of the N sample ROIs. The weights w_1 and w_2 were both set to the value 1.0 in the cited paper; however, alternate weights could be used. Figure 6.2 shows the D-metric versus the number of model components (latent variables) for the prediction of fat, protein, and carbohydrate in cheese. In this case two PLS components were found to be an optimal number for all three ingredients. This new D-metric diagnostic combines both accuracy and precision variance measured throughout the hyperspectral image and is quite robust to variations due to sample presentation errors. This example will be further discussed later in the following section.

6.6   REGRESSION EXAMPLES

Three example datasets are used in this chapter to illustrate a full range of sample source and analyte measurement complexity for hyperspectral imaging. The three example datasets include first, a collection of

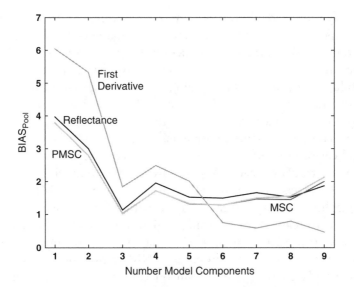

Figure 6.3 $BIAS_{pool}$ PRESS plot for the combined results of ten replicate images containing nearly equal amounts of three ingredients (sugar, citric acid and salicylic acid). The results displayed are for reflectance spectra and three pretreatments: first derivative, MSC, and PMSC. Each image contained approximately 20 000 spectra

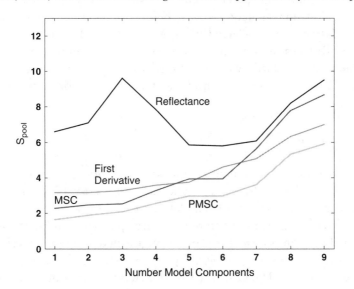

Figure 6.4 S_{pool} PRESS plot for the combined results of ten replicate images containing nearly equal amounts of three ingredients. The results displayed are for reflectance spectra and three pretreatments: first derivative, MSC, and PMSC. The spectral pretreatments reduced the S_{pool} value significantly, and suggest a fewer number of model components required for modeling

artificially produced laboratory samples of mixtures of three pure chemical components. Second, a collection of commercial cheese products used to demonstrate the measurement of fat, protein, and carbohydrate concentrations. Finally, a set of wheat straw materials is used to test the ability of hyperspectral imaging to determine naturally occurring wax content. These datasets demonstrate the advantages and limitations of hyperspectral image regression analysis, and the possible benefits of spectral data conditioning treatments.

In all cases, near infrared (NIR) hyperspectral images were acquired using a Spectral Dimensions MatrixNIR camera, producing images with 256×320 pixels at 118 wavelength channels, 960 – 1662 nm, with 6 nm resolution. Images were digitized with 12 bit resolution, with an average of 10 scans recorded using a 64 ms integration time. The image FOV was $50 \times 62 \, \text{mm}^2$ with approximately $0.2 \times 0.2 \, \text{mm}^2$ pixels. All samples were either placed in circular NIR sample holders with glass covers, or placed on pieces of silicon carbide sandpaper. The sandpaper is inexpensive, disposable, and produces a very low reflectance background.

All sample hyperspectral images were transformed to reflectance units using wavelength and pixel specific transform functions based on sets of Spectralon images acquired just before the sample images. Sample spectra datasets were selected using either object thresholding or user specified ROIs, and cleaned of all hardware and sample class outliers. The internal standards were examined for instrument system consistency, but were not used for instrument standardization corrections.

6.6.1 Artificial Ternary Mixture

An artificial three-component sample set was made from reagent grade citric acid (CAS 77-92-9), salicylic acid (CAS 69-72-7) and sugar (consumer grade Dan Sukker). All three pure chemicals were first screened with a 25 μm sieve to remove large particles, and then blended using a three-component third order augmented simplex experimental design yielding 13 total samples: 3 pure component, 6 binary (1/3, 2/3) and 4 ternary (1/3, 1/3, 1/3; 2/3, 1/6, 1/6; etc.) mixture samples. Samples of each mixture were placed in circular Foss NIRSystems sample holders and imaged. The 3 pure, 6 binary, and 3 augmented ternary (2/3, 1/6, 1/6) mixtures were imaged in duplicate (24 calibration samples). For determining modeling statistics the center point ternary mixture (1/3, 1/3, 1/3) was imaged a total of 10 times (10 test samples). A circular mask with radius of 80 pixels was used to identify the sample ROI, providing over 20 000 spectral samples from each hyperspectral image.

To first examine the effectiveness of PLS modeling and the spectroscopic performance of the instrumentation, the imaging system was tested as a bulk measurement instrument. That is, a single spectrum was first computed as the overall mean spectrum of the spectral dataset from each of the 34 hyperspectral images. The resulting 24 calibration spectra and 10 test set spectra were then subjected to spectral pretreatments to assess their effect on calibration model performance. PLS-I models were computed for percent concentration of each of the three chemical ingredients. The resulting RMSEP values from the 10 samples were pooled using a pooling equation similar to Equation (6.9) or (6.10), and are listed in Table 6.1.

Table 6.1 Pooled RMSEP values for spectral pretreatments of mean spectra from each image. (Bulk analysis)

Spectral treatment	Pooled RMSEP	Number of PLS components
Reflectance	1.0	3
Absorbance	1.6	4
First Derivative	1.7	3
Second Derivative	2.3	3
Kubelka-Munk	4.1	3

These results suggest that a three- to four-component PLS-I model performs adequately well regardless of the spectral pretreatment, yielding RMSEP values of 1–2 %. It should be cautioned however that these results are based on only 10 test spectra. Nonetheless, as a bulk analyzer, the imaging system proved to be very effective.

What happens when the same PLS modeling approach is applied to the individual spectra within each hyperspectral image? The same PLS models based on the 24 mean calibration spectra were used to predict the individual spectra in each of the 10 hyperspectral images. Figure 6.5 indicates the prediction distributions for sugar in just one of the test images, containing roughly 20 000 spectra. The histograms are ordered in increasing number (1–9) of PLS latent variables used, with a vertical line indicating the target concentration of 33 %. The same scaling is used on all histograms.

This series of histograms provides a powerful visual diagnostic for both the determination of calibration model selection and for understanding sample composition. It is apparent that as the number of latent variables is increased, the center of the prediction distribution shifts closer towards the target value, indicating a general decrease in bias.

It is also apparent that the width of the distribution begins to increase significantly when too many latent variables are chosen in the model. These general trends were motivation for the D-metric proposed in Equation (6.11). This metric attempts to balance the decrease in bias with the increase in standard deviation. In this single image example, it appears that a three- or four-component PLS model yields an optimal prediction distribution.

Another important observation in Figure 6.5 is the shape and magnitude of the width of the prediction distribution. The test set of samples were carefully formulated and ideally should have resulted in a narrow distribution centered about the 33 % target value. The observed predicted values generate nice Gaussian like distributions as expected with random noise contributions. Heterogeneous samples or sample impurities would lead to bimodal or complex distribution profiles, clearly not seen here. However the range in the observed values appears quite large, ranging between 20 % and 45 %. This is most likely due to a combination of diffuse reflectance light scattering, incomplete sample mixing, and overly large particles creating heterogeneous pixel regions.

What else can hyperspectral images provide in extending interpretation of these anomalies? Figure 6.6 demonstrates an additional

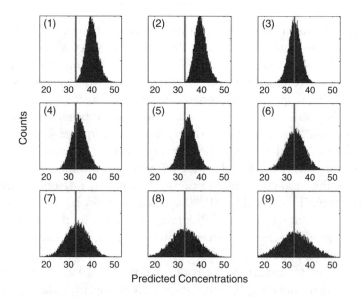

Figure 6.5 Sugar concentrations for one- to nine-component PLS models. Summary histograms represent the results from ~20 000 first derivative spectra contained within a single image ROI. The vertical line indicates the target expected concentration of 33 %

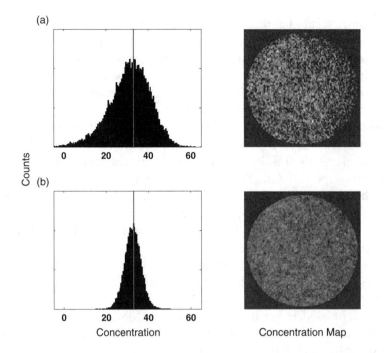

Figure 6.6 Mixture predictions for citric acid with an expected concentration of 33.4 %. Histograms and concentration maps are for ~20 000 spectra selected with a circular ROI mask with an 80 pixel radius. Prediction results are for (a) absorbance and (b) first derivative spectra

exploratory tool, spatial maps of the predicted values. In addition to the histogram indicating the population distribution, the predicted values can be mapped according to their spatial pixel coordinates. The 'speckled' texture effect observed in Figure 6.6 is uniformly distributed throughout the image, at the scaling available with the current pixel resolution. This is important, suggesting that no gross ingredient agglomeration is observed. The application of spectral pretreatments can provide additional information. Figure 6.6(b) indicates the effect of utilizing a first derivative transform on the calibration and test image spectra. The distribution of predicted values is significantly narrower, and the granularity of the 'texture' indicated in the spatial concentration map is much smoother. The derivative transformation appears to compensate or remove a varying baseline offset.

Further diagnostics can be used to compare effectiveness of spectral pretreatments. In this example, the artificial mixtures contain only three ingredients, each with unique spectral features in the measured wavelength range. It can be assumed that the same number of PLS components can be used for each ingredient, since systematic and physical noise contributions should be similar for each ingredient. Consequently the mean bias, RMSEP, and standard deviation values for each ingredient in each test image can be pooled. Figures 6.3 and 6.4 represent PRESS plots for $BIAS_{pool}$ and S_{pool} for models and predictions utilizing different spectral pretreatments.

Two things are important to observe in Figure 6.3. First, the first derivative, MSC, and PMSC pretreatments have little effect on the pooled bias – the results are very similar to the those of the original reflectance spectral data. Secondly, for all four treatment plots, a significant break point can be observed suggesting that a three-component model is sufficient for modeling. This diagnostic expresses a goodness of fit similar to what might be expected from a bulk measurement, independent of spectral pretreatment.

The PRESS plots for S_{pool} in Figure 6.4 indicate substantially different results. The original reflectance data requires a five-component model; however, fairly consistent values are observed for the one- to five-component models for the first derivative, MSC, and PMSC pretreatments. Additionally it can be seen that the pretreatments have reduced the standard deviation significantly.

The trends in pooled bias and pooled standard deviation can be combined to create the D-metric plot indicated in Figure 6.2. In this plot, it is evident that a five-component model is required for the reflectance data. All spectral pretreated data values show a general decrease for the first three components, indicative of the reduction in $BIAS_{pool}$ but with relative constant S_{pool}. The D-metric plot suggests a fewer number of components (three) are sufficient for the pretreated data, compared with five for the reflectance data. It also suggests a general trend in improved performance from reflectance to first derivative to MSC to PMSC pretreatments.

For this set of artificial samples, in can be concluded that the spectral pretreatments do very little to improve the overall or bulk prediction error (pooled bias) but the pixel based heterogeneity is reduced by the first derivative, MSC, and PMSC transforms.

6.6.2 Commercial Cheese Samples

A second dataset was obtained using an assortment of twelve commercial cheese products, specifically selected to span as large a range as possible in concentrations of protein, fat, and carbohydrate. Initially the published values on the packaging labels were used as standard reference values. A parallel set of values for protein and fat content was determined using standard techniques AOAC 976.05 (Kjeldahl protein determination) and BS EN ISO 1735:2004 (fat extraction). The measured values were in close agreement with the values indicated on package labeling.

For imaging purposes, cheese slices 2–4 mm thick were positioned on top of black silicon carbide sandpaper to minimize background contributions. The cheese is a more difficult material to image than fine powders. Initial attempts failed due to exposure to the intense heat generated by light sources – the cheese melted! A small computer box fan was used to cool the cheese surface reducing melting and fat segregation problems, but this air flow may have contributed to sample drying.

In this example, twelve images were used as both calibration and test sets to demonstrate a unique aspect of hyperspectral image regression analysis. For calibration purposes, individual reference values are not available for every pixel location. However bulk sample values are available. A rectangular ROI was first specified for each hyperspectral image, followed by cleaning for class outliers due to holes in the cheese or high specular reflection problems with the cheese surface. A mean spectrum was then computed from the resulting spectral subsets for each cheese, and used with the bulk reference values to compute a calibration model. This model was then used in turn, to predict values at each of the pixel locations indicated in the clean ROIs. This is a very important aspect of hyperspectral image regression: the ability to create a calibration model at one spatial scale, and then apply it for predictions at other scales.

Figure 6.7 shows a D-metric plot for the pooled results of the predictions from twelve cheese images. In this example dataset there was a clear break in the fat modeling errors with only two PLS components. However, with other sets of the same cheeses measured on alternate days, or with different spectral pretreatments applied, no clear break point was obtained. It was difficult to assign a PLS component number consistent with all datasets based on diagnostic values. The D-metric did however seem to appear more stable than RMSEP plots.

From this dataset example it is clear that visual exploratory analysis of complex hyperspectral images and all available diagnostics is critical. After visual inspection of many cheese datasets, it appeared that a four-component model of first derivative transformed data provided the most usable results. Figure 6.8 indicates the results for one set of the 12 cheeses. False color images have been created, mapping the fat, protein,

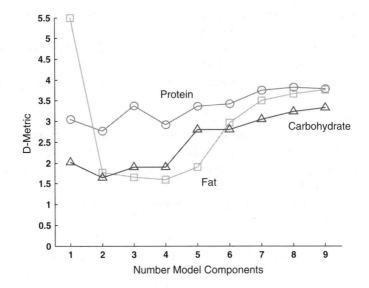

Figure 6.7 D-metric combining $BIAS_{pool}$ and S_{pool} prediction errors in cheese. Protein (circle), fat (square), carbohydrate (triangle) versus the number of components in the calibration model

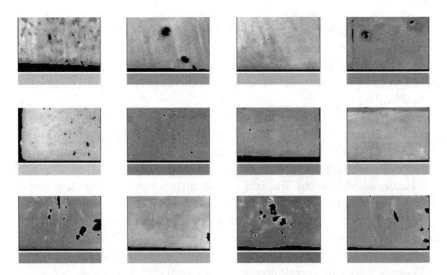

Figure 6.8 False color images indicate the fat (red), protein (green), and carbohydrate (blue) content of 12 cheeses. Predictions are based on a two-component PLS model of first derivative spectra. The small rectangle underneath each image represents the false color target composition expected for each cheese (see Plate 16)

and carbohydrate concentrations into red, green, and blue color spaces. The small rectangle at the bottom of each image indicated the target 'color' expected for that particular cheese. The overall match of color of the predictions and target region is quite good. Additional information can be gained by looking at the spatial 'discolorations'. Cutting marks and drying or agglomeration of ingredients near holes can be seen. This is an extremely useful tool for examining product uniformity or contamination, which would be difficult to detect with a spot probe or bulk measurement instrument.

6.6.3 Wheat Straw Wax

The last example provides an extreme example of the application of hyperspectral image regression. Fifteen samples of wheat plant parts were obtained which had been previously characterized for wax content utilizing gas chromatography - mass spectrometry. As with the cheese samples, only bulk values were available for the different samples. However, unlike the cheese samples with relatively large reference values (as high as 33% fat and 32% protein) the wax content was very small, ranging between 0 % and 1.3 % of total mass. Can hyperspectral image regression succeed with such a large background signal?

Each of the 15 samples was imaged as an individual hyperspectral image, filling the FOV as much as possible with sample material covering black silicon carbide sandpaper. A simple thresholding mask was used to first isolate sample spectra in each image from the background. These datasets were then cleaned of class outliers, and subsequently used for computation of mean spectra for calibration models, and prediction of individual pixel location concentrations. As with the cheese samples, this was a difficult dataset to establish a general consensus of best spectral pretreatment transformation or number of model components. First derivative transformations yielded the most consistent RMSEP for all samples, with a pooled value of 0.40 for three- or four-component PLS models.

Figure 6.9 shows the spatial mapping of the predicted wax concentrations using a three-component model. The gray scale on the bottom indicates the wax values ranging between 0.2 % and 2.0 %, in 0.2 % steps. The rectangles on the right edge of each image indicate the expected wax target values, 0.8 and 0.5 %. The resulting concentration maps indicate a strong segregation of wax levels; some areas of sample contained much higher wax content levels than others. Is this real, or possibly an artifact

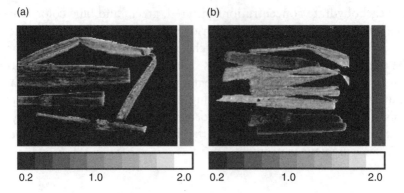

Figure 6.9 Wax content prediction images for two wheat samples, based on three-component models of first derivative spectra. The gray scale on the bottom of each image indicates the 0.2 – 2.0 % range in wax content. The rectangles on the right indicate the target wax content expected: (a) 0.6 % and (b) 0.5 %

of illumination shadow effects? Additional experimentation is needed to verify these results. However, this example indicates the potential of hyperspectral image regression to provide detailed spatial results of ingredient concentration in a noninvasive way.

6.7   CONCLUSIONS

The success of hyperspectral image regression is dependent on the quality of the spectral data contained within the dataset. Data quality can be achieved through two types of processes: data subset selection and/or cleaning, and data conditioning or pretreatment. Because of the quantity of data produced in hyperspectral imaging, automated techniques for data subset selection and the detection and removal of outliers are essential. The validation or assessment of these steps as well as the effectiveness of spectral pretreatments is dependent on both numerical and visual diagnostics.

The very large sample population sizes afforded by hyperspectral images mean that accurate population statistics can be estimated. A new metric based on pooled bias and pooled standard deviation estimates effectively combines both accuracy and precision into a robust diagnostic for the determination of model pseudorank. This metric as well as other population based statistics can provide numerical diagnostics for model interpretation. The large number of prediction results can also be visualized as histograms or concentration maps. Histograms and statistical estimates of these population distributions provide information regarding the uniformity of constituent concentration. Averaged quantitative results can be computed which are comparable to those obtained with spot probe spectroscopic techniques. Two-dimensional mapping of these same concentration values to pixel coordinates adds spatial information, allowing even greater insight into sample uniformity. This may be further accentuated by use of false color mapping visualization techniques. Both histogram and spatial mapping can be equally applied to analyte concentration predictions as well as model residuals to further assist with model and sample interpretation.

In the first two example datasets examined in this chapter, the use of a first derivative transform did little to change the average predicted concentration value of constituents (bias), however, it provided a major reduction in the range in predicted values (RMSEP). This was clearly observed as a peak narrowing in the histograms of predicted values and as a noticeable smoothing in the concentration image maps. General conclusions regarding the advantages or selection of spectral pretreatments cannot be made. Such choices, as with integrating spectrometer data, are sample specific and require careful exploratory analysis to determine their proper application.

One striking property of the second and third examples in this paper is that hyperspectral images often offer large spectral populations but only small sets of reference values. If known spatial regions of interest within an image are mapped to a single (bulk) reference value, the collection of spectra in that region of interest can be reduced or compressed and represented by a single mean spectrum. Sets of these mean spectra can then be used to build calibration models which in turn can be applied to large populations of individual pixel based test set spectra to obtain large populations of predicted values and the uniformity and spatial information previously ascribed. Such information is difficult to obtain with bulk measurement instruments.

The use of internal standards (SRMs) included within the image FOV had little effect on constituent prediction but it is still recommended to enable additional instrument and experiment diagnostics. This is especially important when a standardized instrument image is needed to make use of cataloged library spectra. Examination of the internal standard spectra of questionable sample images can help to confirm whether the instrument is performing properly or whether other sources of variance need to be investigated further.

The massive amount of raw data contained in a single hyperspectral image can appear overwhelming, yet with the proper processing and analysis tools prove to be very beneficial. Multivariate regression of hyperspectral image data provides a powerful tool for increasing the understanding of sample constituent concentrations and their distribution or spatial variation throughout the sample matrix. Effective calibration modeling and prediction is dependent on quality datasets combined with the proper use of chemometric analysis tools.

ACKNOWLEDGEMENTS

The authors thank SLUP and NIRCE, an EU Unizon-Kwarken project for financial support of the research presented in this chapter. Additionally Kempestiftelserna grant SMK-2062 provided funding for hyperspectral imaging instrumentation equipment. Also thanks to Steven Choquette at NIST for providing the glass standard reference material, and Jan Sjögren, TestElek Svenska AB, for providing the Spectralon internal standard gray scale strip material. We also acknowledge Carina Jonsson, Swedish University of Agricultural Sciences, Umeå Sweden for performing additional cheese fat and protein analysis, and Fabien Deswarte, Clean Technology Centre, University of York, York, UK, for providing the wheat straw materials and for the wax content analysis.

REFERENCES

Barnes, R., Dhanoa, M. and Lister, S. (1989) Standard normal variate transformation and detrending of near infrared diffuse reflectance. *Applied Spectroscopy*, **43**, 772–777.

Beebe, K., Pell, R. and Seasholtz, M. (1998) *Chemometrics, A Practical Guide*. John Wiley & Sons, Ltd, New York.

Belsley, D., Kuh, E. and Welsch, R. (1980) *Regression Diagnostics: Identifying Influential Data and Sources of Collinearity*. John Wiley & Sons, Ltd, New York.

Brown, P. (1993) *Measurement, Regression and Calibration*. Oxford Science, Oxford.

Burger, J. and Geladi, P. (2005) Hyperspectral NIR image regression. Part I: calibration and correction. *Journal of Chemometrics*, **19**, 355–363.

Burger, J. and Geladi, P. (2006) Hyperspectral NIR image regression. Part II: preprocessing diagnostics. *Journal of Chemometrics*, **20**, 106–119.

Cook, D. and Weisberg, S. (1982) *Residuals and Influence in Regression*. Chapman & Hall, New York.

Dhanoa, M., Lister, S., Sanderson, R. and Barnes, R. (1994) The link between Multiplicative Scatter Correction (MSC) and Standard Normal Variate (SNV) transformations of NIR spectra. *Journal of Near Infrared Spectroscopy*, **2**, 43–47.

Fox, J. (1990) *Regression Diagnostics: an Introduction*. Sage Publications, Newbury Park, CA.

Geladi, P., McDougall, D. and Martens, H. (1985) Linearization and scatter-correction for near-infrared reflectance spectra of meat. *Applied Spectroscopy*, **39**, 491–500.

Giesbrecht, F., McClure, W. and Hamid, A. (1981) The use of trigonometric polynomials to approximate visible and near infrared spectra of agricultural products. *Applied Spectroscopy*, **35**, 210–214.

Hopkins, D. (2001) Derivatives in spectroscopy. *Near Infrared Analysis*, **2**, 1–13.

Isaksson, T. and Kowalski, B. (1993) Piece-wise multiplicative scatter correction applied to near-infrared diffuse transmittance data from meat products. *Applied Spectroscopy*, **47**, 702–709.

Katsumoto, Y., Jiang, J., Berry, J. and Ozaki, Y. (2001) Modern pretreatment methods in NIR spectroscopy. *Near Infrared Analysis*, **2**, 29–36.

Kubelka, P. and Munk, F. (1931) Ein Beitrag zur Optik der Farbanstriche. *Zeitschrift für Technische Physik*, **12**, 593–604.

Madden, H. (1978) Comments on smoothing and differentiation of data by simplified least squares procedures. *Analytical Chemistry*, **50**, 1383–1386.

Martens, H. and Martens, M. (2001) *Multivariate Analysis of Quality. An Introduction.* John Wiley & Sons, Ltd, Chichester.

Martens, H. and Næs, T. (1989) *Multivariate Calibration.* John Wiley & Son, Ltd, Chichester.

Martens, H. and Stark, E. (1991) Extended multiplicative signal correction and spectral interference subtraction: new preprocessing methods for near infrared spectroscopy. *Journal of Pharmaceutical and Biomedical Analysis*, **9**, 625–635.

Martens, H., Jensen, S. and Geladi, P. (1983) Multivariate linearity transformations for near-infrared spectrometry. In: *Proceedings of the Nordic Symposium on Applied Statistics* (Ed. O. Christie). Stokkand Forlag, Stavanger, pp. 205–233.

Martens, H., Nielsen, J. and Engelsen, S. (2003) Light scattering and light absorbance separated by extended multiplicative signal correction. Application to near-infrared transmission analysis of powder mixtures. *Analytical Chemistry*, **75**, 394–404.

Meloun, M., Militky, J. and Forina, M. (1992) *Chemometrics for Analytical Chemistry Volume 2: PC-aided Regression and Related Methods.* Ellis Horwood, Chichester.

Meloun, M., Bordovska, S., Syrovy, T. and Vrana, A. (2006) Tutorial on a chemical model building by least-squares non-linear regression of multiwavelength spectrophotometric pH-titration data. *Analytica Chimica Acta*, **580**, 107–121.

Næs, T., Isaksson, T., Fearn, T. and Davies, T. (2002) *A User Friendly Guide to Multivariate Calibration and Classification.* NIR Publications, Chichester.

Norris, K. and Williams, P. (1984) Optimization of mathematical treatments of raw near-infrared signal in the measurement of protein in hard red spring wheat. I. Influence of particle size. *Cereal Chemistry*, **61**, 158–165.

Osborne, B., Fearn, T. and Hindle, P. (1993) *Practical NIR Spectroscopy with Applications in Food and Beverage Analysis*, 2nd Edn. Longman Scientific & Technical, Essex.

Savitzky, A. and Golay, M. (1964) Smoothing and differentiation of data by simplified least squares procedures. *Analytical Chemistry*, **36**, 1627–1639.

Steiner, J., Termonia, Y. and Deltour, J. (1972) Comments on smoothing and differentiation of data by simplified least squares procedures. *Analytical Chemistry*, **44**, 1906–1909.

Thennadil, S. and Martin, E. (2005) Empirical preprocessing methods and their impact on NIR calibrations: a simulation study. *Journal of Chemometrics*, **19**, 77–89.

7

Principles of Image Cross-validation (ICV): Representative Segmentation of Image Data Structures

Kim H. Esbensen and Thorbjørn T. Lied

7.1 INTRODUCTION

In this chapter a novel segmentation approach for validation purposes in relation to multivariate image regression (MIR) is presented. Multivariate images contain many significant, large proportions of pixels which are highly redundant. When several thousands to several hundreds of thousands of pixels (or more) represent the same object, special considerations are required for representative validation.

A new approach for user-delineated segmentation is introduced, specifically guided by the covariance data structure as delineated in the multivariate image analysis (MIA) score space.

The practice of 'blind', automated segmentation, which is dominating conventional 2-way cross-validation, is useless in the 3-way MIA/MIR regimen. Problems related to which components to use for the segmentation delineation are illustrated and the necessary conditions for optimal application of the new image cross-validation (ICV)

Techniques and Applications of Hyperspectral Image Analysis   Edited by H. F. Grahn and P. Geladi
© 2007 John Wiley & Sons, Ltd

approach are discussed. The general solution, called high-order components guided random sampling, is described in detail, which also sheds
new light on current 2-way chemometric cross-validation practices. ICV
is illustrated by multivariate image datasets well-known from the multivariate image literature.

This chapter deals with validation issues surrounding MIR, the
design purpose of which is to create regression models between X- and
Y-multivariate images (Esbensen *et al.*, 1992). A general introduction
to MIR is given in Tønnesen-Lied and Esbensen (Tønnesen-Lied and
Esbensen, 2001) in which the phenomenology of the three principal cases
of MIR is given in full detail. See also Chapter 6 for useful equations.

A multivariate image is a 3-D matrix (Geladi and Grahn, 1996;
Esbensen *et al.*, 1988), i.e. two of the 3-D data array ways are object
ways (O) – pixels in rows and columns – while the third is a variable
way (V), comprised by the pertinent set of channels or wavelengths.
This type of array can by categorized as an OOV array (Esbensen
et al., 1988). There are fundamental and distinct differences between
this 3-way domain and the complementary OVV domain, the latter of
which is well-known as 3-way decomposition (Esbensen *et al.*, 1988;
Bro, 1997; Smilde *et al.*, 2004). These two domains do not in general
make use of the same data modeling methods; a comprehensive comparison can be found in Huang *et al.*, (Huang *et al.*, 2003). Here we treat
the standard image analysis array OOV (MIA, MIR) exclusively.

7.2   VALIDATION ISSUES

7.2.1   2-way

For any multivariate regression model to be used for prediction, it is important to know the pertinent predicting power. This is usually done by estimating a measure of the prediction error as a function of the deviations
between reference and predicted values. The standard statistical measure
is the root mean square error of prediction (RMSEP), which is defined as:

$$\text{RMSEP} = \sqrt{\frac{\sum_{i=1}^{n} (\hat{y}_i - y_{i,\text{ref}})^2}{n}} \qquad (7.1)$$

where $\hat{y}_i$ refers to the predicted value, $y_{i,\text{ref}}$ is the reference value and n
is the number of test objects (Martens and Næs, 1996).

The procedure of testing prediction performance is known as validation. To perform this in an optimal fashion, at least two nonoverlapping datasets are required; one for calibration and one for validation (Esbensen, 2001). When a model has been established using the calibration set, the validation set is subsequently used for predicting the $\hat{y}$-values of the validation set for comparison, e.g. according to Equation (7.1). At least two variations of this type of validation exist; one is known as 'test set validation', the other as 'cross-validation'. In test set validation a completely new, independently sampled dataset is acquired in addition to the calibration set. This demands that an identical sampling procedure is used for both datasets – but of course not identical datasets. Esbensen (Esbensen, 2001) argues that this is the only realistic basis upon which to validate the future prediction performance. Only an independently sampled test validation set involves a future sampling event as opposed to the case of cross-validation where a portion of the singular calibration set must play this role.

Thus, only if it demonstrably is not feasible to acquire a proper test set, a different, suboptimal approach will have to suffice (Martens and Næs, 1996; Esbensen, 2001). Cross-validation extracts a pseudo-validation set from the total dataset available, before building the model on the remaining complement; the extracted dataset is used for validation. This approach may take several different forms, which are all closely related however, in that they must correspond to a number of so-called segments in the list: [2,3, 4,5, ... , N], where N stands for the total number of objects in the original dataset. After the prediction error has been estimated based on prediction for the first left-out segment, this is recycled into the modeling database and a new model is created in which a different segment is kept out, hence the term cross-validation (Martens and Næs, 1996). This is continued until every segment has been used once for validation with no overlaps allowed. A special case is the so-called 'full cross-validation', in which each segment consists of only one object (where N is the number of segments). Many schools of thought exist surrounding cross-validation, most of which are based on nothing but myths. Suffice to say, that an important simplification of all possible cross-validation variants can be obtained by treating all variants under the generic principle of segmented cross-validation with the number of segments running the interval [2, 3, ... , N].

To get realistic validation estimates, reliable for the future prediction context of the model, it is essential that the calibration and validation datasets both represent an independent sampling operation from the target (parent) population. The degree of difference between these two nonoverlapping datasets is the only guarantee for capturing the same type of variations, which can be expected to be associated with the future measurement situation in which the regression model is to be used for prediction purposes. It is easy to appreciate that test set validation is the only approach which fully honors all these requirements (Esbensen, 2001).

In 2-way chemometrics there are steadfast, different opinions regarding how exactly to divide the original dataset into cross-validation segments (Martens and Næs, 1996), spanning the gamut from full cross-validation (leave-one-object-out) on the one hand, to two-segment validation, so-called 'test set switch', on the other (the latter represents a singularly unsatisfactory choice of terminology as there is no 'test set' present at all). It is always possible to use *any* intermediate number of segments from the list: [2, 3, 4, . . . , N]. The relationship between test set validation and this cross-validation systematics remains an area of considerable confusion in conventional 2-way multivariate calibration (Martens and Næs, 1996; Esbensen, 2001). These issues are not on the agenda below, except that the present image cross-validation findings shall be shown also to constitute an important caveat for conventional 2-way calibration/prediction.

7.2.2   MIA/MIR

In MIA and MIR, special validation considerations are required to which this chapter is dedicated. There are two major characteristics in image data that are not found in 2-way data.

First, and most dramatic, the number of 'objects', i.e. pixels is astronomical. In a conventional digital image one will find a total number of pixels spanning, say, 500 000 to 5 000 000 'objects', i.e. pixels all in the radiometric range [0, . . . , 255]. Full cross-validation, i.e. removing one single object or one single pixel only from this amount of data seems plainly ridiculous; this is not going to change the model based on the remaining complement of pixels. Also, calculating 0.5 to 5.0 million submodels in full cross-validation is not very tempting, even allowing for the ever increasing computer power available. This state of affairs would constitute one argument for not following the standard reasons for segmented cross-validation, but this issue is not persuasive enough by itself.

Second, and much more important, is the large redundancy that exists in image data. Pixels lying close together in scene space are often likely to represent the same image feature(s), and will therefore tend to have closely similar score values. Simple interleaved two-block datasets, say in which every second pixel is to be used for calibration and validation, respectively, can in this context be seen simply to produce two almost identical images, leading to obvious inferior validation assessments (resulting in over-optimistic prediction error estimates) (Esbensen, 2001). This example forms a case of spatial segmentation in the image, or the scene, space. Below we shall see that spatial delineation in score space forms an absolute necessity in order to comprehend what actually goes on in ICV.

An alternative procedure could be to use 'random sampling'. This corresponds to a notion of a fair, 'blind', automated segmentation strategy, unrelated to the spatial data structure in the original image. At a first glance, this would appear appealing to many data analysts – with reference to the conventional 2-way realm, where this approach forms the very foundation for random cross-validation. Considering the extremely large number of 'objects' behind MIA/MIR, sampling 50 % randomly out of 0.5 to 5.0 million objects will simply likewise produce two practically identical datasets. Using such a 50 % subset will not test any model other than in a hopelessly overoptimistic fashion. A last refuge from mounting frustration when trying to generalize from the well-known 2-way regimen into the 3-way MIA/MIR realm will in all likelihood be: 'Use a larger number of segments, 10 or more . . . '.

Below it is shown that all 'blind' segmentation strategies are doomed to certain failure in the multivariate image regimen, irrespective of the number of segments chosen – if not carefully related to the empirical covariance structure in the multivariate image. In fact, MIA requires a complete reconsideration of strategies for selecting relevant datasets for validation. A new strategy called 'guided random sampling' is suggested. In guided random sampling the user decides how the data are to be divided into the pertinent segments, but this is specifically not done in the traditional sense of random selection however, nor by a pre-specified 'blind' number of segments. It is also important to point out that neither will this constitute a subjective approach, as shall be carefully documented below, since it is based on deliberate and total respect for the empirical data covariance structure present in feature space. A new, different approach from which to attack the image data segmentation problem is required. Following the general MIA experiences this new viewpoint is to be found in score space.

The following notation is used:

X   matrix of predictor variables (2-way, 3-way arrays);
Y   matrix of dependent variables (2-way, 3-way arrays);
y   y-vector (column vector, 1-way array);
T   matrix of X-scores (2-way array);
U   matrix of Y-scores (2-way array).

7.3   CASE STUDIES

For illustration purposes, several examples based on already published multivariate image datasets will be used (Esbensen and Geladi, 1989; Esbensen *et al.*, 1993; Tønnesen-Lied and Esbensen, 2001).

The master dataset consisting of an $512 \times 512 \times 8$ image, the Montmorency Forest experimental dataset, described previously in full detail (Esbensen and Geladi, 1989; Esbensen *et al.*, 1993), is shown in Figure 7.1. For the purpose of illustrating image validation issues, the channel with lowest wavelength is here selected to serve as the Y-image in a regression context. Making use of one of eight channels to be predicted from the remaining seven (partially correlated) channels serves to illustrate the image regression case of full spectral image data coverage [X,Y] for all pixels in the original image (cf. Tønnesen-Lied and Esbensen, 2001). While this particular set-up is a solution for recovering, or repairing, a 'corrupted' channel, which is often met with in remote sensing imagery, we are not interested here in the specific prediction objective as such, only its associated validation.

In Figure 7.2 the pertinent T1–U1 score plot from this MIR application is shown. The cross-validation challenge is to divide this plot into, say, two sets (segments) that are equally representative of the covariance structure present. A score space division following this objective would entail dividing the data structure by a single straight line. However, any straight line in this plot will of necessity give rise to a significant difference between the resulting subsets on either side of this line if the data structure does not comply with a perfect joint multivariate normal distribution – which in all remote sensing imagery is but a fiction. Some features in one set will not be equally represented (if at all) in the other, and the resulting validation will on this account be unbalanced. Any score space two-splitting by a straight line will always be in danger of being unbalanced in this sense. The solution proposed here takes a novel approach to resolve this dilemma.

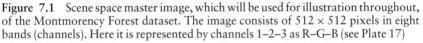

Figure 7.1 Scene space master image, which will be used for illustration throughout, of the Montmorency Forest dataset. The image consists of 512×512 pixels in eight bands (channels). Here it is represented by channels 1–2–3 as R–G–B (see Plate 17)

To resolve this issue, it is here suggested that all score space datasets are generically divided into at least eight segments, sampling both along as well as across the dominant covariance data structures, to be fully explained.

Initially the score data is split into two halves with a cut along the main covariance direction. In Figure 7.2 this would be a line passing through the two modes of the highest frequencies of pixels with similar score signatures, i.e. two topographic 'peaks' (cf. Geladi and Grahn, 1996; Esbensen and Geladi, 1989), which are colored red and orange in Figure 7.3. Each of these parts should now contain approximately 50 % of the objects, and all main image objects should be represented – but only those classes which contribute to the dominating elongated covariance trend.

Then, intersecting the first line, a new line should be drawn representing the second most important covariance direction, as judged from the pertinent MIA score cross-plot(s). It is important that this direction corresponds to the second most representative part of the overall covariance structure (more examples to be given below). Thus there are

Figure 7.2   T1–U1 score plot from MIR analysis of the image in Figure 7.1, based on all 7 + 1 channels [X,Y]. Note a few 'corrupted' artifacts appearing as horizontal lines out of correspondence with the general covariance structure, caused by channel no. 7; these artifacts are of no consequence for the theme in the present work (see Plate 18)

no requirements for orthogonality of these two user-delineated covariance directions. This gives four segments, which each ideally should contain approximately 25 % of the objects (there is no demand for precisely 25 % due to the high redundancy present) – barring whatever 'surprises' may be in waiting in the higher-order components not captured in the first, low(er) order score space renditions. This first (generally oblique) axis-cross delineation is all the user has to supply in order for the new cross-validation procedure to take over. The following description relates to Figure 7.3.

After the user has drawn the second covariance trend line, the ICV algorithm automatically delineates four additional bisecting lines. This approach only involves locating the intersection point between the user-supplied first two intersecting lines, after which it is an easy task to calculate also the midpoints between the corners of an enveloping outer frame. Bisecting lines are drawn between these midpoints and the central intersection point. An example of the resulting eight-segment mask is also shown in Figure 7.3. This configuration illustrates a generic eight-segment mask. It is the user's obligation to delineate the first

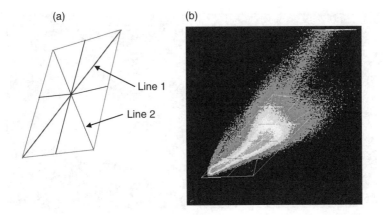

Figure 7.3 (a) ICV segmentation as initiated by two intersecting master lines drawn by the user (Line 1 and Line 2). (b) Example of subsequent automatic eight cross-validation segments defined in score plot T1–U1. Note that outlying pixels can be *excluded* already when delineating this mask (if decided by an *informed* analyst) (see Plate 19)

axis-cross specifically following the T–U (or T–T) score plot disposition as given by the MIR (or MIA) decomposition.

With this type of mask, there are now three functional combinations of subsets available, consisting of eight, four or two validation segments, respectively. When selecting and/or combining sets, the essential requirement is that segments must be mirror opposites with regard to the center point. This is a crucial feature of the proposed ICV approach.

Figure 7.4 shows two compounded sets used as a basis for two-segment cross-validation. In general all such nonoverlapping, two-fold divisions of image covariance structure take the form of a Maltese cross (also known as the Templar cross), as illustrated in Figure 7.4. Notice in Figures 7.3 and 7.4 how some obvious, outlying features can been excluded already in this first stage of cross-validation segmentation (by the informed, reflecting image analyst). Should the image analysts so desire, the first axis cross may alternatively include any part of the score structure. ICV has been designed to allow the user complete freedom with respect to the central score space masking operations, following the general MIA philosophy (Esbensen and Geladi, 1989; Geladi and Grahn, 1996).

7.3.1 Case 1: Full Y-image

The first case is a study of what was found in Tønnesen-Lied and Esbensen (Tønnesen-Lied and Esbensen, 2001), to be a comparatively

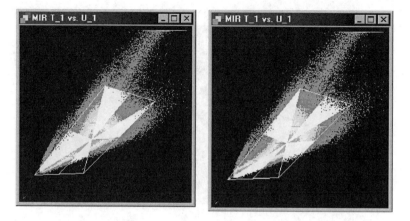

Figure 7.4 Two *complementary* 'Maltese cross' validation datasets selected in the T1–U1 score plot shown in Figure 7.2. Note how both achieve good data structure representation (except for the deliberately left out *outlier* region) (see Plate 20)

rare situation in image analysis; the full Y-image. In this situation, each object in X, each pixel, has a one-to-one corresponding representation in Y. This furnishes a particularly illustrative example of ICV to be outlined in full detail below. A more common situation is studied in case 3.

While Figure 7.4 shows the two validation datasets in the score plot, Figure 7.5 displays the same data in image space. Pixels marked in white in the scene space correspond to the active set of four segments in Figure 7.4.

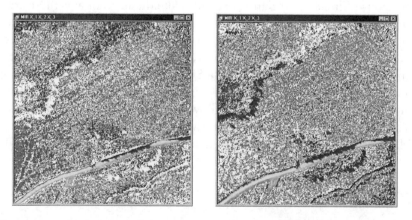

Figure 7.5 The two complementary validation segments selected in Figure 7.4 projected into the original image space. Note how both achieve satisfactory *general coverage* and spatial representation, except for minor *outlier* parts (see Plate 21)

Some marginal parts of the data were left out of the validation set entirely, because these pixels were identified as outliers already when delineating the problem-specified Maltese cross region of interest. Alternatively outlier removal can be refined by making a local model (Geladi and Grahn, 1996; Esbensen and Geladi, 1989) prior to cross-validation, allowing only the specific, problem-dependent features to be represented in the score plot (standard MIA issues).

Studying the images in Figure 7.5, it is fair to say that these two datasets both are good general representations of the original image, covering fairly well the same features at the scale of the full image, with only a small difference at the most detailed levels.

The user has the full ability to iterate his or her first tentative delineations of the Maltese cross configuration by careful inspection of the resulting disposition of the two nonoverlapping scene space renditions (Figure 7.5), until a fully satisfactory and interpretable result has been achieved.

Figure 7.6 shows what happens if eight segments were to be used independently as in a conventional eight-segment cross-validation. Obviously there are extreme differences between these eight datasets, in fact there is an absolute certainty that these submodels will be almost totally incommensurable with each other. This is also a dramatic illustration of the general cross-validation 'problem' when the relationship between the X- and the Y-space is complex. In the present image analysis example, it is visually evident what goes wrong, when using a 'blind' eight-segment (12.5 %) cross-validation scheme in which the data structure is not observed.

It cannot be emphasized too strongly, that Figure 7.6 constitutes a veritable death sentence over conventional 'flying blind' segmented cross-validation in the standard 2-way context.

7.3.2   Case 2: Critical Segmentation Issues

It is possible to run into various intrinsic problems with this new approach if proper insight is not applied in the Maltese cross segmentation step. If the segments are too small, they will very likely not be able to be representative for the entire dataset as is illustrated in Figures 7.7 and 7.8. Another possibility, as will be shown in case 2, is related to not drawing the optimal guiding lines for the Maltese cross.

Figures 7.7 and 7.8 show what happens when the guiding lines split the data in an off-centered fashion relative to the objective, most

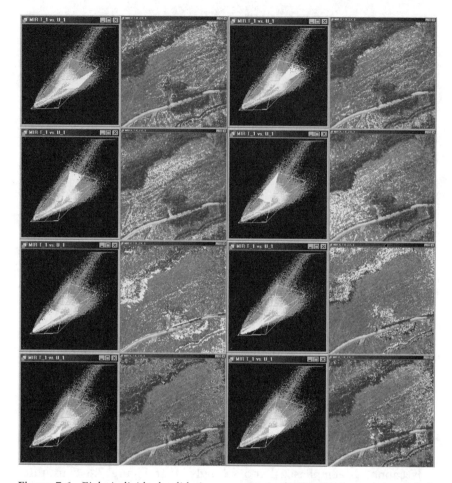

Figure 7.6 Eight individual validation segments in the T1–U1 score plot and their corresponding image space renditions. None of these segments achieve either data structure or spatial representativity, from which a 'blind' cross-validation would result in nonsensical, nonrepresentative performance (see Plate 22)

prominent mode in the score space. Clearly the two complementary Maltese cross configurations are not making up a balanced (i.e. roughly 50/50%) cross-validation base. As can be seen, even a small off-centered delineation has a dramatic effect on the coverage and the relative size of the two datasets because of the very high number of similar pixels making up the backbone of the covariance data structure. When one dataset is significantly different from the other, the result is obvious: a poor, nonrepresentative validation result is unavoidable.

While the new approach is sensitive to the relevance of, and the understanding behind, the critical user interaction and first delineation

(a)   (b)

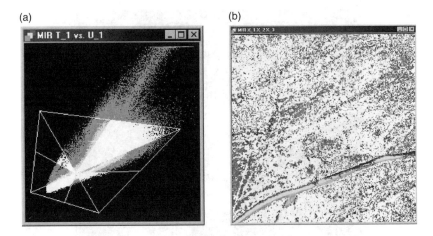

Figure 7.7 Corresponding (a) score plot (T1–U1) and (b) scene space image for an *off-centered* Maltese Cross. The alternative 50 % validation set is shown in Figure 7.8. Off-centering is the result of ill-informed, or outright sloppy acknowledgement of the data structure information by the image analyst (see Plate 23)

(a)   (b)

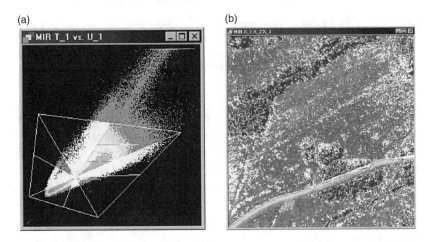

Figure 7.8 Corresponding (a) score plot and (b) scene space image for an *off-centered* Maltese cross. The complementing segments are shown in Figure 7.7 (see Plate 24)

of the Maltese cross, this is far from fatal however. Far from being a weakness, this user-dependency actually opens up very powerful data analysis and cross-validation.

Another potential problem is when the modes (the 'peaks') in the score plot do not lie on a straight line. If there are more than two peaks of interest, drawing a representative two-split line through them can of

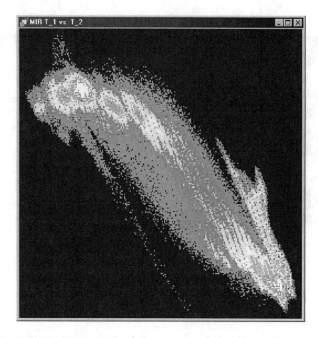

Figure 7.9 MIA T1–T2 score plot from a very complex dataset showing an 8-mode ('8-peaked'), curved covariance data structure. Observe how it is almost impossible to apply a Maltese Cross segmentation on a data structure as complex as this. See text for details on how to circumvent this type of problem (see Plate 25)

course be difficult, if not impossible. This problem is illustrated well by a score plot from a different representative dataset (Tønnesen-Lied and Esbensen, 2001) (Figure 7.9). This example illustrates clearly why multivariate image analytical endeavors usually are an order of magnitude more complex than in the ordinary 2-way regimen. Observe how it is almost impossible to apply the simple Maltese cross segmentation on a data structure as complex as this. However, there is more than one way to resolve how to segment even such complex covariance data structures (Esbensen and Geladi, 1989; Esbensen *et al.*, 1992; Esbensen *et al.*, 1993; Geladi and Grahn, 1996; Tønnesen-Lied and Esbensen, 2001).

7.3.3   Case 3: Y-composite

More common than having access to the full Y-image, is the case when the X and Y images are constructed as composites from several smaller images. This is a useful approach when making a reference dataset as a basis for a regression model. A typical 'composite image'

(a)   (b)

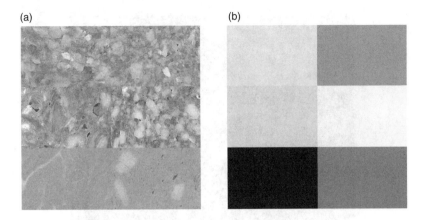

Figure 7.10 Illustration of the composite Y-image case. (a) Six sausages (X-image, RGB) and (b) corresponding fat content (Y-image). The fat content has been rescaled to [0,255] grey-levels

is shown in Figure 7.10. This image consists of six smaller images of different foodstuff (sausages in this case), imaged for analytic textural purposes. For the present purpose, the corresponding Y-image contains the averaged analytical results for overall fat content for each subimage only. Thus fat content is represented as a common grey-level for all pixels in the relevant subimages (Figure 7.10). This data set-up was discussed extensively in Tønnesen-Lied and Esbensen (Tønnesen-Lied and Esbensen, 2001), where it was used as a vehicle for explaining the concepts of MIR. In the present context the objective for the MIR set-up is to be able to predict the average fat content in the six rather heterogeneous sausages.

As can be seen from the Y-image in Figure 7.10, there is no longer a unique Y-value for each pixel in the X-image. This phenomenon occurs when an average value is to be predicted from an X-image. This situation has a very clear destructive effect on the conventional T–U score-plot, shown dramatically in Figure 7.11.

In Figure 7.12 it is demonstrated that applying the eight-segment Maltese cross scheme on a T–U plot, as the one in Figure 7.11, is not straightforward. The nature of the T–U plot in the composite Y-image case will force an uneven distribution in the image-space, quite irrespective of the details of how the eight-fold segmentation mask is delineated, leading to extremely unbalanced delineations in the image scene space (Tønnesen-Lied and Esbensen, 2001).

In Figure 7.12 it is evident that especially the two lower and the right center X subimages are very poorly represented in the complementary

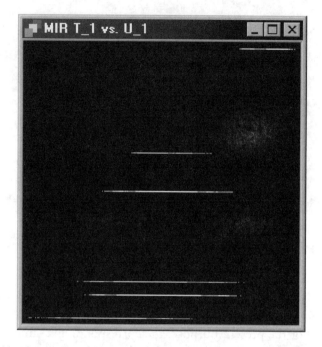

Figure 7.11 Conventional T1–U1 score plot from the sausage fat prediction case 2. Each horizontal line represents a specific Y-value, or subimage [cf. Figure 7.10(b)]

validation segments. This is even more evident if the eight individual segments were to be illustrated sequentially as shown in the first example in Figure 7.6. To save space, this is not repeated for the current case.

Thus what initially seemed to be a good idea, i.e. the 'Maltese cross' eight-fold cross-validation segmentation in the T–U score space, on further inspection has proved to be, at best, a very sensitive approach. It will be shown however, that this is merely a question of application.

The critical issue is not so much how the lines are drawn in the plot(s), it is in which plot the lines are drawn. So far, the procedure has been applied to plots where there are strong correlation in the data, the familiar low-order score plot(s), e.g. T1–U1, etc., which all play a dominating role in conventional 2-way multivariate calibration (Martens and Næs, 1996; Esbensen, 2001). Chemometricians will be familiar with the fact that in the score space, the first dimensions contain the most structured parts of the data, while for the higher-order components there is bound to be less and less variance.

With this in mind, the next steps will surely appear surprising. In the present image exploration context it is nevertheless proposed to focus explicitly on this higher-order score space.

Plate 1

Plate 2

Plate 3

Plate 4

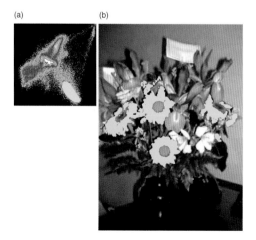

Plate 5

(a) (b) (c)

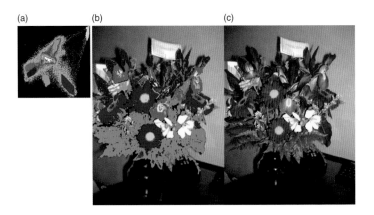

Plate 6

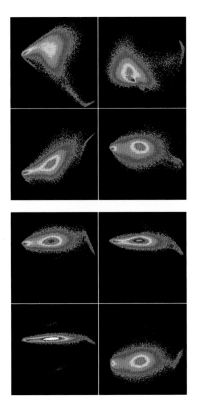

Plate 7

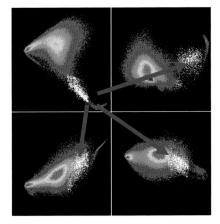

Plate 8

Plate 9

(a) (b)

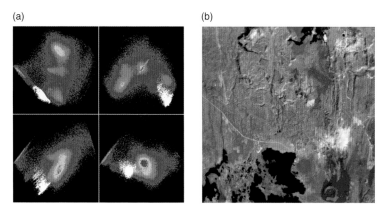

Plate 10

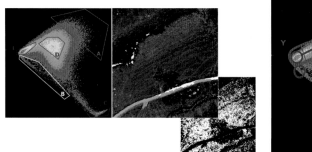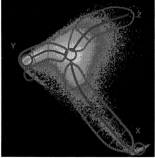

Plate 11   Plate 12

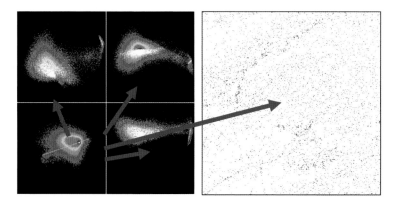

Plate 13

(a)

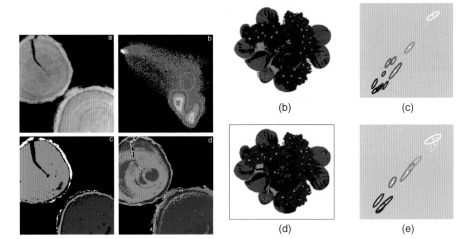

Plate 14

Plate 15

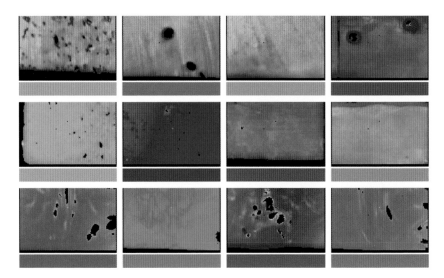

Plate 16

Plate 17

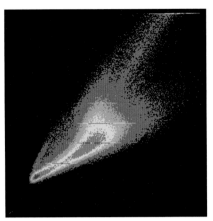

Plate 18

(a)

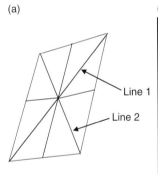

Line 1

Line 2

(b)

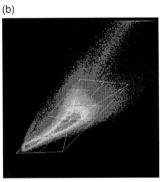

Plate 19

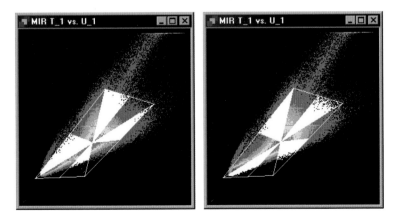

Plate 20

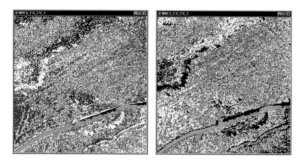

Plate 21

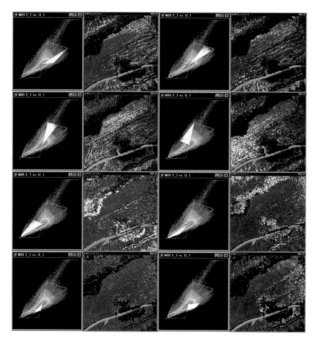

Plate 22

(a) (b)

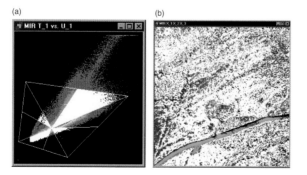

Plate 23

(a)
(b)

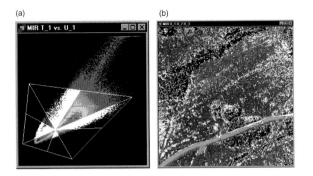

Plate 24

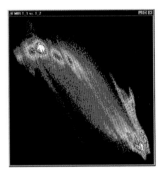

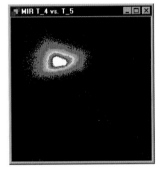

Plate 25   Plate 26

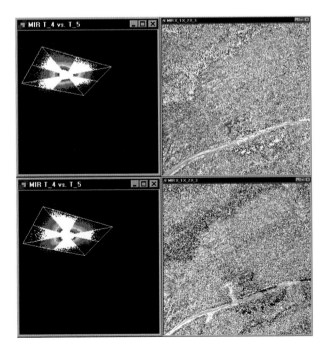

Plate 27

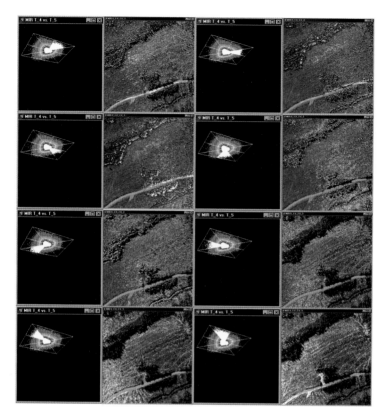

Plate 28

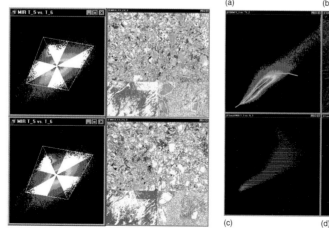

Plate 29

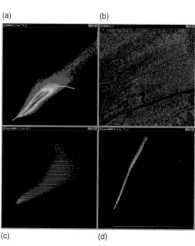

Plate 30

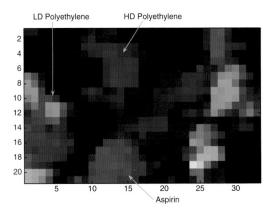

Plate 31

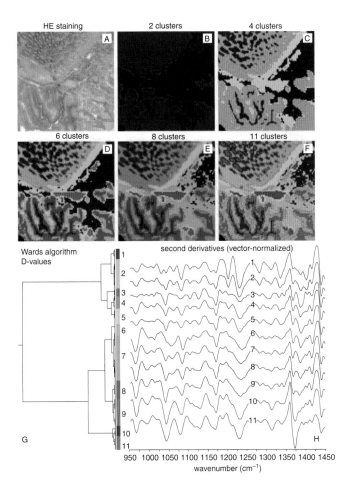

Plate 32

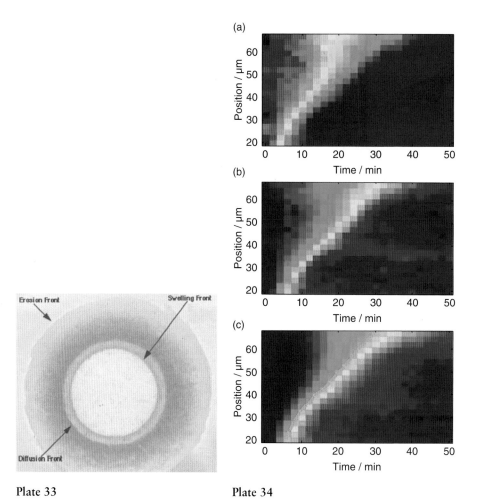

(a)

(b)

(c)

Plate 34

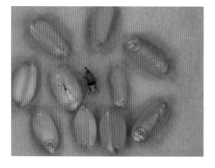

Plate 33

Plate 35

Figure 7.12 Composite Y-image case with two-block Maltese cross segmentation from selection in T1–U1 plot in Figure 7.11. Note extensive *imbalance* in the image space (right)

Figure 7.13 shows such high-order T scores (T4 versus T5) from the Montmorency Forest example. What is interesting in this plot is that most of the structural information is now orthogonal to the covariance data structure delineated in this particular score cross-plot. Notice also that even the outlying data parts have been subsumed into this quasi-normal ('hyper-spherical') distribution. This indicates that the current plot is well suited as a starting point for comprehensive cross-validation segmentation of the type under development.

Applying the Maltese cross-validation segmentation to the score plot shown in Figure 7.13, results in Figure 7.14, which shows the resulting two, nonoverlapping segments both in score space and image space. There is now a satisfactory, very even coverage in these two 50/50 segments (only very close investigation revealed some minor differences between the complementary image-space representations).

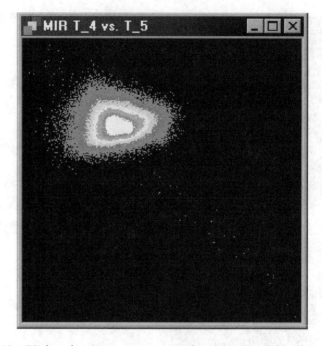

Figure 7.13 'High-order' components score plot (T4 versus T5 in this case) from the Montmorency Forest dataset (see Figure 7.1) (see Plate 26)

The conclusions from Figures 7.13–7.15 are obvious: when delineating the new image analytical eight-fold cross-validation segmentation in an appropriate high-order score space rendition, in which the overwhelming parts of the data structure are orthogonal, the sampling sensitivity associated with the first components is now satisfactorily controlled, indeed almost completely eliminated.

Figure 7.15 illustrates the coverage displayed by the eight individual segments. It is evident that eight individual score-space segments, even when delineated in the optimal high-order component basis, lead to distinctly unbalanced pixel divisions in the corresponding image space.

From this exposition it is concluded that many such segments always must be combined to form larger fractions of the entire field of view, e.g. compounded to constitute two 50 % segments as shown in Figure 7.14.

Stepping back to the difficult Y-grid example in case 2 (prediction of sausage fat content), it will be interesting to see how this high-order components approach fares. Using T-scores 5 versus 6 and drawing the two lines that split this dataset in as equally representative fashion as possible produces the segments shown in Figure 7.16.

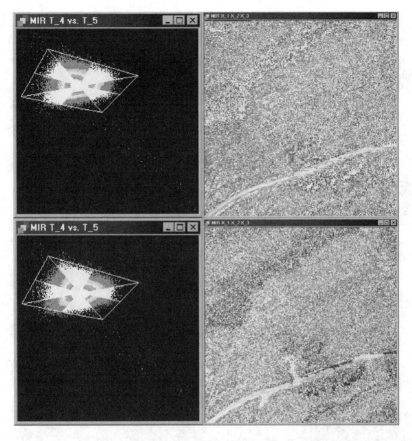

Figure 7.14 Maltese cross-validation segments selected in higher-order T4–T5 score plot from Figure 7.13. Note equal representative data structure coverage in *both* score and image space (see Plate 27)

Compared with Figure 7.12, Figure 7.16 shows a strikingly much more uniform coverage of the one composite validation segment in the image with respect to the complementary set – there are indeed only a very few, minor differences. Overall, this partition will lead to a realistic validation of the prediction model performance even for this very complex data structure.

7.3.4   Case 4: Image Data Structure Sampling

One of the key features of image analysis is the huge redundancy. Having between 500 000 and 500 000 objects describing typically up to 10–20

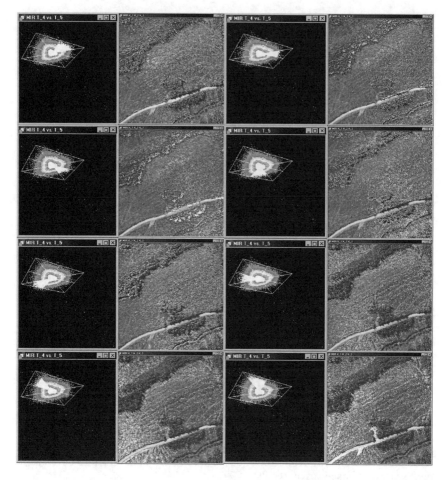

Figure 7.15 Eight segments from the T4–T5 score plot cross-validation split shown in both score and image space. Even with the optimal high-order segmentation, there is still an unsatisfactory unbalanced pixel coverage in each *individual* segment (see Plate 28)

classes (often substantially fewer though) is obviously class overkill. In MIR cases where reducing this redundancy is essential, it is possible to reduce the number of objects dramatically by a simple new image sampling procedure [cf. Esbensen and Geladi (1989) in which this case was described for MIA].

This new image sampling facility suggestion is shown in Figure 7.17 in the form of a curved (hand-drawn) line, where the number of pixels has been reduced to a small fraction of the original, yet deliberately covering all the important classes of interest in the image. This is so

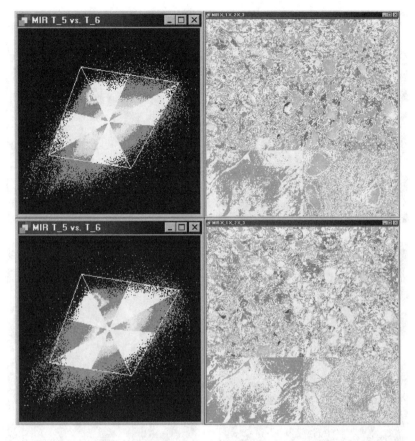

Figure 7.16 Cross-validation segments in [T5,T6] score and the [1,2,3] image space for the sausage prediction example. Note that an acceptable representation has been achieved in both score and image space (cf. Figure 7.12) (see Plate 29)

because the line has been drawn specifically to 'cover' the most dominating global covariance trend of the image feature space. Since this mask is positioned directly along the 'topographic' highs (cf. Esbensen and Geladi, 1989), it will, of necessity, be maximally representative for the essential data structure present while at the same time allowing for the exclusion of all similar pixels lying outside its width (typically one – three pixels wide) without risk of losing any of the essential representative structures. Observe how we have made use of this feature in the so-called 'Pred.–Meas.' plot (predicted versus measured), well-known from conventional 2-way multivariate calibration validation.

(a)   (b)

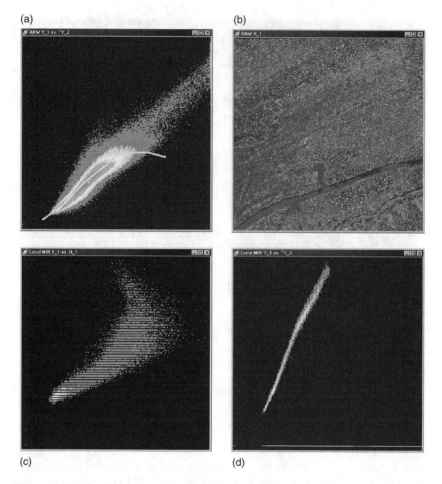

(c)   (d)

Figure 7.17 Freehand 'sampling line' (blue) covering the essential covariance data structure in a Pred.–Meas. plot (a). After a local MIR model has been created on this basis, the corresponding new T1–U1 plot (c) and the complementing new Pred.–Meas. plot (d) are clearly a valid representative sampling of the original MIA/MIR data. The image space coverage is shown in (b) (see Plate 30)

Starting out in the strongly correlated Pred.–Meas. plot, a one-pixel-wide line is drawn covering all the main data structure of interest. This line follows the global covariance trend backbone as best as possible. All pixels covered by this line are then used as objects in a new, local model (Esbensen and Geladi, 1989; Geladi and Grahn, 1996).

The local model will contain far fewer objects – the redundancy in the original data will be strongly reduced. In Figure 7.17(c) a T1–U1 plot for this local model is shown. This can now be used as a starting point for the Maltese cross-validation segmentation for example. The local model Pred.–Meas. plot is shown in Figure 7.17(d).

Technical comment. The points and line that can be observed in the lower part of the Pred. – Meas. plot represent the objects that have been *left out* of the model (outliers). In the calibration procedure, they have been removed from the data modeling, but for image displaying purposes, it is necessary to include these pixels. To avoid them from interfering with the image, they are set to zero value, and are displayed black in the image. The lower-left point is hence the (0,0) coordinate, as all score values are scaled in the range [0–255] to optimize the display.

Figure 7.17(a) and (b) represents a novel, powerful way to subsample a multivariate image. Because of its design purpose, the manually drawn 'freehand' line can be made to follow all desired data structure as closely as desired (barring outliers). This feature makes for an unprecedented fidelity with respect to the degree of representativeness of the selected pixel subset. A comparison between Figure 7.17 and Figures 7.6–7.8 and 7.16 testifies to a highly acceptable coverage in Figure 7.17, clearly representing all the major pixel classes present (Tønnesen-Lied and Esbensen, 2001). If the number of pixels selected by this procedure happens to be too big, it is trivial to set up scene space pruning on a random selection basis.

The main issue here is that the most meaningful subsampling of the image is not set up in the scene space, but in score space, indeed following MIA original intentions to the letter (Esbensen and Geladi, 1989; Esbensen *et al.*, 1992; Geladi and Grahn, 1996; Tønnesen-Lied and Esbensen, 2001).

7.4 DISCUSSION AND CONCLUSIONS

A significant level of informed experience is needed when employing the Maltese cross segmentation facility on low-order component plots (e.g. T1–U1), in which significant, indeed severe, 'off-center' sensitivity was demonstrated.

A new approach for image validation is proposed in which segmentation takes place in an orthogonal data representation in higher-order score components. This is a powerful addition to the MIA/MIR regimen, which enables a realistic two-split cross-validation foundation (acceptably close to 50/50 coverage in all scenarios). The Maltese cross as applied to high-order component score plots will always result in two segmentation composites which are nonoverlapping 'mirror' configurations of each other.

This starting point enables an optimal, representative split of any image training dataset (calibration set) representing all important covariance data structures, making it ideal for image cross-validation purposes. In general it is not recommended to use cross-validation in MIA with a number of segments higher than two, and then only in the form of the composite Maltese cross due to much higher complexity in multivariate image data relative to experiences from the conventional 2-way realm.

In MIA, there is usually a high degree of redundancy in the data structures expressed. In such cases with relatively few physical objects (classes), data reduction with local modeling should be considered prior to validation. We have delineated a simple approach for this – the one-pixel-wide sampling swath closely following the backbone of the dominating data covariance structure(s). This is a novel image sampling facility with many applications because of its simplicity and its guaranteed representativiness in image space, following the MIA general score space imperative (Geladi and Grahn, 1996; Tønnesen-Lied and Esbensen, 2001).

7.5 REFLECTIONS ON 2-WAY CROSS-VALIDATION

The above presentation of ICV also serves to put a spotlight on conventional 2-way validation, the issue of the vastly different number of pixels notwithstanding.

There is superior information value in T–U plots. The T–U plot furnishes the only insight into the empirical (spectral) data structure, necessary to delineate the most relevant segmentation for validation purposes. Simple carrying-over from conventional 2-way data analysis into the image realm, will lead to overemphasis on the image space far too early in the image analysis process.

The T–U plot is critical for delineation of the spectral [X,Y] data structures [scores, loading weights]. This is in contrast with the fact that some revered PLS algorithm implementations do not make use of U-scores for decomposing the Y-space. Historically, within chemometrics, there has been a marked focus on modeling issues relative to the prediction/validation issues, resulting in a wealth of multivariate modeling approaches emphasizing algorithmic issues alone. This has resulted in a very popular focus on certain algorithmic PLS solutions only using T-scores for both X- as well as Y-space, due to the fact that prediction results are similar.

While this is true for the prediction values in the narrow sense, the price to be paid includes any insight into the interrelated T–U score relationships – a very high price indeed, and to these authors an unacceptable price. Precluding even the possibility for insight into the specific X–Y interrelationships (the very heart of the matter for any regression model), is exactly that which unavoidably leads to a preference for 'blind' segmentation. This in turn leads to a dependence on "schools of thought" regarding what constitutes a proper validation approach, e.g. a preference for a certain number of segments, etc., as the universal weapon to be directed at all data structures regardless of their internal [X,Y] relationships.

There is here a clear and present danger for losing out on all data structure irregularities that do not correspond to an archetypical assumption of a joint multi-normal distribution, which in the image field has been shown to be a total fiction (Esbensen and Geladi, 1989; Esbensen et al., 1992; Esbensen et al., 1993; Geladi and Grahn, 1996; Tønnesen-Lied and Esbensen, 2001; Huang et al., 2003). ICV puts the traditional use of 'blind' segmentation on notice – especially in combination with not using U-scores for the Y-space.

Chemometric data analysts in the 2-way realm are familiar with non-normally distributed T–U data structure distributions however. Any such deviation will be picked up by the T–U score plot – but no information will ever be present as to this issue when not employing both T- and U-scores. There are many well-known PLS-modeling/prediction examples from chemometrics which use no U-scores at all, while simultaneously steadfastly recommending only using leave-one-out cross-validation – there is here a fatal inherent danger, indeed certainty, of inferior 'blind' validation results.

The present ICV illustrations speak very clearly against this state of affairs:

1. 'Blind' cross-validation is unscientific. It is based on an untenable assumption of multivariate joint normal distribution which almost never holds up in real world datasets. It is quite unacceptable to generalize from the very few counterexamples available.
2. Leave-one-out cross-validation is actually relevant, indeed only acceptable, in one scenario only – the proverbial 'small sample' case of a very small number of objects, say below 20. Any other use of this cross-validation option should be discontinued.

3. Test set validation is the only scientific option in view of the manifest objective of validation: assessment of the future prediction performance.

REFERENCES

Bro, R. (1997) PARAFAC. Tutorial and applications. *Chemometrics and Intelligent Laboratory Systems*, **38**, 149–171.

Esbensen, K. (2001) Multivariate *Data Analysis – in Practice*, 5th Edn. CAMO, Oslo.

Esbensen, K. H. and Geladi, P. (1989) Strategy of multivariate image analysis (MIA). *Chemometrics and Intelligent Laboratory Systems*, **7**, 67–86.

Esbensen, K. H., Wold, S. and Geladi, P. (1988) Relationships between higher-order data array configurations and problem formulations in multivariate data analysis. *Journal of Chemometrics*, **3**, 33–48.

Esbensen, K., Geladi, P. and Grahn, H. (1992) Strategies for multivariate image regression (MIR). *Chemometrics and Intelligent Laboratory Systems*, **14**, 357–374.

Esbensen, K. H., Edwards, G. and Eldridge, G. (1993) Multivariate image analysis in forestry involving high resolution airborne imagery. In: *Proceedings of the 8th Scandinavian Conference in Image Analysis*, NOBIM, Tromsø, pp. 953–963.

Geladi, P. and Grahn, H. (1996) Multivariate Image Analysis. John Wiley & Sons, Ltd, Chichester.

Huang, J., Wium, H., Qvist, K. B. and Esbensen, K. H. (2003) Multi-way methods in image analysis–relationships and applications. *Chemometrics and Intelligent Laboratory Systems*, **66**, 141–158.

Martens, H. and Næs, T. (1996) *Multivariate Calibration*. John Wiley & Sons, Ltd, Chichester.

Smilde, A., Bro, R. and Geladi, P. (2004) *Multi-way Analysis: Applications in the Chemical Sciences*. John Wiley & Sons, Ltd, Chichester.

Tønnesen-Lied, T. and Esbensen, K. H. (2001) Principles of MIR, multivariate image regression. I: Regression typology and representative application studies. *Chemometrics and Intelligent Laboratory Systems*, **58**, 213–226.

8

Detection, Classification, and Quantification in Hyperspectral Images Using Classical Least Squares Models

Neal B. Gallagher

8.1 INTRODUCTION

The classical least squares (CLS) model is most often used in spectroscopy and is directly applicable to multivariate image (MI) analysis. (The acronym MI can be applied to hyperspectral, omnispectral, gigaspectral, etc., images.) CLS is extremely useful because it is easily interpretable from a chemical perspective; it estimates concentrations from spectra. However, the model generally requires a 'spectrum' for every source contributing to the measured signal. This is because CLS includes interferences in the model explicitly. Interferences include spectra from both chemical and physical sources and, in many cases, obtaining spectra for the interferences can be a difficult task. In contrast, inverse least squares (ILS) models, such as partial least squares (PLS) and principal component regression (PCR), model interferences implicitly. This ability to implicitly account for interferences is often given as one of the biggest advantages of ILS over CLS. However, variations of CLS that use the extended mixture model and generalized least squares

Techniques and Applications of Hyperspectral Image Analysis Edited by H. F. Grahn and P. Geladi
© 2007 John Wiley & Sons, Ltd

can be used to explicitly account for interference in the signal. It is the objective of this chapter to discuss CLS-based models for use in hyperspectral image analysis (HIA).

The reason for the interest in CLS-based models has already been stated; the models are easily interpretable from a chemical and physical perspective. For example, CLS results in concentration images rather than scores images from ILS models. Concentration images are gray-scale images (they might be colored) that have an estimate of the concentration of each analyte in each pixel. In contrast, scores images tend to be linear combinations of all the analytes present in each pixel. The result is that concentration images are far easier to interpret than scores images. Concentration images of known interferences as well as target analytes can also be developed. All these pieces are useful information provided by CLS models in HIA. For this reason, and perhaps also in part due to the gaining popularity of multivariate curve resolution and remote sensing applications, CLS models are being 'rediscovered'. Further discussion of the differences between CLS and ILS models can be found in Sanchez and Kowalski (Sanchez and Kowalski, 1988).

This chapter discusses the practical aspects of using CLS models with the objective of stating model assumptions and potential problems for practical application of CLS to HIA. It is not the intention to provide all the detailed statistical properties of each model. Section 8.2 introduces CLS, the extended mixture model [resulting in extended least squares (ELS)], and finally the generalized least squares (GLS) model. Section 8.3 uses an example to show how the GLS model can be used for detection, classification, and quantification for MIs. The chapter finishes by discussing possible future directions.

8.2   CLS MODELS

Three CLS-based models are discussed for application to HIA: CLS, ELS (based on the extended mixture model) and GLS. The latter two models are designed to account for interferences in the measurements. For additional detail on the models see Martens and Næs (Martens and Næs, 1989).

The three models are linear mixture models, however it should be noted that some systems are nonlinear. For example, reflectance spectroscopy can result in a nonlinear response (Gallagher *et al*, 2003a; Nascimento and Bioucas Dias, 2005) and the model can be quite complex (Hapke, 1993; Bänninger and Flühler, 2004). In some cases, data pre-processing such as Kubelka–Munk or log transforms can be used to help linearize the data (Griffiths, 1996), and other methods can be used to help account for scattering in the data (Martens et al., 2003; Gallagher et al., 2005). Linear response with concentration would be best if it could be obtained, but even when the system is nonlinear, linear models can provide insight into HIA. Therefore, data pre-processing will not be discussed and the focus will be on linear models.

8.2.1 CLS

CLS is an extremely useful model for spectroscopic analysis because it provides quantitative and interpretable chemical information (Martens and Næs, 1989). The CLS model can be considered a multi-component Beer's law model as given in Equation (8.1):

$$X = CS^T + E \qquad (8.1)$$

where X is a $M \times N$ matrix of measured spectra with each row considered an object and each column a spectral channel, C is a $M \times K$ matrix of concentrations, S is a $N \times K$ matrix of spectra, and E is a $M \times N$ matrix of residuals. The columns of C and S correspond to K different analytes, the rows of C correspond to individual objects (e.g. pixels), and each column of S is an analyte spectrum. This model can be considered a force-fit through zero, i.e. when all the concentrations are zero, the measured signal should be zero. If the spectra S are known, the least squares estimate of the unknown concentrations C is given by:

$$\hat{C} = XS(S^TS)^{-1} \qquad (8.2)$$

where the ˆ indicates an estimate.

For a given set of spectra S, the CLS model can be applied to images. To do this, image matricizing is introduced (Smilde et al., 2004). For example, for a $M_x \times M_y \times N$ image $\underline{X}$, i.e. each image at each spectral channel $n = 1, \ldots, N$ has $M_x \times M_y$ pixels, the matricized form of the image is a $M \times N$ matrix X where $M = M_x M_y$. The matricized form of the measured image X is used in Equation (8.2). The next step is the inverse matricizing operation, i.e. the $M \times K$ matrix $\hat{C}$ of estimated concentrations is rearranged to a $M_x \times M_y \times K$ image $\underline{\hat{C}}$. Each of the $k = 1, \ldots, K$ layers then corresponds to a $M_x \times M_y$ 'concentration image' for the kth analyte. (For ease of reading, the $\hat{\ }$ is dropped from the estimates in the remainder of this chapter.)

Associated statistics can also be estimated for the CLS model. Definitions of Hotelling's T^2 and Q residuals used here follow the analogous statistics used in principal components analysis (Jackson, 1991). The residual for the mth pixel, e_m, is a $1 \times N$ vector given by:

$$e_m = x_m - c_m S^T = x_m - x_m S \left(S^T S\right)^{-1} S^T = x_m \left(I - S \left(S^T S\right)^{-1} S^T\right) \quad (8.3)$$

The residual e_m is also called the contribution to Q for the mth pixel, where Q_m is the sum-of-squared residuals given by:

$$Q_m = e_m e_m^T = x_m \left(I - S \left(S^T S\right)^{-1} S^T\right) x_m^T \quad (8.4)$$

Q images can be used to find pixels that do not lie in the subspace spanned by S, i.e. pixels that have contributions to the signal that are not well modeled by the set of spectra S. The source of this unusual variation can be due to new analytes not included in S, 'bad' pixels, instrument malfunction or other sources that lead to variation not captured by the model spectra. Plots of the residual contributions vector e_m can provide insight as to the source of the unusual variation.

Hotelling's T^2 statistic (Anderson, 1971) for concentrations in the mth pixel is be given by:

$$T_{C,m}^2 = (c_m - \bar{c}) \left(\frac{(C - 1\bar{c})^T (C - 1\bar{c})}{M - 1}\right)^{-1} (c_m - \bar{c})$$

$$= (c_m - \bar{c}) \left[\mathrm{cov}\,(C)\right]^{-1} (c_m - \bar{c}) \quad (8.5)$$

where c_m is a $1 \times K$ vector of the estimated concentrations for the mth pixel, $\bar{c}$ is a $1 \times K$ vector column mean of C, and 1 is a $M \times 1$ vector of ones. T_C^2 images can be used to identify pixels with unusual concentrations. T_C^2 contributions can be obtained by examining Equation (8.5). But first it is useful to discuss estimation of the inverse of the covariance matrix of C, $\mathrm{cov}\,(C)$ as it will be revisited later in this chapter. The inverse can be easily obtained by first decomposing $\mathrm{cov}\,(C)$ using the singular value decomposition:

$$V S_{v,C} V^T = \mathrm{cov}\,(C) \quad (8.6)$$

where $\mathbf{V}$ is a $K \times K$ matrix with orthonormal columns corresponding to the right singular vectors, and $\mathbf{S}_{v,C}$ is a diagonal matrix containing the singular values of $\mathrm{cov}(\mathbf{C})$. The subscript v indicates that $\mathbf{S}_{v,C}$ is a matrix of singular values and is not a matrix of spectra. Note that since $\mathrm{cov}(\mathbf{C})$ is symmetric the matrices of left and right singular vectors are equal, and only the right singular vectors are used. The inverse of $\mathrm{cov}(\mathbf{C})$ is obtained by simply inverting $\mathbf{S}_{v,C}$ and Equation (8.5) can then be written as:

$$T_{C,m}^2 = (\mathbf{c}_m - \overline{\mathbf{c}})\, \mathbf{V}\mathbf{S}_{v,C}^{-1}\mathbf{V}^T\,(\mathbf{c}_m - \overline{\mathbf{c}}) = (\mathbf{c}_m - \overline{\mathbf{c}})\, \mathbf{V}\mathbf{S}_{v,C}^{-1/2}\mathbf{V}^T\mathbf{V}\mathbf{S}_{v,C}^{-1/2}\mathbf{V}^T\,(\mathbf{c}_m - \overline{\mathbf{c}})$$
(8.7)

The right-hand expression for $T_{C,m}^2$ provides a definition for the contributions to Hotelling's T^2, $\mathbf{t}_{C,con}$. For the mth pixel this is:

$$\mathbf{t}_{C,con,m} = \mathbf{V}_1\mathbf{S}_{v,1}^{-1/2}\mathbf{V}_1^T\,(\mathbf{c}_m - \overline{\mathbf{c}})$$
(8.8)

Similar to contributions to Q, this definition for the T^2 contribution has the property that the sum-squared contributions for the mth pixel $\mathbf{t}_{C,con,m}$ yields the original $T_{C,m}^2$. Inspection of plots of $\mathbf{t}_{C,con}$ can help determine the source of unusual variation observed in Hotelling's T^2 statistic, e.g. it can help identify which analyte concentration(s) are unusual.

The concept of net analyte signal (NAS) is useful and is briefly reviewed here. NAS is the portion of an analyte spectrum that is unique to that analyte (Lorber, 1986). If we define $\mathbf{S}_{-j}$ as a $N \times K - 1$ matrix of analyte spectra with the jth column removed, i.e. spectrum $\mathbf{s}_j$ removed, then the NAS for the jth analyte, $\mathbf{n}_j$, is given as:

$$\mathbf{n}_j = \left(\mathbf{I} - \mathbf{S}_{-j}\left(\mathbf{S}_{-j}^T\mathbf{S}_{-j}\right)^{-1}\mathbf{S}_{-j}^T\right)\mathbf{s}_j$$
(8.9)

Equation (8.9) says that $\mathbf{n}_j$ is the portion of $\mathbf{s}_j$ that is orthogonal to $\mathbf{S}_{-j}$. Therefore, it might be expected that the larger the number of analytes in an image K, that NAS for any analyte will generally decrease; the chance that a portion of an analyte's spectrum to project onto other spectra increases. And, as Lorber showed, as NAS decreases, the error in the estimated concentration will generally increase.

There is a potential problem with the CLS model. The matrix $\mathbf{S}$ is an oblique basis for the row space of $\mathbf{X}$. Therefore, to obtain accurate estimates of $\mathbf{C}$ all spectra associated with every source of signal generally needs to be included in $\mathbf{S}$. Signal from sources not included in $\mathbf{S}$

can seriously bias the estimates of **C**. And, obtaining good estimates of spectra for all the sources of variability can be difficult when there are many sources of signal, or if the sources are difficult to characterize, e.g. due to temperature shifts (Wülfert *et al.*, 1998) and other clutter. (The word 'clutter' is borrowed from the radar community and is defined as signal that is not of interest.) Clutter usually includes random noise associated with the measurement plus any systematic signal not included in the model (e.g. spectra not included in **S**). Statistics such as Q and T^2 can help identify pixels for which predictions may be suspect, however the CLS model does not directly correct for unmodeled signal. The difficulty associated with identifying all sources of variance in a signal is one reason that ILS models are often used. However, a few methods for handling these interferences are discussed below.

Obtaining good estimates of **S** can be difficult. This can be a problem even when the number of sources of signal K is small. At times, library spectra are available. At other times, the spectra must be estimated from the data directly. Multivariate curve resolution (MCR) (also known as self-modeling curve resolution and end-member extraction) is a class of methods for estimating **S** from the data (Andrew and Hancewicz, 1988; Winter, 1999; Cipar *et al.*, 2002; Budevska *et al.*, 2003; Gallagher *et al.*, 2003b). The advantage of using MCR is that spectra estimated from the data tend to be relevant for the specific problem of interest, e.g. measurement matrix effects are included in the estimates that might not be included when using spectra from pure analytes. However, practitioners need to be aware of the limitations of MCR that can limit its usefulness in estimating spectra including multiplicative and rotational ambiguities (Manne, 1995; Gemperline, 1999; Tauler, 2001).

Even if estimated spectra are available (e.g. from a library), it may be that differences between the instrument that measured the spectra and the instrument that measures the image may result in spectral missmatch (and therefore, biased predictions). Instrument-based pixel-to-pixel variability can also lead to lowering the net analyte signal. Fortunately, there are methods related to instrument standardization and calibration transfer that can be used to account for this problem (Wang *et al.*, 1992; Feudale *et al.*, 2002), and newer papers for imaging spectrometers (Burger and Geladi, 2005). The importance of the image-based standardization is that it removes pixel-to-pixel variability within an image thus making each pixel appear as if it had the same response (gain and offset) as all the other pixels in the image. The result is that there is more net analyte signal available and higher signal-to-noise that can be used for detection, classification, and quantification.

8.2.2 ELS

There are typically two approaches to modifying the CLS model that are used to account for unknown interferences or clutter. Both methods rely on measuring multiple spectra associated with the clutter. The first to be discussed is based on the extended mixture model (Martens and Næs, 1989; Haaland and Melgaard, 2002) that yields an ELS model. The ELS model was the basis of orthogonal background suppression used by Hayden *et al.* (Hayden *et al.*, 1996) for remote sensing applications. The approach was found to be quite useful for accounting for clutter.

Suppose M_c measurements associated with clutter are made. The result is a $M_c \times N$ matrix of clutter spectra X_c. It is required that these measurements (a) do not include the analytes of interest or (b) that the analytes of interest are present but do not vary in the clutter measurements. The clutter measurements X_c can be estimated from a portion of an image under analysis or from other images. However, if instrument drift or differences are not accounted for using some form of instrument standardization discussed above, it would be best if these measurements were taken from the same image as that to be analyzed. The clutter pixels need not be in close spatial proximity, but should be representative of clutter in pixels that contain signal of interest. This requires a reasonable approximation of where in the image signal is present and absent. Iterative application of the ELS approach can be used to modify the selection of pixels included in X_c. The extended mixture model can then be given by:

$$X - 1\bar{x}_c^T = \left[C\, T_u \right]\left[S\, P_c \right]^T + E \qquad (8.10)$$

In Equation (8.10), X is a $M \times N$ matrix of measured spectra (e.g. a matricized image), $\bar{x}_c$ is the $N \times 1$ mean of the clutter matrix X_c, 1 is a $M \times 1$ vector of ones, C is a $M \times K$ matrix of concentrations to be estimated, T_u is a $M \times K_c$ matrix of unknown coefficients to be estimated, S is a $N \times K$ matrix of known spectra, P_c is a $N \times K_c$ subspace that spans the systematic variation in the centered matrix $X_c - 1_c\bar{x}_c$ and E is a $M \times N$ matrix of residuals. This model assumes that each measured spectrum in X can include signal from analyte spectra and clutter. The clutter subspace P_c can be obtained using principal components analysis (PCA) as shown in Equation (8.11) (Jackson, 1991).

$$X_c - 1_c \bar{x}_c^T = T_c P_c^T + E \qquad (8.11)$$

Here, 1_c is a $M_c \times 1$ vector of ones, T_c is a $M_c \times K_c$ matrix of scores ($T_c \neq T_u$), P_c is a matrix with columns corresponding to the PCA loadings, and E is a $M_c \times N$ matrix of residuals [different from the residuals matrix in Equation (8.10)]. For S known, the ELS estimated concentrations are obtained from:

$$[C T_u] = (X - 1\bar{x}_c) [S P_c] \left([S P_c]^T [S P_c] \right)^{-1} \qquad (8.12)$$

Matrix decomposition techniques other than PCA can be used to obtain a satisfactory estimate for P_c [e.g. using MCR (Gallagher et al., 2003c)]. Concentration images can be obtained from C and interference images can be obtained from T_u. Estimates of Hotelling's T^2 and Q residuals can be estimated for the ELS model in a manner analogous to that used for CLS discussed above.

One difficulty with the ELS approach, is determining the correct number of factors to keep in the PCA model of clutter K_c. This can often be done using cross-validation or inspecting plots of the clutter eigenvalues, and K_c is typically $\leq \min(M_c, N)$. However, to get a feel for what changing the size of K_c does, consider one extreme where $M_c \geq N$, and all principal components are kept to model $X_c - 1_c \bar{x}_c$, i.e. $K_c = N$. In this case, P_c would span the entire N-space, S could be described by linear combinations of the columns of P_c, and the inverse in Equation (8.12) would not exist. In general, it can be expected that as $K_c \to N$ the inverse in Equation (8.12) would become more ill-conditioned. Fortunately, in practice it is typical that $K_c \leq N$. However, it should be noted that increasing K_c will generally decrease the net analyte signal as can be seen by examining the net analyte signal for the ELS model given in Equation (8.13):

$$n_j = \left(I - [S_{-j} P_c] \left([S_{-j} P_c]_{-j}^T [S_{-j} P_c]_{-j} \right)^{-1} [S_{-j} P_c]_{-j}^T \right) s_j \qquad (8.13)$$

This equation shows that, in general, the more clutter the lower the net analyte signal. This is consistent with the observations by Funk et al. (Funk et al., 2001). They showed that using local models of the clutter (determined using clustering) in effect lowered the number of clutter components K_c and improved the predictive performance of their model.

8.2.3   GLS

The second method typically used to account for interferences in CLS models is referred to as GLS (Mardia *et al.*, 1979; Martens and Næs, 1989). A nice derivation of the GLS model is given in Draper and Smith (Draper and Smith, 1981). GLS can be considered a weighted least squares compliment of the ELS model and is the basis of the matched filter (Funk *et al.*, 2001). The estimator is given by:

$$C = (X - 1\bar{x}_c) W_c^{-1} S (S^T W_c^{-1} S)^{-1} \qquad (8.14)$$

where X is a $M \times N$ matrix of measured spectra (e.g. a matricized image), C is a $M \times K$ matrix of estimated concentrations, W_c is a $N \times N$ clutter covariance matrix $\text{cov}(X_c)$, and S is a $N \times K$ matrix of analyte spectra. The clutter covariance is estimated from the $M_c \times N$ matrix of measured clutter X_c using Equation (8.15):

$$W_c = \frac{1}{M_c - 1} (X_c - 1_c \bar{x}_c^T)^T (X_c - 1_c \bar{x}_c^T) \qquad (8.15)$$

where 1_c is a $M_c \times 1$ vector of ones, and $\bar{x}_c$ is the $1 \times N$ mean of the clutter matrix.

The inverse of W_c can be obtained using the singular value decomposition (SVD) in a manner similar to that described above for the concentration covariance matrix. The SVD gives:

$$V_c S_{v,c} V_c^T = W_c \Rightarrow W_c^{-1} = V_c S_{v,c}^{-1} V_c^T \qquad (8.16)$$

where V_c is a $N \times N$ matrix with orthonormal columns corresponding to the right singular vectors, and $S_{v,c}$ is a diagonal matrix containing the singular values (eigenvalues) of W_c. The subscript v indicates that $S_{v,c}$ is a matrix of singular values and is not a matrix of spectra, and subscript, c, indicates clutter. The inverse of W_c is obtained by simply inverting $S_{v,c}$ as shown on the right-hand-side of Equation (8.16).

The Hotelling's T^2 statistic for concentrations can be obtained using Equation (8.17) and residuals from Equations (8.3) and (8.4) that were used for the CLS model. However, since the model was minimized using a weighted residual it is more informative to use the weighted residual. The weighted residual for the *m*th pixel, $e_{w,m}$, is a $1 \times N$ vector given by:

$$e_{w,m} = (x_m - c_m S^T) W_c^{-1/2} \qquad (8.17)$$

The weighted sum-of-squared residuals for the mth pixel, $Q_{w,m}$, is given by:

$$Q_{w,m} = \mathbf{e}_{w,m}\mathbf{e}_{w,m}^T = \left(\mathbf{x}_m - c_m \mathbf{S}^T\right) \mathbf{W}_c^{-1} \left(\mathbf{x}_m - c_m \mathbf{S}^T\right)^T \qquad (8.18)$$

As with the CLS model, Q_w and T^2 images can be used to find unusual pixels. However, there is another T^2 statistic that can be used when $\mathbf{S}$ is unavailable. The T^2 statistic given in Equation (8.7) provides a measure of variability of estimated concentrations relative to a representative set of concentrations. In contrast, the following T^2 statistic for the mth pixel, $T_{x,m}^2$ gives a measure of the variability of the measured spectra relative to the variability in the clutter:

$$T_{x,m}^2 = \left(\mathbf{x}_m - \bar{\mathbf{x}}_c\right) \mathbf{W}_c^{-1} \left(\mathbf{x}_m - \bar{\mathbf{x}}_c\right)^T \qquad (8.19)$$

$T_{x,m}^2$ can be used as a first cut at detecting unusual pixels, i.e. pixels that have unusual signal relative to the clutter. Some of these pixels may include signal of interest, but the T_x^2 does not discriminate between interesting and noninteresting variation.

The GLS model may not seem intuitive. Therefore, a short discussion of an error analysis and net analyte signal is given here with the intention of exploring the model in a little detail. It is a relatively straightforward procedure to show that the error covariance for the GLS model is given by (Draper and Smith, 1981):

$$E\left(d\mathbf{c}^T d\mathbf{c}\right) = \left(\mathbf{S}^T \mathbf{W}_c^{-1} \mathbf{S}\right)^{-1} \qquad (8.20)$$

where $d\mathbf{c}$ is a $1 \times K$ vector of the differential for a single row of $\mathbf{C}$ given in Equation (8.14) and $E()$ is the expectation operator. This equation assumes that $E(\mathbf{e}_w) = 0$, i.e. the GLS estimator is unbiased. To see what Equation (8.20) means, the univariate case can be examined. For example, assume that there is only one analyte and one spectral channel (i.e. $\mathbf{S} \rightarrow s_{nk}$ and $\mathbf{W}_c \rightarrow w_{c,nn}$) then Equation (8.20) reduces to:

$$E\left(dc_k^2\right) = \left(s_{nk} w_{c,nn}^{-1} s_{nk}\right)^{-1} = w_{c,nn} / s_{nk}^2 \qquad (8.21)$$

where

$$w_{c,nn} = \frac{1}{M_c - 1} \sum_{m=1}^{M_c} \left(x_{c,mn} - \bar{x}_{c,n}\right)^2 \qquad (8.22)$$

This says that the error increases as the clutter variance at channel n increases relative to the signal s_{nk}. Equation (8.21), and therefore Equation (8.20), can be read as noise-to-signal.

The diagonal elements of the term $(S^T W_c^{-1} S)$ are Hotelling T^2 statistics for the analyte spectra projected onto the inverse clutter matrix. Therefore, we would expect that the larger the T^2, the smaller the error. The contributions to this T^2 statistic, the columns of the $N \times K$ matrix $T_{con,S}$ is given by:

$$T_{con,S} = V_c S_{v,c}^{-1/2} V_c^T S \qquad (8.23)$$

The columns of $T_{con,S}$ can be considered spectra weighted by the inverse of the noise, and might be considered signal-to-noise spectra. These weighted spectra can be used as the basis for variable selection methods [e.g. by maximizing $\det(S^T W_c^{-1} S)$ for selected channels].

Equation (8.23) shows that directions in W_c associated with large eigenvalues (i.e. large entries in the diagonal matrix $S_{v,c}$) will be de-weighted. These directions are associated with large clutter, and the more of these directions that have high clutter the larger the estimation error. The consequence is that pixels selected to represent clutter in the matrix X_c should not contain more variability (more sources of clutter) than is necessary. Ideally, X_c will only contain sources of variability observed in each pixel for which concentration estimates are being made, i.e. the clutter model is local to the pixels being tested. This minimizes estimation error and is consistent with observations by Funk et al. (Funk et al., 2001). For example, in one extreme, systematic variation is absent in the clutter, the noise is white, and W_c is diagonal with elements corresponding to the noise in each spectral channel. If the noise is the same in all the channels for this example, then the GLS model reduces to the CLS model.

Some of the above observations can be seen by examining the net analyte signal for the GLS model:

$$n_j = W_c^{-1/2} \left(I - W_c^{-1} S_{-j} \left(S_{-j}^T W_c^{-1} S_{-j} \right)^{-1} S_{-j}^T \right) s_j \qquad (8.24)$$

Equation (8.24) is a slightly modified version of the equation given in Brown (Brown, 2004). What this equation implies are two things. First, the more analytes included in S_{-k} the lower the expected net analyte signal n_k. And second, the larger the magnitude of the clutter covariance W_c, the lower the expected net analyte signal. This again suggests that local clutter models will result in better prediction performance.

A practical problem that must be addressed when using a GLS model is estimating the inverse of $\mathbf{W}_c$, $\mathbf{W}_c^{-1}$. When the clutter covariance matrix is full rank the inverse can be easily obtained using the procedure outlined in Equation (8.16). However, if the number of clutter measurements is less than the number of spectral channels ($M_c < N$) or $\mathbf{W}_c$ is not full rank then $\mathbf{W}_c^{-1}$ will not exist and something must be done to allow estimation of the concentrations. Even if $\mathbf{W}_c$ is of full rank, but is ill-conditioned, the following procedures for obtaining the inverse can be useful.

One approach is to use a subset of spectral channels, N_{sub} such that $N_{sub} \leq M_c$. The specific channels and N_{sub} might differ for each real-ization of $\mathbf{X}_c$ (e.g. each local clutter model). However, this procedure runs the risk of losing signal averaging that is available when using all the channels. Additionally, selecting a subset of channels may not be a trivial matter.

Another approach that has been suggested is to throw out the vectors in $\mathbf{V}_c$ (directions) associated with small eigenvalues in $\mathbf{S}_{v,c}$ similar to PCR. However, this is precisely the wrong thing to do because the small eigenvalues are associated with low levels of clutter. It would be better to throw out the directions associated with the large eigenvalues in $\mathbf{S}_{v,c}$. If the selected directions corresponding to large variance are removed, the result is that the target spectra will be projected onto the null-space of the clutter. The result is a biased prediction, however good results can be obtained. The problem with this approach (analogous to estimating the number of factors K_c) is estimating which directions should be eliminated.

Two forms of ridging have also been proposed for use in estimating $\mathbf{W}_c^{-1}$. The first method simply adds a small constant θ to all the eigen-values in $\mathbf{S}_{v,c}$, and the second used in this chapter saturates eigenvalues that are less than some critical level $s_{v,c,crit}$ to be a small fraction of the largest eigenvalue $s_{v,c,1}$ (Funk et al., 2001). This can be viewed as setting the condition number to a desired level. The difficulty then is estimating $s_{v,c,crit}$ and the desired condition number. However, values that result in a condition number for $\mathbf{W}_c$ around 10^3 to 10^6 are common.

8.3   DETECTION, CLASSIFICATION, AND QUANTIFICATION

The GLS estimator provides a basis for detection, classification, and quantification in MIs. It should be noted that there quite a bit of overlap between these three objectives. The tasks are demonstrated using a Raman image discussed in Gallagher *et al.* (Gallagher *et al.*, 2003b), and briefly described here.

The image is 21×33 pixels $\times 501$ Raman spectra of an aspirin/polyethylene mixture. The sample is aspirin tablet scrapings (Bayer Corp.) and granulated high density (HD) polyethylene (Dow Chemical, Inc.) dispersed on a silica-glass slide. The image was collected on a fiber-coupled Raman microscope (Hololab 5000, Kaiser Optical Systems, Inc.) over a $105 \times 165\,\mu m^2$ region at $5\,\mu m$ steps using a $10 \times$ Olympus objective. The laser wavelength was $785\,nm$ and the image spectra are in the range $600 - 1660\,cm^{-1}$ at $2\,cm^{-1}$ increments. The objective and glass slide resulted in significant luminescence viewed in the present context as clutter. Three targets of interest are low density (LD) and (HD) polyethylene, and aspirin. The normalized target spectra, S, are shown in Figure 8.1. [Target spectra were estimated from the image using MCR (Gallagher *et al.*, 2003b)].

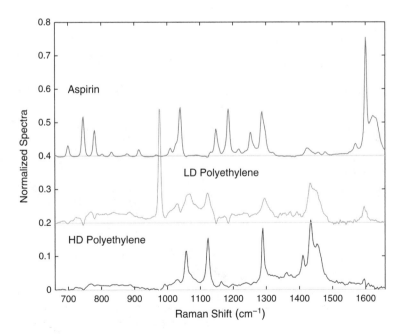

Figure 8.1 Normalized target spectra for the Raman image

8.3.1 Detection

Detection of any of the three targets might be performed using PCA of the entire image and looking for unusual pixels (Chang and Chiang, 2002). However, in this approach, variance of the targets is included in the PCA model making it less clear when a detection event has occurred. Instead, Equation (8.19) and the T_x^2 statistic is used here to identify unusual pixels. First an estimate of clutter $\mathbf{X}_c$ and clutter covariance $\mathbf{W}_c$ are required. Signal-free pixels can be found by exploratory analysis, a priori knowledge, or applying Equation (8.14) and iteratively removing pixels with high target concentration. For this example, window target factor analysis (WTFA) (Lohnes *et al.*, 1999) adapted to images was used with the three target spectra. (WTFA is suited to just this type of problem but is outside the scope of the present discussion.)

Several pixels were identified to have little or no target signal and were fairly representative of the clutter. The measured spectra for these 40 target-free pixels are characteristic of luminescence from the glass slide and are shown in Figure 8.2. $\mathbf{W}_c^{-1}$ was estimated using Equation (8.16) where values in $\mathbf{S}_{v,c}$ less than $10^{-5} s_{v,c,1}$ were set to $10^{-5} s_{v,c,1}$. Therefore, the condition number of $\mathbf{W}_c$ and $\mathbf{W}_c^{-1}$ was 10^5. Figure 8.3 shows a plot

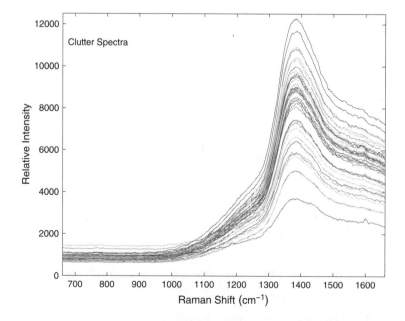

Figure 8.2 Measured clutter spectra (glass luminescence)

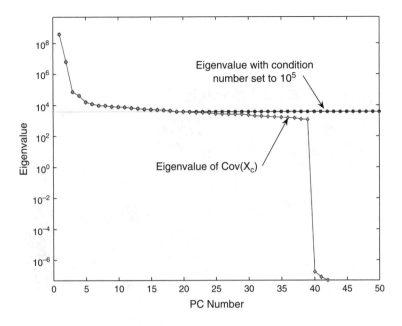

Figure 8.3 Eigenvalues of the clutter covariance $\mathbf{W}_c$

of the eigenvalues of $\mathbf{W}_c$. This plot shows that the rank of the 501×501 matrix $\mathbf{W}_c$, prior to saturating the eigenvalues, is approximately 39; eigenvalues for factors >39 are zero to within machine precision.

The T_x^2 image is shown in Figure 8.4. Because a few high values can use up most of the dynamic range, the image is shown for $\ln\left(T_x^2\right)$. The 95th percentile for the clutter pixels is 36.5. Therefore, values > $\ln(36.5) \approx 3.6$ might be considered a detection event, this includes most of the pixels. Three pixels are highlighted and the measured spectra for these pixels are plotted in Figure 8.5. Pixel (20,31) has a low T_x^2 and Figure 8.5 shows that its spectrum is similar to the clutter spectra in Figure 8.2.

Pixel (6,10) has the highest T_x^2. The corresponding contribution plot and a comparison of the measured spectrum in Figure 8.5 to clutter spectra in Figure 8.2 suggest that the reason it is unusual is the slope in the spectrum at shifts <1000 cm^{-1}. Pixel (19,12) has a high T_x^2. Comparison of its measured spectrum in Figure 8.5 to target spectra in Figure 8.1 suggest that the reason it is unusual is the presence of aspirin. The T_x^2 detection approach finds unusual pixels due to variation not present in the clutter. The unusual variation could be due to bad spectra [e.g. Pixel (6,10)] or presence of target [e.g. Pixel (19,12)].

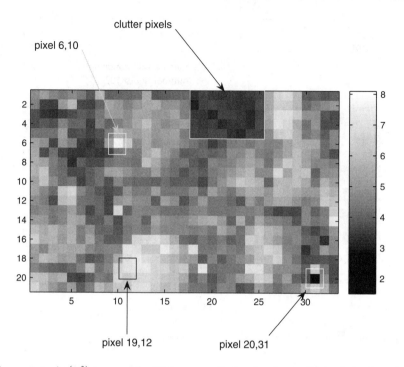

Figure 8.4 $\ln\left(T_x^2\right)$ image. The 95th percentile for the clutter T_x^2 statistics is ≈ 3.6

8.3.2 Classification

Classification can be performed by simply examining the concentration estimates from Equation (8.14) and finding concentrations that are greater than some limit. For example, the 95th or 99th percentile of predicted concentrations on the clutter pixels could be used. Concentrations higher than this limit are considered present in a pixel. This approach can be considered a detection, classification and quantification algorithm. The problem is determining the limit for detection.

Stepwise procedures can also be used. A reverse stepwise procedure employs Equation (8.14) iteratively for each pixel. Target analytes are removed from the basis **S** until the fit residual given by Equation (8.18) increases significantly using an F-test. Alternatively, a forward stepwise procedure starts by including only a single analyte spectrum in the basis **S** and estimating the fit residual given in Equation (8.18). If the fit improves significantly, the analyte with the lowest residual is added to the basis. The procedure is repeated with remaining analytes until the fit does not improve significantly. The problems with this approach are

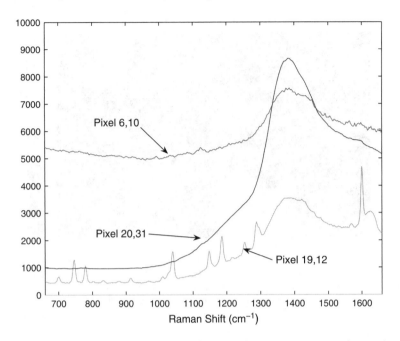

Figure 8.5 Measured spectra for highlighted pixels in Figure 8.4. Pixels (6,10) and (19,12) have high T_x^2. Pixel (20,31) has low T_x^2

defining the appropriate significance level for the F-test and applying it one pixel at a time. The latter problem can be alleviated by using a weighted average of a subset of pixels. For example, a single pixel could be obtained by using a T_x^2 weighted average. Another potential problem occurs when analyte spectra in the basis S are linear combinations of each other or nearly so. In this case, the inverse of $(S^T W_c^{-1} S)$ will not exist or will be ill-conditioned resulting in poor concentration estimates. Therefore, caution is advised when trying stepwise procedures with large databases of target spectra.

8.3.3 Quantification

Figure 8.6 shows the concentration image for aspirin, where the concentrations have been scaled to the 95th percentile of aspirin concentrations in the clutter pixels. The same equation used for classification was used here for quantification [Equation (8.14)]. Note that aspirin is apparent in three locations in the image; bottom center, bottom right, and top right.

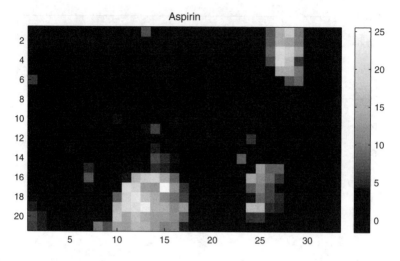

Figure 8.6 Estimated concentration of aspirin. Concentrations have been normalized to the 95th percentile of estimated concentrations in the clutter pixels. Dark colors are low concentrations and light colors are high concentrations

Recall that the target spectra are normalized, i.e. they do not necessarily correspond to unit concentration. This is a result of a multiplicative ambiguity in the MCR estimates of the target spectra. Therefore, concentrations in each pixel relative to other pixels can be compared for each analyte but not between analytes. Ideally, standards could be used to resolve the multiplicative ambiguity. However, practitioner should be aware that it is typical that reflectance measurements do not have an absolute concentration scale due to sampling artifacts (e.g. sample matrix effects, changes in packing density, and changes in light penetration depth) and some pre-processing may be required; for example, see Martens *et al.* (Martens *et al.*, 2003). Even after preprocessing some ambiguity may exist. For example, the problem may be due to an ambiguity between the concentration and the path length (Gallagher *et al.*, 2003a). No pre-processing of the measured spectra was performed for the present example, therefore quantification is relative on a pixel-to-pixel basis for each analyte.

Figure 8.7 shows a concentration image for all three targets. In this case red corresponds to aspirin, green is LD polyethylene, and blue is HD polyethylene. Dark areas are relatively low in all three analytes and whiter areas have all three analytes present. Estimated concentrations have been scaled to 0–255 and converted to unsigned 8-bit data prior to displaying the color image.

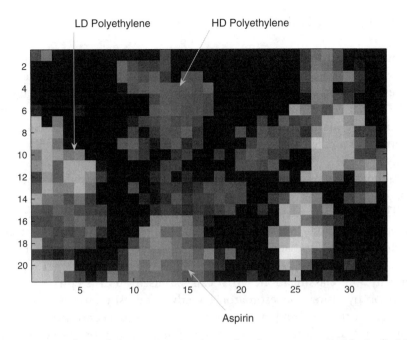

Figure 8.7 Color-coded concentration image for three targets: aspirin (red); LD polyethylene (green); HD polyethylene (blue). The image has been contrast enhanced (see Plate 31)

Often it is found that there are a few pixels with very high and/or very low concentration with most of the pixels constrained to a small concentration range. Therefore, the bulk of the pixels have nearly the same color while only a few pixels are given the extremes of the color range. This is the problem with visualizing concentration images in the present example. For this reason, contrasting was used in Figure 8.7. Contrasting is performed by setting upper and lower limits on the concentrations. Concentrations above the upper limit are artificially set to equal the upper limit, and those below the lower limit are set to the lower limit. The effect is to spread out the estimated concentrations in the bulk of the pixels over a wider range of colors. However, contrasting sacrifices some quantification information in an image for one that is easier to discriminate by eye.

Although, it is not shown, the Q_w image should be examined along with concentration images. In this example, the residuals image shows that pixels (6,10) [very high], (17,15), (19,19), and (14,33) have relatively high Q_w statistics.

8.4   CONCLUSIONS

This chapter discussed three CLS-based models and their application to hyperspectral image analysis; CLS, ELS and GLS. ELS and GLS are used to explicitly model clutter (noise and interferences). The advantage of ELS is that contributions for interferences are estimated resulting in the ability to visualize images corresponding to interferences. In contrast, clutter is modeled with a weighting matrix in GLS; GLS is not typically used to estimate contributions of interferences. However, of the three methods discussed GLS is probably the most popular approach for HIA.

The tasks of detection, classification, and quantification are intertwined. For example, GLS can be used for detection using a Hotelling T^2 statistic based on the clutter (strict anomaly detection) or by using the concentration estimator, Equation (8.14), directly. In classification, GLS can be used in a stepwise regression or by using the estimator directly. The estimator is also used for quantification and a significant result is concentration images. These are an advantage over scores images often produced by ILS models, because concentration images provide chemical information directly. This makes interpretation easier and more relevant for many applications.

GLS performs well for HIA and new algorithms based on its backbone are presently being developed. For example, detection algorithms that use Hotelling T^2 statistic with cross-validation are expected to be more sensitive than present approaches. Estimators that automate identification of local clutter models are also expected to provide ease of use and better signal-to-noise properties. Spatial information was not used in the approaches outlined here, and it is possible that explicit use of spatial information will improve performance by enhancing signal averaging in an analysis. The advantage of CLS-based approaches discussed, and those still being researched, is that they provide relevant chemical and physical information from hyperspectral image measurements.

ACKNOWLEDGEMENTS

Kaiser Optical Systems and Jeremy M. Shaver provided the Raman aspirin image.

REFERENCES

Anderson, T. W. (1971) *An Introduction to Multivariate Statistical Analysis*, 2nd Edn, John Wiley & Sons, Ltd, New York.

Andrew, J. J. and Hancewicz, T. M. (1998) Rapid Analysis of Raman Image Data Using Two-Way Multivariate Curve Resolution, *Appl. Spectrosc.*, 52(6), 797–807.

Bänninger, D. and Flühler, H. (2004) Modeling Light Scattering at Soil Surfaces, *IEEE Trans. Geosci. Remote Sens.*, 42(7), 1462–1471.

Brown, C. D. (2004) Discordance between Net Analyte Signal Theory and Practical Multivariate Calibration, *Anal. Chem.*, 76(15), 4364–4373.

Budevska, B. O., Sum, S. T. and Jones, T. J. (2003) Application of Multivariate Curve Resolution for Analysis of FT-IR Microspectroscopic Images of in Situ Plant Tissue, *Appl. Spectrosc.*, 57(2), 124–131.

Burger, J. and Geladi, P. (2005) Hyperspectral NIR Image Regression. Part I: Calibration and Correction, Journal of Chemometrics, 19(5–7), 355–363.

Chang, C. I. and Chiang, S. S. (2002) Anomaly Detection and Classification for Hyperspectral Imagery, *IEEE Trans. Geosci. Remote Sens.*, 40(6), 1314–1325.

Cipar J. J., Meidunas, E. and Bassett, E. (2002) A Comparison of End Member Extraction Techniques, *Proc. SPIE*, 4725, 1–9.

Draper, N. and Smith, H. (1981) *Applied Regression Analysis*, 2nd Edn, John Wiley & Sons, Ltd, New York.

Feudale, R. N., Woody, N. A., Tan, H., Myles, A. J., Brown, S. D. and Joan Ferre', J. (2002) Transfer of Multivariate Calibration Models: a Review, *Chem. Intell. Lab. Sys.*, 64(2), 181–192.

Funk C. C., Theiler, J., Roberts, D. A. and Borel, C. C. (2001) Clustering to Improve Matched Filter Detection of Weak Gas Plumes in Hyperspectral Thermal Imagery, *IEEE Trans. Geosci. Remote Sens.*, 39(7), 1410–1420.

Gallagher, N. B., Wise, B. M. and Sheen. D. M. (2003a) Estimation of Trace Vapor Concentration–Pathlength in Plumes for Remote Sensing Applications from Hyperspectral Images, *Anal. Chim. Acta*, 490, 139–152.

Gallagher, N. B., Shaver, J. M., Martin, E. B., Morris, J., Wise, B. M. and Windig, W. (2003b) Curve Resolution for Images with Applications to TOF-SIMS and Raman, *Chem. Intell. Lab. Sys.*, 77(1), 105–117.

Gallagher, N. B., Sheen, D. M., Shaver, J. M., Wise, B. M. and Shultz, J. F. (2003c) Estimation of Trace Vapor Concentration–Pathlength in Plumes for Remote Sensing Applications from Hyperspectral Images, *SPIE Proc.*, 5093, 184–194.

Gallagher, N. B, Blake, T. A. and Gassman, P. L. (2005) Application of Extended Multiplicative Scatter Correction to mid-Infrared Reflectance Spectroscopy of Soil, *J. Chem.*, 19(5–7), 271–281.

Gemperline, P. J. (1999) A Method for Computing the Range of Feasible Profiles Estimated by Self-modeling Curve Resolution, *Anal. Chem.*, 71, 5398–5404.

Griffiths, P. R. (1996), Letter: Practical Consequences of Math Pre-treatment of Near Infrared Reflectance Data: log(1/R) vs F(R), *J. Near Infrared Spectrosc.*, 3, 60–62.

Haaland, D. M. and Melgaard, D. K. (2002) New Augmented Classical Least Squares Methods for Improved Quantitative Spectral Analyses, *Vib. Spectrosc.*, 29, 171–175.

Hapke, B. (1993) *Theory of Reflectance and Emittance Spectroscopy*, Cambridge University Press, New York.

Hayden, A., Niple, E. and Boyce, B. (1996) Determination of Trace-Gas Amounts in Plumes by the Use of Orthogonal Digital Filtering of Thermal Emission Spectra, *Appl. Opt.*, **35**(16), 2802–2809.

Jackson, J. E. (1991) *A User's Guide to Principal Components*, John Wiley & Sons, Ltd, New York.

Lohnes, M. T., Guy, R. D. and Wentzell, P. D. (1999) Window Target-testing Factor Analysis: Theory and Application to the Chromatographic Analysis of Complex Mixtures with Multiwavelength Fluorescence Detection, *Anal. Chim. Acta*, **389**, 95–113.

Lorber A. (1986) Error Propagation and Figures of Merit for Quantification by Solving matrix Equations, *Anal. Chem.*, **58**, 1167–1172.

Manne, R. (1995) On the Resolution Problem in Hyphenated Chromatography, *Chem. Intell. Lab. Sys.*, **27**, 89–94.

Mardia, K. V., Kent, J. T. and Bibby, J. M. (1979) *Multivariate Analysis*, Academic Press, San Diego.

Martens, H. and Næs, T. (1989) *Multivariate Calibration*, John Wiley & Sons, Ltd, New York.

Martens, H., Nielsen, J. P. and Engelsen, S. B. (2003) Light Scattering and Light Absorbance Separated by Extended Multiplicative Signal Correction. Application to Near-infrared Transmission Analysis of Powder Mixtures, *Anal. Chem.*, **75**, 394–404.

Nascimento, J. M. P. and Bioucas Dias, J. M. (2005) Does Independent Component Analysis Play a Role in Unmixing Hyperspectral Data? *IEEE Trans. Geosci. Remote Sens.*, **43**(1), 175–187.

Sanchez, E. and Kowalski, B. R. (1988) Tensorial Calibration: I. First Order Calibration, *J. Chemom.*, **2**, 247–263.

Smilde, A., Bro, R. and Geladi, P. (2004) *Multi-way Analysis with Applications in the Chemical Sciences*, John Wiley & Sons, Ltd, New York.

Tauler, R. (2001) Calculation of Maximum and Minimum Band Boundaries of Feasible Solutions for Species Profiles Obtained by Multivariate Curve Resolution, *J. Chemom.*, **15**, 627–646.

Wang, Y., Lysaght, M. J. and Kowalski, B. R. (1992) Improvement of Multivariate Calibration Through Instrument Standardization, *Anal. Chem.*, **64**, 562–564.

Winter, M. E. (1999) Fast Autonomous Spectral End-member Determination in Hyper-spectral Data, *Proceedings of the Thirteenth International Conference on Applied Geologic Remote Sensing*, Altarum Geologic Conferences, ERIM International, Inc., Ann Arbor, MI, Vol. II, pp. 337–344.

Wülfert, F., Kok, W. T. and Smilde, A. K. (1998) Influence of Temperature on Vibrational Spectra and Consequences for the Predictive Ability of Multivariate Models, *Anal. Chem.*, **70**, 1761–1767.

9

Calibration Standards and Image Calibration

Paul L. M. Geladi

9.1 INTRODUCTION

This chapter introduces standards for calibration in hyperspectral imaging. It starts with describing the need for calibration in general and for image calibration in particular. Then some definitions are given of hyperspectral image resolution and spectroscopic nomenclature. Calibration standards for visible and near infrared (NIR) imaging are introduced and an example is given of the calibration of an imaging camera for the NIR spectral region.

9.2 THE NEED FOR CALIBRATION IN GENERAL

All scientific activities need calibration. This is to make sure that the reported measurement results are reliable and to make an estimate of accuracy and precision. The calibration activity in a laboratory pertains to a wide range of actions. Examples are: obtaining pure water at the correct temperature and pure dry chemicals to make standard solutions with the aid of an analytical balance (calibrated and with all necessary corrections made) and calibrated flasks; calibrating a pH meter with standard buffer solutions; calibrating a calorimeter by burning a known

Techniques and Applications of Hyperspectral Image Analysis   Edited by H. F. Grahn and P. Geladi
© 2007 John Wiley & Sons, Ltd

amount of a pure chemical, etc. Most introductory books on chemistry contain information on calibration methods and where they are needed (Harris, 1997; Christian, 1980). Standards and calibration are basic building blocks of all activities in experimental chemistry and physics. A good overview is given by Stoeppler *et al.* (Stoeppler *et al.*, 2001). The remote sensing community has a long tradition of calibration and correction (Lillesand *et al.*, 2004).

When one thinks carefully about this, any reported analytical result is conditional on the validity of a large number of calibrations. All calibrations should as much as possible refer to SI or IUPAC standards (Page and Vigoureux, 1975; Mills *et al*, 1993; Taylor, 2001). All results that are produced should be expressed in SI units or other appropriate units and the chain of relationshsip of the actual result to the official units should be transparent. Millimole liter^{-1} should mean the most accurate definition of mole and the most accurate definition of liter that is achievable. An estimate of inaccuracy is a necessity. Luckily, manufacturers often have good estimates of accuracy and precision available for their equipment.

9.3 THE NEED FOR IMAGE CALIBRATION

Images contain visual information and most of us are used to the subjective nature of this visual information. The human perception pattern recognizer detects things like: a horse, a still-life with flowers, a standing nude, a mountain landscape, a battle scene, a windmill, etc. The human pattern recognizer also corrects for size, illumination, aging and other factors. A windmill is recognized as a windmill whether the picture is a miniature or covers a whole wall of a palace. Flowers are recognized as flowers in naïve, impressionistic or romantic paintings and this irrespective of lighting conditions: shade, bright sunlight, spotlight or fluorescent tubes.

Scientific imaging, e.g. in medical applications or in microscopy, brings with it a need for objective measurement. The result of observing the image is not an esthetic impression, but a measurement result. Also here, calibration and a transparent path to SI units are necessities and an estimate of inaccuracy with respect to the SI units is required. One important measure is size of objects in the imaged sample. See Figure 9.1 for how to report size of an image. It is equally important to have intensity calibration. Terms like white, black, light gray, medium gray, dark gray, etc., are subjective and not at all accurate. The intensities

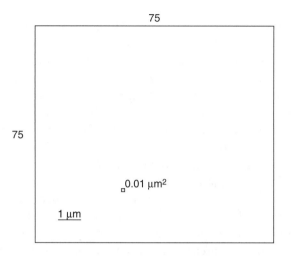

Figure 9.1 To express size in an image, a size bar (here 1 μm) can be added. It is also possible to give pixel size (here 0.01 μm²) and the number of rows and columns (here 75 × 75) or it is possible to give the total image size (here 7.5 μm × 7.5 μm)

in the image should be expressed in a more accurate manner and this requires a calibration standard or an absolute physical measurement. In astronomy, radiation sources are not controllable by the experimenter and units like radiation, radiant flux, radiant flux density, lux and lumen are used. More general and reasonably robust spectroscopic definitions for the laboratory are given in a following section.

The calibration of images, multivariate images and hyperspectral images is related to the different types of resolution in the images. For all calculation purposes an image is digitized (Duff, 1983) and the resolution resulting from this digitization is always a limiting factor for the accuracy and precision of measured results.

9.4   RESOLUTION IN HYPERSPECTRAL IMAGES

A digitized black and white image has a fixed size, e.g. 512 × 512 pixels. Most digitized images are square or rectangular. A pixel has a 2-D size and a distance between two pixels A and B in the image is also a size. The SI unit for distance is the meter and a square pixel has a size expressed in square meters. This is good for satellite imaging, but in microscopy smaller units are used: micrometer, nanometer and their squared variant for areas. This is easy, provided that there are no nonlinearities. Horizontal and vertical scale in an image are often different and if perspective is involved, different parts of the image may have different pixel sizes. The intensity (grayvalue) in each pixel is also a digitized value. Old satellite images have only 128 (2^7) possible grayvalues. Newer satellite images and some microscopic images have 256 (2^8) grayvalues. In radiology, 4096 (2^{12}) grayvalues are usual and also some microscopic techniques have this resolution. A more ideal resolution is 2^{16} grayvalues but that is still rare. The resolution in intensity, grayvalues is also called radiometric resolution (Kramer, 1992). For spectroscopic purposes, digitized count units have to be transformed to other units. This is explained further on in the text.

The human eye stops seeing differences somewhere between 32 and 64 grayvalues, so for subjective visualization, only 64 grayvalues are needed in a black and white image. For color images, color information replaces some of the intensity information, but the way this is done subjectively is very complicated. Color perception depends on the size of the colored area and the needed number of grayvalues is different for different colors (Johnson and Fairchild, 2003).

When an image is published, some size indication is needed. Sometimes a bar of a fixed length (1mm, 1 μm or any other) is added. An alternative is to indicate the complete image size (width and height in mm or μm). It is also possible to indicate the pixel size. As an example, one may report the image size as 512×512 pixels and each pixel as $1 \, \mu m^2$. In the geographical and remote sensing literature a map scale factor is often given. Some techniques can make 3-D images and then there are three size factors: width, height and depth. A pixel then becomes a voxel and has a volume expressed in m^3.

For hyperspectral images, wavelength resolution is important. A typical example would be 128 wavelength bands between 908 and 1670 nm with 6 nm resolution. The SI unit is the nanometer, a fraction of a meter. In this wavelength range, the resolution of 6 nm is quite adequate for solids and liquids. In the gas phase, a Fourier transform spectrum with $2 \, cm^{-1}$ resolution would be the right choice. The SI unit for wavenumber is actually m^{-1}, but cm^{-1} is considered more practical for infrared and NIR spectroscopy.

Spectral resolution is a difficult topic because it depends on at least two factors. The first one is the resolution of the monochromator when fixed at one wavelength. The output of the monochromator is an intensity distribution with a maximum. The nominal wavelength is usually given as this maximum value. A bandwidth can be given as full width at half maximum (FWHM). This avoids the need to know the distribution in detail or to handle skew distributions. The second factor is the spacing of the wavelegnths at which the signal is digitized. Ideally the distance between two adjacent wavelengths should be equal to the FWHM, but this is not always the case. Sometimes the spacing of the wavelengths is less than the FWHM and then the spectrum consists of overlapping bands. If the wavelength spacing is more than the FWHM there are gaps between the bands. Similar arguments hold for interferometric spectrometers. See Figure 9.2 for a graphical explanation.

Intensity

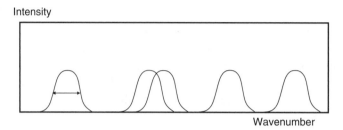

Wavenumber

Figure 9.2 The spectral bandwidth of a monochromator is given as the FWHM in the intensity or energy distribution. Wavelength spacing can be such that the bands overlap or that there are gaps between them

In some applications, time resolved or temporal images are made. In this case, the time interval determines the temporal resolution. The SI unit for time is the second. Time intervals can be regularly or irregularly spaced depending on the application.

In conclusion, hyperspectral images have size, wavelength and sometimes time resolution and each pixel (voxel) also has a radiometric (intensity) resolution (Kramer, 1992). The images have to be made in such a way that SI or IUPAC units and definitions can be used and that an estimate of accuracy is possible.

9.5   SPECTROSCOPIC DEFINITIONS

The intensity value in a pixel or voxel is expressed as a number between 0 and the maximum: 255, 4095 or some other value. The range depends on the resolution of the A/D convertor 2^8, 2^{12} or some other value. In spectroscopy other units are required (Mills *et al.*, 1993). For transmission imaging, transmittance is used:

$$T = I/I_0 \qquad \text{or} \qquad T\% = 100(I/I_0) \qquad (9.1)$$

I_0 is the intensity of the radiation without the object under study in the image and I is the radiation transmitted through the object under study.[1] In microscopic imaging, one would first make an image of an empty microscope slide G_0 and then an image of the microscope slide with the sample under study G.[2] The element by element ratio of the two matrices would be the transmission matrix or image T. Transmission has no units.

$$T = G /^* G_0 \tag{9.2}$$

where $/^*$ means element by element division.

The division by G_0 also compensates for uneven illumination, because the intensity distribution of the illumination source is a part of G_0.

According to Beer's law (or Lambert–Beer–Bouguer's law) concentrations of chemicals are responsible for the absorption of radiation:

$$A = klc \tag{9.3}$$

$$A = -\log_{10}(T) \tag{9.4}$$

where A is the absorption, l is pathlength and c is the concentration. k is the molar absorption coefficient of the chemical. According to this equation it would be useful sometimes to express images as absorption images:

$$A = -\log_{10}(T) \tag{9.5}$$

The logarithm is taken pixelwise. In such an image, the intensity in A would be proportional to the concentration of the chemical under study. A and T are in accordance with the IUPAC definition. They both have no units.

Images are also made of nontransparent materials. Then the image G is a reflection image, also called a sample image. As a reference, an image of a highly reflecting material is made and called G_0, also called a calibration image. This gives the reflectance image as a pixelwise ratio:

$$R = G /^* G_0 \tag{9.6}$$

---

[1] Instead of absolute measures of radiation (radiant flux, radiant flux density, etc.) used in astronomy and remote sensing, relative measures are used. This is more practical in laboratory situations when using artificial and controllable radiation sources.

[2] In matrix notation I is the identity matrix, so G is used for images in order to avoid confusion.

Pseudoabsorbance is then calculated as:

$$A = -\log_{10}(\mathbf{R}) \qquad (9.7)$$

Pseudoabsorbance is a typical word used in NIR spectroscopy. It has not been accepted by the organizations for spectroscopic standardization.

An interesting question is whether $\mathbf{G}$ and $\mathbf{G}_0$ are stable in time and space. Even though a ratio is calculated in Equations (9.2) and (9.6), too much unevenness of $\mathbf{G}_0$ in space can give low accuracy in parts of the image. Another question is whether the detector and digitization process are linear. Calibration and measurement of accuracy and precision have to take into account such matters. Interesting books on the relationship between radiation properties and imaging are by Kopeika (Kopeika, 1998) and Callet (Callet, 1998).

As a conclusion, arbitrary intensity is not an accepted spectroscopic unit. In order to get to spectroscopic units, transmittance, reflectance, absorbance or pseudoabsorbance have to be calculated. All these have no SI units. They are unit-free.

9.6   CALIBRATION STANDARDS

The history of calibration standards for microscopy, photography and microphotography is very old. Many standard materials exist for determining and adjusting all kinds of properties of images. Also television and video have produced numerous standards. The most wellknown is the test image that many television stations and cable companies transmit during the night and early morning. Such images are used for testing geometrical errors (using a circle, vertical and horizontal lines, sometimes a grid), intensity errors (a grayscale) and color errors (color bars). There are also bars for testing resolution degradation in the chain from transmitter to receiver.

For photography, the test images exist on printed cards. For intensity calibration, standard gray cards are used. Also standard color bars exist for testing color reproduction. One of the most sold standards is a gray card that reflects 18 % of the incoming light.

There also exist test cards for adjusting scanners.

All these standards are valid for the visual wavelength range 400–780 nm. Geometrical standards also work in the NIR and ultraviolet (UV) if they remain visible. Grayscale and color standards are however seldom made for wavelengths above 800 nm, so for reflectance and wavelength accuracy specific products have to be used. Figures 9.3 and 9.4 show some examples of geometrical calibration standards. The

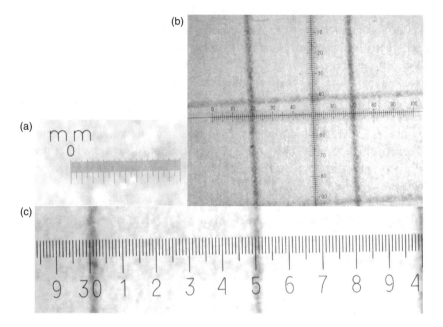

Figure 9.3 (a) 1 mm divided in to $10 \mu m$ portions. (b) Crossed rulers with $100 \mu m$ divisions. (c) Part of a 50 mm line divided in to 1 mm portions

standards are made by etching glass or by evaporating metal on glass. Table 9.1 gives a listing of some commercially available testing targets for photography and video cameras. Some secondary standards are for use in microscope oculars (reticles).

For diffuse reflectance, materials such as PTFE (commonly known as Teflon), finely powdered gold and $BaSO_4$ can be used. Some of these materials are sold as complete flat tiles and some are available as coatings to be applied to a surface of choice. For UV, NIR and visible spectroscopy there exist Spectralon reflection standards supplied with NIST calibration certificates. These are actually made for spectrometers with integrating spheres, but can be used for hyperspectral imaging too. They exhibit the highest known diffuse reflection for the visible and NIR regions. The measured values of tiles with nominal reflectance values of 2 %, 25 %, 50 %, 75 % and 99 % are shown in Figure 9.6. The reflectance values are actually 0.02, 0.25, 0.50, 0.75 and 0.99, but the percentage values are often used. The values are only given every 50 nm,

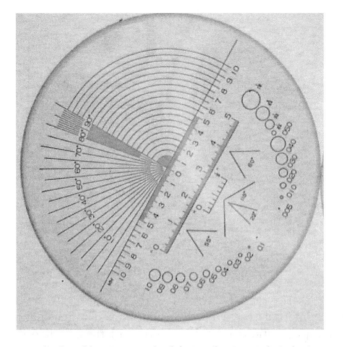

Figure 9.4 Multiple calibration standard for angle, size and circle size

Table 9.1 Image quality standards for photography and video cameras

Linear size (see Figure 9.3)
Area
Disk size (see Figure 9.4)
Angle (see Figure 9.4)
Resolution (USAF, NBS 1963A, IEEE and ISO targets)
Geometrical correctness / distortion
Depth of field
Modulation transfer function
Grayscale (see Figure 9.5)
Astigmatism (Star target)
Color (Macbeth color checker)

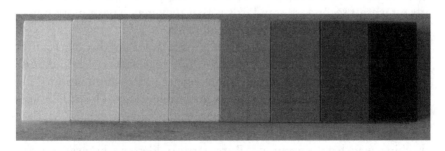

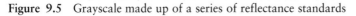

Figure 9.5 Grayscale made up of a series of reflectance standards

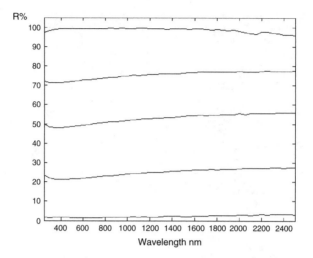

Figure 9.6   Reflectance curves of Spectralon reflectance standards

but can be interpolated by curve fitting. In remote sensing, there have been some publications on the properties of these standard reference materials. (Baba and Suzuki, 1999; Bruegge *et al.*, 2001; Secker *et al.*, 2001). Figure 9.5 shows a series of reflectance standards going from white to black. Figure 9.7 shows a magnification illustrating how mixing carbon black in white PTFE can give predetermined bulk reflectance values. It also shows that these standards cannot be used for high magnifications unless some averaging is used. Averaging can be done by mathematically averaging spectra over an area, by blurring the image (defocusing) or by moving the standard while images are averaged.

Any material that is homogeneous, flat, stable in time and has fixed and known reflectance values can be used as a secondary standard. Secondary standards can be calibrated against the primary Spectralon standards. Materials that exhibit diffuse reflection such as unglazed ceramic tiles would make good secondary standards. One of the possible advantages of secondary standards is that they may have a more homogeneous structure at higher magnifications.

Tiles with a fixed reflectance over a wide wavelength range can be used as reflectance standards to measure G_0 in Equations (9.2) and (9.6). They can also be used for testing homogeneity of the illumination. They do not supply information about wavelength accuracy. This can be seen by looking at Figure 9.6, where the spectra are almost flat.

Standards with sharp peaks in the NIR region are needed for checking wavelength accuracy. Rare earth oxides can be used for this

Figure 9.7 Detail of a grayvalue reflectance standard. The material is made for bulk reflectance measurement, so high magnification shows a nonhomogeneous structure. The scale below the picture has 1 mm marks

purpose. The NIST wavelength calibration standard is a lanthanum zirconium borosilicate glass doped with samarium oxide, ytterbium oxide, neodymium oxide and holmium oxide. The standards are transmission filters. SRM 2035 is for the NIR and SRM 2065 is for UV-visible-NIR.

The absorption bands of SRM 2035 are centered around 5140 (Ho), 6804 (Sm), 7313 (Sm), 8179 (Sm), 8682 (Ho), 9294 (Sm) and 10245 (Yb) cm^{-1} (Choquettte *et al.*, 2003; Duewer *et al.*, 2003). In principle, lasers can also be used to provide wavelength standards. It should be mentioned that all wavelength standards are temperature-sensitive and that also water vapor can give errors.

9.7 CALIBRATION IN HYPERSPECTRAL IMAGES

The example given here is from NIR hyperspectral imaging using a camera and LCTF filter (Geladi *et al.*, 2004). Many aspects of the example are also relevant for other hyperspectral imaging techniques.

The instrument is a MatrixNIR from Spectral Dimensions. It consists of an InGaAs array based camera sensitive between 850 and 1720 nm and a liquid crystal tunable filter (LCTF) between sensor and objective. The set-up uses four quartz halogen lamps for illumination of samples under study. The measurement is done in reflectance (Figure 9.8). The LCTF can be computer controlled, so the camera only detects intensities in a narrow wavelength band. The signal from the camera is digitized in 12 bits (intensities from 0 to 4095). The final result is a hyperspectral image of 256×320 pixels (the camera resolution) in up to 128 wavelength bands (the wavelength resolution). Each element of the hypercube is a measured and digitized intensity. The instrument allows averaging to reduce noise, so the intensities are not always integers, but can have decimal values.

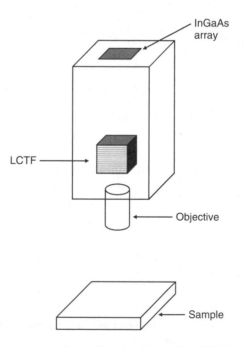

Figure 9.8   Schematic view of an InGaAs camera with LCTF filter

The result of one image collection is a hyperspectral image $\underline{G}$. This is a 3-way array with indices $i = 1, \ldots, I$ and $j = 1, \ldots, J$ for the pixels and $k = 1, \ldots, K$ for the variables.

The internal intensity calibration of the camera consists of setting the integration time of the InGaAs camera so that the maximum intensity of 4095 is not reached for any wavelength. With this calibration, no image will suffer from saturation, but the obtained intensities are still of arbitrary size. Some images will be very dark because the sensitivity of the InGaAs array is low close to 900 and 1700 nm. Figure 9.9 shows a typical example of an intensity curve. The example in Figure 9.9 is between 900 nm and 1700 nm with 10 nm steps and all intensities are averages of five measured values.

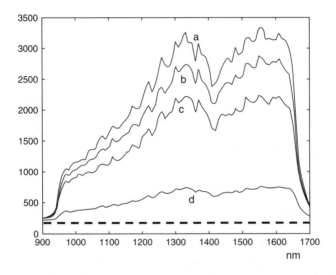

Figure 9.9 Intensity in counts for a few pixels in a hyperspectral image. Dashed line: approximate location of the camera dark current

Besides wavelength dependencies, there is also the effect of nonhomogeneous spatial distribution of intensity as shown in Figure 9.10. This effect may come from bad illumination, lens errors or InGaAs array errors. The figure also shows some dark pixels that are the result of dead pixels in the camera. In this example it turns out that less than ideal lamp placement is responsible for most of the intensity errors, while lens errors and InGaAs array errors are below detector noise. With a good illumination set-up, quite homogeneous images can be achieved and it is more important to correct for wavelength-dependent intensity differences than for spatial ones.

It is useful to point out a number of features of Figure 9.9. The dashed line is the approximate size of the minimum camera dark current. This is the signal coming form the InGaAs array without light coming through the objective. It usually gives a minimum signal of approximately 200 in

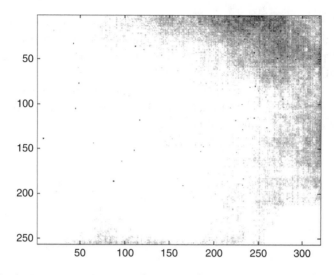

Figure 9.10 Image at 1700 nm of a 2 % reflectance standard showing camera errors (dead pixels) and nonhomogeneous illumination. The contrast is exaggerated. Most pixels have intensities between 225 and 235 counts

intensity. Curve a is one pixel of a 99 % reflectance Spectralon standard. The signal is safely below the saturation maximum of 4095. The shape of the curve is a combination of three effects: (1) the intensity distribution of the incoming light is unknown and probably not even; (2) the camera is not equally sensitive for all wavelengths; and (3) the LCTF does not have the same transmission for all wavelengths. Especially below 950 nm and above 1650 nm the InGaAs array loses sensitivity.[3] The small peaks in curve a are from LCTF instability. Curve c represents a pixel of a 50 % reflectance standard. Curve b is a pixel from a 75 % reflectance standard. Curve d is a pixel from a 2 % reflectance standard. The curves in Figure 9.9 show clearly that there is a need for transformation from intensity to IUPAC definitions, in this case reflectance.

With the availability of a 100 % reflectance standard, reflectance can be calculated as:

$$R = (I - d) \, (I_0 - d)^{-1} \qquad (9.8)$$

---

[3] It is possible to redo the imaging with longer integration times for these less sensitive wavelengths. In that case the more sensitive wavelengths would give saturation, but they can be skipped.

where d is the dark current. Equation (9.8) is valid for each pixel (i, j) in an image and each wavelengh (k = 1, . . . , K) so it can be rewritten as:

$$R_{ijk} = (I_{ijk} - d_{ijk})(I_{0ijk} - d_{ijk})^{-1} \qquad (9.9)$$

Matrix notation avoids the cumbersome use of indices:

$$\underline{R} = (\underline{G} - \underline{D}) /^* (\underline{G}_0 - \underline{D}) \qquad (9.10)$$

This is comparable with Equation (9.6), with /*signifying a pixel by pixel division. $\underline{D}$ is the dark current image. (The symbols are underlined because they signify 3-way arrays and not matrices.)

Equations (9.8)–(9.10) are for the use of only one reflectance standard. More standards can be used to increase the accuracy of the calculated reflectances. Using more reflectance standards gives a number of advantages: (1) there is an averaging effect; (2) the sample reflectance may be closer to that of a standard (bracketing); and (3) nonlinearities in the detector can be observed and corrected for.

(1) When using a single reflectance standard, all errors in the standard are propagated in the results. If a 100 % reflectance standard has become contaminated and only reflects 98 % then all calculations according to Equation (9.9) are in error. When using many standards, there may still be errors, but statistically there should be as many on the positive side as on the negative side, so the errors cancel out more or less.
(2) It is always a good idea to have bracketing. If an image area has a reflectance of 30 % then using standards close to 30 % (e.g. 25 % and 35 %) gives good results for linear interpolation.
(3) Equations such as (9.9) can neither detect nor correct for nonlinearities. Using more standards allows this.

Some useful calibration equations are:

$$r = b_1 g + b_0 + e \qquad (9.11)$$

$$r = b_0 + b_1 g + b_2 g^2 + e \qquad (9.12)$$

where r is a vector containing the nominal reflectances (see Figure 9.4) for the used standards, so the size of the vector is (number of standards $\times$ 1), g is a vector containing the measured intensities of the used standards, g^2 means squaring element by element, b_p are regression coefficients ($p = 0$, 1 or 2) and e is a residual vector.

In these equations r is known and finding the b_p is the goal. The coefficient b_1 in Equation (9.11) is a least squares averaging over the used standards. b_0 is an offset and replaces the dark current of Equation (9.9). b_2 in Equation (9.12) is zero if the detector is linear. Once the coefficients in Equations (9.11) and (9.12) are found, the same equations can be used to calculate reflectances for the sample images. An alternative to this is bracketing by linear interpolation between two close standards, one slightly above and one slightly below.[4]

Equations (9.11) and (9.12) should be established for each wavelength in a hyperspectral image. They should also be established for each pixel in the image if the illumination is not homogeneous. This would give the vectors g and e indices i, j and k. For the example of Equation (9.11):

$$r_k = b_{1ijk}g_{ijk} + b_{0ijk} + e_{ijk} \qquad (9.13)$$

b_{1ijk} and b_{0ijk} become hyperspectral images of coefficients. The multiway array equation would become very complicated and is left out. Once the coefficients b_p are found, the A/D converter counts for the sample image can be filled in in Equation (9.11) or (9.12) to find reflectances:

$$r_{ijk} = b_{1ijk}g_{ijk} + b_{0ijk} \qquad (9.14)$$

$$r_{ijk} = b_{0ijk} + b_{1ijk}g_{ijk} + b_{2ijk}g_{ijk}^2 \qquad (9.15)$$

In these equations the b_{pijk} ($p = 0$, 1 or 2) are known and the goal is to calculate an r_{ijk}.

It was shown in Geladi et al. (Geladi et al., 2004) how Equation (9.12) gives the best fit and Equation (9.9) gives the worst fit. In the wavelengh ranges 900–950 nm and 1620–1700nm, no good fit is found because the data are too noisy. Even the best prediction gives a bias of 1–2 % reflectance and the worst cases can give up to 6 % reflectance bias. In Burger and Geladi (Burger and Geladi, 2005) a more complete description of how to use the equations is given.

---

[4] If all possible reflectances are present in the sample image, standards that span the whole range 0–100 % should be used. Many images only contain limited reflectances and then a limited number of standards matching these reflectances may be more useful.

All of the previous material is about internal calibration of each sample image. The goal is to get the most correct reflectance value for each pixel and wavelength in the image array.

Between-image stability is also an issue. The camera, filter and illumination source can fluctuate or drift when a series of images is taken over a longer time. Using an internal standard can be used to detect for drift and fluctuations. The internal standard can also be used to correct a series of images. Standards such as the one in Figure 9.6 can be included as part of the image and equations similar to (9.11), (9.12), (9.14) and (9.15) may be used to compensate for drift. In Burger and Geladi (Burger and Geladi, 2005) it is shown how an artificial lamp drift giving a bias of more than a factor two in the A/D count images can be compensated to a bias of only 2–3 % reflectance in this way. In order to compensate for wavelength errors, a wavelength internal standard has to be included. In the same paper it is also shown that calibration using reflectance standards improves classification and regression models.

9.8   CONCLUSIONS

It is important to have control over all aspects of a hyperspectral image. In order to express all results in SI units or according to IUPAC definitions, reference standards are available or should be made and tested. Size and distortion standards control the spatial aspect of the hypercube. Grayscale standards control the intensity aspect, preferably expressed in transmittance, reflectance or absorbance. Wavelength standards should be used to check wavelength accuracy.

The example describes the use of reflectance standards to:

- allow transformation of arbitrary signals to reflectance;
- allow correction of internal (within an image) errors: mainly nonhomogeneous illumination;
- allow correction of external (between images) errors: lamp ageing, temperature effects, etc. This is done by including internal standards in each sample image.

REFERENCES

Baba, G. and Suzuki, K. (1999) Gonio-spectrophotometric analysis of white and chromatic reference materials, *Analytica Chimica Acta*, **380**, 173–182.

Burger, J. and Geladi, P. (2005) Hyperspectral NIR image regression. Part I : Calibration and correction, *Journal of Chemometrics*, **19**, 355–363.

Bruegge, C., Chrien, C. and Haner, D. (2001) A Spectralon BRF data base for MISR calibration applications, *Remote Sensing of Environment*, **76**, 354–366.

Callet, P. (1998) *Couleur-Lumière Couleur-Matière. Interaction Lumière-Matière et Synthèse d'Images*, Diderot, Paris.

Choquette, S., O'Neal, L. and Duewer, D. (2003) Rare-earth glass reference materials for near-infrared spectrometry. Correcting and exploiting temperature dependencies, *Analytical Chemistry*, **75**, 961–966.

Christian, G. (1980) *Analytical Chemistry*, 3rd Edn, John Wiley & Sons, Ltd, New York.

Duewer, D., Choquette, S., O'Neal, L. and Filliben, J. (2003) Rare-earth glass reference materials for near-infrared spectrometry. Sources of X-axis location variability, *Analytica Chimica Acta*, **490**, 85–98.

Duff, M. (Ed.) (1983) *Computing Structures for Image Processing*, Academic Press, London.

Geladi, P., Burger, J. and Lestander, T. (2004) Hyperspectral imaging: calibration problems and solutions, *Chemometrics and Intelligent Laboratory Systems*, **72**, 209–217.

Harris, D. (1997) *Quantitative Chemical Analysis*, 5th Edn, W.H. Freeman and Company, New York.

Johnson, G. and Fairchild, M. (2003) Visual psychophysics and color appearance, in: Sharma, G. (Ed.), *Digital Color Imaging Handbook*, CRC Press, Boca Raton, pp. 116–171.

Kopeika, N. (1998) *A Systems Engineering Approach to Imaging*, SPIE Optical Engineering Press, Bellingham.

Kramer, H. (1992) *Observation of the Earth and its Environment. Survey of Missions and Sensors*, 2nd Edn, Springer, Berlin.

Lillesand, T., Kiefer, R. and Chipman, J. (2004) *Remote Sensing and Image Interpretation*, 5th Edn, John Wiley & Sons, Ltd, New York.

Mills, I., Cvitas, T., Homann, K., Kallay, N. and Kuchitsu, K. (1993) *Quantities, Units and Symbols in Physical Chemistry*, Blackwell Science, Oxford.

Page, C. and Vigoureux, P. (Eds) (1975) *The International Bureau of Weights and Measures 1875–1975*, NBS Special Publication 420, NIST, Gaithersburg.

Secker, J., Staenz, K., Gauthier, R. and Budkewitsch, P. (2001) Vicarious calibration of airborne hyperspectral sensors in operational environments, *Remote Sensing of Environment*, **76**, 81–92.

Stoeppler, M., Wolf, W. and Jenks, P. (Eds) (2001) *Reference Materials for Chemical Analysis*, Wiley- VCH, Weinheim.

Taylor, B. (Ed.) (2001) *The International System of Units (SI,)* NIST Special Publication 330, NIST, Gaithersburg.

10

Multivariate Movies and their Applications in Pharmaceutical and Polymer Dissolution Studies

Jaap van der Weerd and Sergei G. Kazarian

10.1 INTRODUCTION

Multivariate imaging aims at the extraction of spectroscopic information from heterogeneous systems, where the spectral properties of different components can be elucidated along with their spatial distribution. In cases where a heterogeneous sample changes with time, a time series of multivariate measurements is needed in order to achieve a complete description of the sample. This chapter will provide an introduction into these time series, or multivariate movies, and discuss their applications in the fields of polymer dissolution and pharmaceutical science.

The introduction will describe the basic format of multivariate movies and suggest several ways to reduce the large amount of data normally obtained. We will not focus on a description of the instruments involved. The second part of this chapter, applications, will review studies of polymer dissolution and pharmaceutical tablet dissolution. The close relation between these two areas will be explained and a short introduction on polymer dissolution will be provided, followed by a more

Techniques and Applications of Hyperspectral Image Analysis   Edited by H. F. Grahn and P. Geladi
© 2007 John Wiley & Sons, Ltd

detailed description of several studies, including optical photography, nuclear magnetic resonance (NMR), Raman and Fourier transform infrared (FTIR) imaging. The focus of these descriptions is on the experimental set-up and the applied methods of data processing rather than on the wider implications of the obtained results.

10.1.1   Introducing the Time Axis

Imaging spectrometers have become available for a large number of analytical techniques in the past decade (Barbillat *et al.*, 1994; Williams *et al.*, 1994; Ebizuka *et al.*, 1995; Lewis *et al.*, 1995; Caprioli *et al.*, 1997; Johansson and Petterson, 1997; Price *et al.*, 1999; Burka and Curbelo, 2000; Pollock *et al.*, 2000; Schueler, 2001; Chan *et al.*, 2003). The imaging approach is a powerful extension of an analytical technique, whenever the distribution of different compounds in a sample needs to be studied. In an analogue way, time series of analyses are the obvious extension when the evolution of a system is important. In a number of cases, both spatial and temporal resolution is required simultaneously and this is where multivariate imaging time series or multivariate movies are needed.

The best known example of an analysis that needs both temporal and spatial resolution is the monitoring of the atmosphere by satellite-borne instruments, e.g. the American total ozone mapping spectrometer (TOMS) (Krueger and Jaross, 1999), and its European counterparts global ozone monitoring experiment (GOME) and Sciamachy (Bovensmann *et al.*, 1999; Kaiser *et al.*, 2004). These imaging spectrometers provide daily updates on the concentrations of ozone and other gases in the world's atmosphere.

Miniaturisation of imaging techniques has enabled more down-to-earth applications of multivariate movies, but there still are only few appearances of these studies in the literature. It is to be expected that the acquisition of time series will become more common along with the broader availability of imaging instruments. This chapter will introduce the basic structure of multivariate movies and will explain the need for data reduction. Several means to achieve reduction will be discussed. After this general introduction, some applications will be highlighted. The discussed systems are solvent intake into polymers, and the drug release from pharmaceutical tablets.

10.1.2   Data Structure and Reduction

The basic coordinate system of a multivariate movie is sketched in Figure 10.1. In this axes system, every row of data contains a single spectrum. A dataset contains many spectra, namely for several points

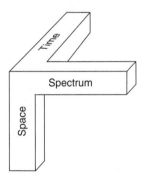

Figure 10.1   The basic coordinate system of a multivariate movie, consisting of a spectrum dimension, a time dimension, and one or more space dimensions

in space (down the columns) and time (in the third dimension, drawn towards the back). Space is represented here in a single dimension, but should generally be extended to two or three dimensions, depending on the analytical technique used. When a plane (2-D) is being analysed, the resulting dataset will be 4-D. As a sheet of paper is restricted to visualisation of only two dimensions, it is hardly possible to visualise a multivariate movie in a single image. Hence, many authors use multiple plots to display their results. Generally, the distribution of n materials at m time points is displayed as a 'matrix' of $n \times m$ images, such as the schematic fading pattern in Figure 10.2. In this fashion, articles have appeared with tens or even hundreds(!) of plots. Visualisation on a screen, where a succession of images can be displayed as a movie, reduces the number of images to be shown simultaneously to the number of materials n. Whether on paper or on a screen, large numbers of images are needed to show the results of a multivariate movie. Movies made in this way are often an invaluable tool in creating an understanding of processes in a sample, based on the impressive pattern recognition capabilities of the human mind. It should however be realised that they contain only a few per cent of the total amount of acquired data: only the concentrations of n ($\sim$2–4) materials are usually displayed rather than one image for every of the hundreds of data points generally present in

Materials

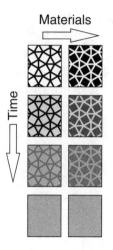

Figure 10.2 General method to visualise a multivariate movie as a 'matrix'. In this series of images, two materials (or factors) are derived from every spectrum dimension, and the time-dependent changes in the distribution of these materials are shown down the columns

the spectrum dimension. Somehow, a way has to be found to summarise these hundreds of data points in only a few numbers (amounts or concentrations of the relevant materials). Such a summary or 'compression' in a dimension of the multivariate movie reduces the size of the dataset, thereby facilitating visualisation and thus interpretation. The quality of the applied data reduction steps has a direct impact on the quality of the interpretation, and sensible compression is thus essential. Most of the first part of this chapter is concerned with various ways of data compression. Compression of the spectral dimension, mentioned above, is an important step in data processing, but certainly not the only step. Different compression steps are available for the different dimensions. In the remaining part of this section, the compression of the spectral dimension, space dimensions, and time dimension will be discussed, respectively. Successive compression of different dimensions is generally possible, but will not be discussed explicitly. Finally the simultaneous compression of multiple axes will be discussed.

10.1.3   Compression of Spectra

The spectrum dimension in a multivariate dataset consists of a large number of values, representing the response of a sample at different wavelengths, wavenumbers, masses, energies, etc., depending on the analytical technique used. In many cases, these spectra are a linear combination of only a few underlying spectra, or factors rather than completely random numbers. If these underlying factors can be resolved, it becomes possible to compress a complete spectrum to a few numbers, namely one number per factor; each number representing the bearing of the corresponding factor in the total spectrum. If a factor represents a specific material, the derived number indicates the concentration or total amount of that material.

Once every spectrum in a multivariate imaging dataset has been summarised in a few numbers, the complete set can be visualised using a few images. A multivariate movie can accordingly be visualised as a few movies on a screen. Several ways have been developed to achieve compression of the spectrum dimensions. All techniques available for multivariate imaging or for datasets consisting of a collection of single point spectra (Beebe *et al.*, 1998; Massart *et al.*, 1988; Esbensen, 2001; Geladi, 2003; Geladi *et al.*, 2004) can in principle be applied to multivariate movies, since the spectra in a multivariate movie are typically processed regardless of their original position or time stamp (generally called an 'unfolding' approach) (Geladi and Grahn, 1996; de Juan *et al.*, 2004). A quick, easy, and regularly used method to achieve compression of the spectrum axis is the selection or integration of a single band (univariate selection). The value of this approach is dependent on the availability of isolated bands for each important component or factor. If a selected band is specific for a component, its intensity or absorbance provides a concentration or total amount of the corresponding component.

Overlap between bands of different compounds seriously affects this simple approach and multivariate techniques have to be applied, such as principal component analysis (PCA), principal component regression (PCR), partial least squares (PLS), classical least squares (CLS), etc. In all cases, the final aim is to replace the complete spectrum by one or a few relevant numbers, which can then be displayed as one or a few movies, similar to integrated band movies. The choice of multivariate techniques is dependent on a number of circumstances (Beebe *et al.*, 1998). If all components in a sample are known, and their spectra are available, then CLS is a fast and easy option. The basic idea of this method is that every sample spectrum (obtained as part of imaging experiments) is a mixture (or linear combination) of the spectra of the pure components involved. A schematic representation of this is shown in Figure 10.3. In this figure, S is a measured (imaging) dataset, consisting of a large number (n) of spectra in rows. Spatial and temporal

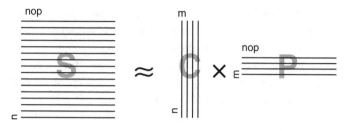

Figure 10.3 Reduction of a 2-D data matrix to smaller matrices. See text for details

information regarding the origin of the spectrum is temporarily removed (the dataset is unfolded) to facilitate calculations. Each spectrum in S contains a number of data points (*nop*), i.e. one per column. Matrix **P** is the set of pure compound spectra, in this case containing the spectra of *m* pure components. Obviously, the *nop* per spectrum should be equal to that in matrix **S**. Both **S** and **P** are determined experimentally. The aim of CLS is to find the concentration matrix **C**, which contains the concentrations of the *m* components in the *n* spectra. Calculation of **C** is achieved by multiplying matrix **S** with the pseudoinverse P^+ of matrix **P**.

$$C = S \cdot P^+ \qquad (10.1)$$

If not all components in the sample are known, but samples with known amounts of the important components can be obtained, inverse techniques, such as inverse least squares (ILS), PCR and PLS can be used. These inverse methods are taught (or calibrated) to calculate the concentration of the important components from the spectra acquired from the known samples. This calibration process determines which data points in the spectrum are relevant to calculate the concentration of important components. As the inverse methods (unlike CLS) can determine which data points are relevant and irrelevant, it becomes possible to ignore interfering components, and it is thus not necessary to completely characterise the sample. A more complete discussion on the merits, limitations, and dangers of inverse methods can be found elsewhere (Beebe *et al.*, 1998; Esbensen, 2001).

Finally, if nothing is known about the dataset, or if it is the choice of the analyst not to share this knowledge with the computer, PCA and multivariate curve resolution (MCR) are useful techniques. PCA uses the dataset to define a number of factors or principal components (PCs). The representation in Figure 10.3 is also valid for PCA. In this case, **P** contains the calculated PCs, the corresponding values in **C** are called scores. The main difference between PCA and CLS is thus the origin of the matrix **P**. In CLS, it is based on actual measurements of pure materials, so the spectra in **P** and accordingly the concentrations in **C** have an actual physical meaning. In PCA, **P** is extracted from **S** via matrix calculations. PCs are chosen in such a way that any number of PCs, when multiplied with a suitable **C** matrix, replicate the original dataset **S** as well as possible. Normally, PCA can describe a dataset using only a few PCs. However, this description should not be confused with an explanation: the physical meaning of **P** (and thus **C**) is not always obvious, and their interpretation should be made cautious. MCR (Tauler and Barcelo, 1993; Andrew and Hanciewicz, 1998; Schoonover *et al.*, 2003) aims to automatically extract physically relevant factors from the datasets by imposing restrictions on the factors to be retrieved, such as a non-negativity of factors and concentrations.

All techniques described above compress the spectrum dimension into one or a few variables, and thereby allow the visualisation of the dataset as a series of images or movies. Movies or images obtained in this way are useful and informative, and suffice in many cases to draw conclusions. The concentration, amount or any other calibrated quantity is known for every position at every time point, provided that a reliable model was used to compress the individual spectra. However, the information along the spatial and temporal dimensions is still qualitative; they lack the quantification essential for many scientific studies. For example, it is easy to see from a movie that a particle is small or big, circular, rod-like, or any other form (spatial domain), or (going to the time domain) that a particle is growing, shrinking, or moving, but these parameters are difficult to quantify. Therefore it might be useful to use additional steps to compress the other dimensions into a few important numbers. Some of these techniques will be highlighted in subsequent sections.

10.1.4 Space Dimensions

The previous section focused on a multivariate movie as a large collection of spectra. Instead, the current section will view the same dataset as a large number of images. A multivariate movie consists of one image for every time step as well as every variable in the spectrum dimension. All these images are basically ordinary greyscale images. Some basic tools for image processing are normally present in the software tools developed for multivariate imaging, such as the use of look up tables (LUTs) for creating fancy false colour images, and image inter-polation, to enhance the apparent spatial resolution. Other image tools become available by loading an image into image-processing software, Photoshop® or one of its competitors, to adjust brightness/contrast settings, sharpening, etc. The increased aesthetics of an image or movie attainable in this way should of course be exploited thoroughly, but do not normally lead to a better understanding or further data reduction. Other ways have to be found for this purpose. Traditionally, these oper-ations are not classified as chemometrics, but rather as image processing. Below we will discuss a number of techniques that might be helpful in the analysis of multivariate movies. Only a few of these techniques have actually been applied to multivariate movies, and in many cases the discussion will be an outlook of the possibilities that lie ahead rather than a review of what has been done. These techniques are far less common to spectroscopists, and are normally only applied after the spectrum dimension has been compressed using multivariate techniques. This procedure is justified by the relatively simple chemical systems studied by multivariate movies thus far. The techniques to compress the image dimensions discussed below may be applied to the raw datasets, or after compression of the spectrum dimension.

A very basic technique to achieve compression of the space axes is binning, in which adjacent spectra are grouped, and the corresponding spectra averaged. Binning directly reduces the spatial resolution of the dataset, and is as such not a very sensible approach. Its main use is the reduction of the size of the dataset in order to reduce calcu-lation times in subsequent data processing. A number of cases have been shown in the literature where an image was radically binned to a single number. In these cases, the spectrum dimension had already been compressed, so the values comprising the image indicated 'concen-trations' or 'amounts' of a material. Summation of all the concen-trations in the FOV resulted in a measure of the total amount of a component in the field of view (FOV). In this way, Snively (Snively and Koenig, 1999) calculated the total amount of solvent absorbed by a polymer from a series of FTIR imaging measurements. In a similar approach, Coutts-Lendon (Coutts-Lendon *et al.*, 2003) and van der Weerd (van der Weerd and Kazarian, 2004a) calculated the total amount of drug released from a pharmaceutical tablet. In these extreme cases of binning, spatial resolution is basically discarded. It is, however, clear that many studies would benefit from a compression that preserves all or at least most of the information in the space dimensions of multivariate movies.

Some of the methods introduced thus far in literature can best be classified as ad hoc approaches, i.e. they critically depend on knowledge (symmetry) of the sample. In a number of studies, the surface of a polymer slab is positioned parallel to one of the image dimensions. The sample and its evolution are thus independent of this image direction, as shown in Figure 10.4(a) (some of these experiments will be discussed in more detail below). When this assumption is justified, it becomes possible to select a single line in the Y dimension, or to take an average of several lines in the Y dimension without losing information. Another example of an ad hoc approach is provided by van der Weerd and Kazarian (van der Weerd and Kazarian, 2004b) who assume that their sample is circular, and thus that the value of angle α [Figure 10.4(b)] is irrelevant. Accordingly, the two image dimensions X and Y can be compressed into a single dimension, namely the radius.

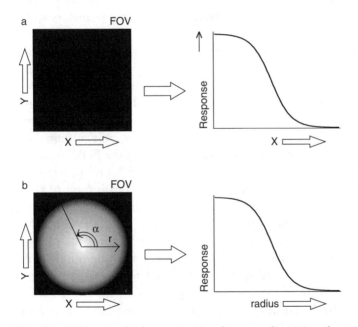

Figure 10.4 Easy ad hoc methods to compress the space dimensions by assuming that (a) the sample is homogeneous in one image dimension (is independent of X) or (b) has radial symmetry (is independent of α)

An infinite number of variations on this theme can be found. These are easy and effective, but their common feature is that their significance critically depends on the validity of assumptions regarding the sample. If these assumptions turn out to be incorrect or inaccurate, the compression will suffer from errors. More objective (unsupervised) approaches would be preferable if such assumptions can not be made *a priori*, or if an unsupervised method is preferred. A number of these objective techniques are available, but their use in multivariate movies is not yet reported.

It is possible to perform a Fourier transformation (FT) of the image (Niblack, 1986; Seul *et al.*, 2000) transforming the space into the frequency dimensions. This procedure will reveal possible patterns in the image. If many particles are visible in the FOV, the particle size distribution will be revealed. The space dimensions are effectively compressed into one or two numbers, if such a particle size distribution is used to calculate the mean particle size, or the mean particle size and its standard deviation. Another interesting, but still unexplored technique for multivariate movies is edge detection (Niblack, 1986; Petrou and Bosdogianni, 1999; Seul *et al.*, 2000), which serves to automatically locate objects in the sample. Once automatic location of objects is achieved, it becomes possible to objectively calculate its parameters such as mean particle size, the ratio area/circumference, and a mean spectrum.

A more common objective approach to compress the space dimensions is clustering of similar spectra. Classification algorithms define a quantification for the similarity or dissimilarity of spectra, and group pixels with similar spectra together (Niblack, 1986; Schowengerdt, 1997; Beebe *et al.*, 1998; Massart *et al.*, 1988; Petrou and Bosdogianni, 1999; Seul *et al.*, 2000; Esbensen, 2001). Normally, only a limited number of classes, representing homogeneous regions in the FOV, replace the large number of original spectra. The regions are found purely on the basis of their spectral similarity, as the original localisation of the different spectra is not taken into account during the classification process (i.e. classification in this way is an unfolding method). Once the clusters are formed, an optimised or averaged spectrum for each cluster can be calculated, reducing the number of spectra to the number of clusters. These spectra can subsequently be used to interpret the chemical or physical nature of the classes. A supervised method of classification, as opposed to the unsupervised method described above, uses a calibration set to define the clusters. Interpretation can thus be deducted from the calibration set or from the knowledge of samples in the calibration set. The different clusters can be displayed in order to locate them in both space and time. Such a visualisation is found in a number of studies on multivariate imaging. Clustering of spectra has been shown for results of remote sensing (Schowengerdt, 1997) and multivariate imaging (Shaw *et al.*, 2000; Lasch *et al.*, 2002, 2004; Noordam and van den Broek, 2002; Noordam *et al.*, 2003; Zhang *et al.*, 2003), mostly to deal with complex spectra. An outstanding example is provided in Figure 10.5. Application of clustering to multivariate movies would add a number of useful possibilities. Plotting the time dependent area occupied by a particular cluster would provide a useful insight in the evolution of a scene; growth of particles or disappearance of a reactant become directly visible. Alternatively, it is possible to base the clustering on the time dimension rather than the spectral dimension. In this way, pixels with a similar trend would be clustered instead of pixels with a homogeneous chemical nature, and it becomes possible to study which components appear in or disappear from the sample simultaneously.

10.1.5   Time Dimension

In this section, a multivariate movie will be interpreted as a large number of simultaneously acquired time series; one for every image position and spectrum point. Compression of the time dimension in a multivariate movie has been shown in a number of ad hoc approaches, requiring initial assumptions about the sample. For example, the speed of solvent intake into polymers has been determined in a number of studies (Ribar *et al.*, 2000, 2001; Gonzalez-Benito and Koenig, 2002; Miller-Chou and Koenig, 2002; Coutts-Lendon *et al.*, 2003; Gupper *et al.*, 2004a; van der Weerd *et al.*, 2004). The speed of the solvent summarises the time dependent location of a solvent front into a single value. If the speed of solvent intake is not constant, it might still be possible to summarise the 'trend' of the speed of solvent in a single value, as will be described below.

A more general method to compress the time dimension is particle tracking, where the position of a particle or multiple particles is followed in time. An elegant implementation of a particle tracking algorithm is found in a study by Adler *et al.* (Adler *et al.*, 1999) (Figure 10.6), who loaded a pharmaceutical tablet with small fluorescing particles. A sequence of fluorescence images was acquired during dissolution of this tablet. The particles were found to move due to the swelling of the tablet. Tracking enabled the calculation of the particle position and speed. The dataset used in this study is not multivariate, as ordinary fluorescence images were acquired. Accordingly, the chemical specificity of these analyses is low, and special markers (in this case fluorescent markers) have to be used. Multivariate techniques could aid here to increase the specificity, as e.g. reported by van der Weerd *et al.* (van der Weerd *et al.*, 2004), who showed the displacement of a caffeine particle in a tablet, as studied by FTIR imaging.

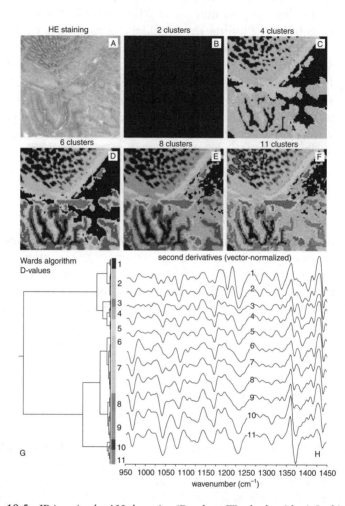

Figure 10.5 IR imaging by AH clustering (D-values, Wards algorithm). In this example of cluster imaging, spectra of a specific cluster are encoded by a specific colour, shown at the coordinates at which each IR spectrum was collected. The images are reassembled by colour, encoding 2, 4, 6, 8 or 11 classes [(B)–(F)]. In the 11-class classification trial, the malign epithelial cells of the adenocarcinoma [red areas of (F)] can be clearly separated from the respective epithelial cells of non-neoplastic crypts [blue and dark blue regions of (F)]. (G) Dendrogram with the respective colour assignments of the 11-class classification trial. The class-mean spectra (second derivatives) are given in (H) (numbering corresponds to the class numbering of the dendrogram) (Lasch *et al.*, 2004). Reprinted from Biochim. Biophys. Acta-Mol. Basis Dis. 1688(2), P. Lasch, W. Haensch, D. Naumann and M. Diem, Imaging of colorectal adenocarcinoma using FT-IR microspectroscopy and cluster analysis, 176–186. 2004 with permission from Elsevier (see Plate 32)

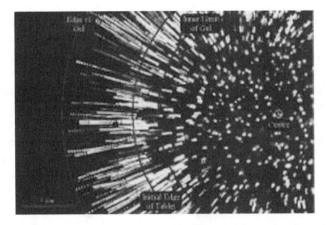

Figure 10.6 Tracks made by fluorescent microspheres in a xanthan tablet hydrating in 0.05 M NaCl over 80 min. Also marked are the initial location of the tablet edge and the final limits, inner and outer, of the gel layer (Adler *et al.*, 1999). Reproduced by permission of John Wiley & Sons Inc.

The processing methods above are efficient and well suited for the circumstances in which they were used, as they were specifically implemented for a single application. More general techniques are readily available, but applications in multivariate imaging have not been found in the literature. These techniques have been developed in the field of time series analysis. This scientific field is vast and comprises applications in fields as diverse as economy, sociology, astronomy, medicine, seismology, physics, marketing, demography, process control and meteorology (Granger and Newbold, 1977; Chatfield, 1984; Harris and Sollis, 2003). The main driving force for most time series analysis is the formulation of models that describe a time series as well as possible. If a fitting model is established, it characterises the behaviour of the measured variable or enables forecasts of future values in the time series.

One simple aid in the analysis of time series of images is to differentiate the series by subtracting the previous point from every time point (Rosenfeld and Kak, 1982; Chatfield, 1984). This procedure highlights the moving or changing features in a time series by ignoring the stable features and trend lines. Another basic technique generally used in time series analysis is the calculation of autocorrelation (self-correlation). In this procedure, the time series is delayed with a specific lag time, and the correlation coefficient of this delayed series with the original time series calculated. If the lag time is zero, the correlation coefficient is by definition one; for longer lag times the coefficient will normally drop to zero. A plot of the correlation coefficient versus the lag time ('correlogram') is an effective way to identify memory effects in the series, i.e. the extent to which a certain point in a time series depends on previous points. Various models established for linear time studies aim to describe the relation between successive points in a time series more specifically. The most important of these (Granger and Newbold, 1977; Chatfield, 1984) are known as random (no correlation), linear cyclic (time series consists of a number of strictly periodic components), random walk (next value is random, but close to the previous value in the series), moving average (which assumes that the time series is a smoothed white noise), and autoregressive (next value is a regression of the previous values in the time series). A combined model containing both autoregressive and moving average elements, the so-called ARMA model, is particularly popular nowadays.

One of the techniques used extensively in time series analysis is 'spectral analysis', in which a time series is converted into the frequency domain by Fourier transformation. In this way, periodicities in the time series can be elucidated. Obvious periodicities, such as the daily or yearly periodicity in outdoor temperature, are easily found in this way, though it should be noted that these obvious periodicities are normally filtered out of the dataset before spectral analysis, so that the emphasis is on 'hidden periodicities' (Chatfield, 1984).

The general applications of time series analysis differ significantly from the situation encountered in most multivariate movies. The time series captured in most multivariate movies are short, only a few tens of data points covering a number of hours or less, compared with many data points covering time spans of weeks to decades. On the other hand, a multivariate movie contains many (number of pixels × number of variables per pixel) simultaneously acquired time series, whereas normal time series analysis is concerned with one or a few variables. A time series is called multivariate if two or more variables are monitored. The main interest in the analysis of multivariate time series is finding the relation between the observed variables. A basic way to investigate this relation is by determining the cross covariance or cross-correlation coefficient. As with autocorrelation, one of the time series is delayed for a specific lag time. Afterwards, the correlation coefficient of the delayed time series with the second time series is calculated. Similarities of the different time series, and the influence of the delay on this similarity can by characterised by plotting the cross-correlation as a function of the chosen lag time in a so-called cross-correlogram (Chatfield, 1984). If the different time series are related, it might be beneficial to use past values of one series in the modelling or prediction of the other series. In many cases, predictions can be improved if used models are based on more than one time series. One specific relation between time series is causality, i.e. one parameter influences the other. 'Granger causality' is a way to determine this. To establish if Granger causality is present, a prediction is made based on a single time study and compared with a prediction based on two time series. If the prediction improves as a result of the inclusion of the second series, it is concluded that the second series 'Granger causes' the first series (Chen *et al.*, 2004). Initially, Granger causality was used to test economic hypotheses, but the theory has been applied extensively outside econometry. An interesting example in imaging is provided by Goebel *et al.*, who applied Granger causality analyses to results obtained by functional magnetic resonance imaging (fMRI) (Goebel *et al.*, 2003). fMRI is used extensively to locate activated (oxygen consuming) areas in the brain. Granger causality was used to reveal the effective connectivity (directed influences) between activated brain areas.

A last global techniques might be interesting for analysis of time series. Firstly, it is possible to compress the time series by means of PCA. The different positions in the images (samples) versus the time points (variables) will now be decomposed into a number of factors that can be seen as different 'trends'. The significance of a specific trend for a certain spectrum can be derived from the obtained scores. PCA of time series has been proposed under the name dynamic factor analysis (Zuur *et al.*, 2003). The technique has been applied in environmental science, where analysts are obviously concerned with time series and trend lines, e.g. to trace climate changes. A similar technique has been developed for cyclical time series, i.e. time series containing periodicities (Paatero and Juntto, 2000).

It is clear that much work has to be done before the achievements of the large field of time series analysis will be fully exploited in the just emerging field of multivariate movies. Higher time resolutions and broader use of multivariate movies, both of which are to be expected, will however require these advanced techniques to exploit the obtained data to the full.

10.1.6   Simultaneous Compression of all Variables

A currently unexplored approach in the analysis of multivariate movies is *N*-way analysis. These techniques are designed to find correlated features in multiple (N) dimensions. The parafac model, which is a multidimensional PCA, is the most comprehensive of these systems (Bro, 1997). A more general model is the Tucker3 model. A schematic representation of these N-way techniques is provided in Figure 10.7. Note

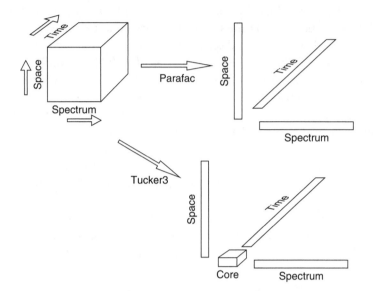

Figure 10.7 Graphical explanation of the Parafac and Tucker3 (after Pravdova *et al.*, 2002). Reprinted from Anal. Chim. Acta 462(2) V. Pravdova, C. Boucon, S. de Jong, B. Walczak and D.L. Massart, Three-way principal component analysis applied to food analysis: an example, 133-148, 2002, with permission from Elsevier

that space in these images is compressed into a single dimension, as in Figure 10.1. Again, this is just a limitation of graphical representation; mathematical processing is not affected by higher dimensionality. Parafac decomposes an N-dimensional dataset to N factors. These factors are calculated in such a way that they can be used to replicate the original in the best possible way, similar to PCA. If these N factors do not suffice to describe the dataset satisfactorily, a second component of N factors (etc.) can be added, until the differences between the original and a reconstructed dataset basically describe noise. In the parafac model, every dimension has the same number of factors. The Tucker3 model is more general, as it allows a different number of factors in different dimensions. This is useful if the complexity of the different dimensions is different. Imagine a dataset where the spectrum dimension is fairly complex; every spectrum represents a random mixture of many components and many factors will thus be needed to describe every spectrum. If the evolution of this complex set over time is simple, e.g. say that all components evaporate from the FOV at the same rate, only one factor will suffice for the time dimension [some more realistic examples can be found in Huang *et al.* (Huang *et al.*, 2003)]. Tucker3 can handle these different complexities by allowing a different number of factors for every dimension. The core matrix (plotted in Figure 10.7) describes the relations between these factors and is needed to reproduce the whole dataset from these factors.

Application of N-way analysis in cyclical time series has been shown by Pateero (Paatero and Juntto, 2000). Other studies (Gurden *et al.*, 2003; Huang *et al.*, 2003) describe the processing of imaging data using N-way techniques. Application of these techniques to multivariate movies has not yet been found in the literature, but should be a straightforward extension. A suitable sample would be one where a species of interest cannot be distinguished on the basis of its spectrum alone, but only on a combination of spectrum and particle size, or spectrum and degradation time. Time will show if this kind of processing to multivariate movies will turn popular, and if applications can be found that benefit from these analyses.

10.2 APPLICATIONS: SOLVENT DIFFUSION AND PHARMACEUTICAL STUDIES

This section will describe some applications of multivariate time series. Most of the studies published up to now discuss the study of solvent diffusion in two closely related fields, namely polymers and drug release from tablets. These applications share a number of features that make them interesting for study by multivariate movies:

- The timescales (minutes to hours) and position changes (μm to mm range) on which these processes take place match with the temporal and spatial resolutions that can be achieved by mid infrared (MIR) and Raman imaging.
- The components in the sample are known and it is easy to acquire reference spectra, which facilitates data processing by means of univariate or multivariate techniques.

Therefore, we will start with a short introduction on diffusion of solvents in polymers. Subsequently, some studies on these topics will be highlighted.

10.2.1   Solvent Diffusion in Polymers

The intake of solvents into polymers is an important process with applications in microlithography, membrane science and plastics recycling (Miller-Chou and Koenig, 2003). As a result, a large number of studies have appeared to characterise solvent diffusion. A common goal for these studies is the determination of the rate of solvent intake. In a simple approximation, this rate can be described by the diffusion coefficient D, defined as the proportionality factor between a concentration gradient $(\delta c/\delta x)$ of a solvent in a polymer and the resulting flow F (Fick's first law)(Crank, 1975):

$$F = D \cdot \frac{\partial c}{\partial x} \tag{10.2}$$

A higher D thus indicates higher flow rates and a faster equilibration of the system (i.e. concentration is constant everywhere).

The diffusion coefficient can be calculated using Fick's first law when the concentration gradient of a solvent and its flow are known. The steady state approach, one of the traditional approaches to determine D, measures the flow of a solvent through a membrane and the solvent gradient from the vapour pressures on either side of the membrane, (Crank and Park, 1968). Experimental challenges of this system, such as its susceptibility to leaks, have made other approaches popular. In these other experimental set-ups, it is however not possible to directly determine the flow of a solvent, and the quantitative description of a diffusion coefficient based on Equation(10.2) is thus hardly possible. Fick's second law solves this by defining D in terms of the partial derivatives of concentration: the flow induced by a concentration gradient in space $(\delta c/\delta x)$ will lead to a changing concentration in time $\delta c/\delta t$, according to:

$$\frac{\partial c}{\partial t} = D \cdot \frac{\partial^2 c}{\partial x^2} \tag{10.3}$$

As in Fick's first law, D is a proportionality factor, which connects the observables. A number of useful and extensively applied relationships, normally called the 'square root relationships' have be derived from solutions of this formula (Crank and Park, 1968), and relate D to easily analysed quantities, such as position and weight:

- The distance of penetration of any given concentration is proportional to the square root of time.
- The time required for any point to reach a given concentration is proportional to the square of its distance from the surface.
- The amount of diffusing substance entering the medium through a unit area of its surface varies as the square root of time.

The last of these relationships is easily exploited by the sorption technique, another traditional method to determine D. In this method, a polymer slab is kept in an atmosphere of constant vapour pressure and the intake of solvent into a polymer is monitored by weight. Single point spectroscopic studies have been used analogous to these gravimetric studies (Brantley *et al.*, 2000; Cotugno *et al.*, 2001; Musto *et al.*, 2002). In this case the chemical specificity of the spectroscopic technique was used to quantify the amounts of polymer and solvent in a polymer slab and so determine the total amount of solvent in a polymer slab.

These approaches have been used extensively, but do have a number of restrictions:

- They are usually restricted to systems in which solvents are supplied from the vapour phase. Solvent supplied as a fluid impedes the determination of concentration or weight. Additionally, dissolution of the polymer might be an issue.
- They provide an integral view over a polymer film and will therefore only work properly if the diffusion coefficient is homogeneous throughout the polymer film. However, the diffusion coefficient is normally a function of the concentration of the solvent. Such a variable D cannot be directly determined by the mentioned approaches. If D is variable, the values found with steady state and sorption techniques will be a sort of mean value.

Obvious ways to circumvent the first restriction is to apply techniques such as attenuated total reflection (ATR) spectroscopy (see Linossier *et al.* (1997) and Balik and Simendinger (1998 and references mentioned therein). A general solution for both restrictions is the application of imaging experiments, starting with optical microscopy (Ueberreiter, 1968). Later imaging experiments, both line (Challa *et al.*, 1996, 1997;

Snaar *et al.*, 1998; Narasimhan *et al.*, 1999; Gupper *et al.*, 2004b) and global chemical imaging experiments (discussed below), were applied as a powerful addition to earlier techniques due to their ability to determine the concentration, mobility, and diffusion of a solvent in a polymer with a spatial resolution rather than providing a general overview of the polymer slab as in optical photography. As a result, imaging techniques present a more direct way to derive a concentration dependent D.

Mathematically (though not experimentally), the simplest way to derive a concentration dependent D is spatially resolved determination of the concentration of a membrane in the steady-state (Barrer, 1946; Crank and Park, 1968), as D is inversely proportional to the concentration gradient $\partial c/\partial x$ [Equation (10.2)]. A single imaging measurement of the spatially resolved concentration of solvent through the polymer film thus suffices to calculate D, even if it is non-constant. The drawbacks of such an experiment (mentioned above) make that non-steady-state experiments, determining the concentration of solvent as a function of both space and time, are more popular. One of the drawbacks of the steady-state approach is that glassy polymers cannot be fully characterised (Crank, 1975). Glassy polymers show two competing processes upon contact with solvent, namely the diffusion of solvent and the relaxation of the glassy polymer to a gel state. Information on polymer relaxation, however, cannot be obtained from steady-state measurements. As the diffusion coefficient normally depends on the conformation of the polymer, solvent intake into glassy polymers cannot generally be described by Fick's laws, even if D is assumed concentration dependent. The divergence from the Fickian behaviour is obviously determined by the relative speeds of solvent intake and polymer relaxation, and a classification of glassy polymers is generally made into Case I (or Fickian) diffusion (solvent intake is slow compared with conformational changes) and Case II diffusion (vice versa). Anything between these extremes is classified as anomalous.

The distinction between Case I and Case II behaviour can be made by fitting the profile of one of the observables $M(t)$, i.e. depth of solvent penetration or total solvent intake (see square root relationships above) to the power law:

$$M(t) = k \cdot t^n \tag{10.4}$$

The value of k determines the actual magnitude of the observed parameter. Here, it is used merely as a scaling factor. Classification can be based on the found value for n, ranging from $1/2$ for Case I (Fickian) behaviour (slab configuration) to 1 for Case II behaviour (Crank, 1975). If intermediate values are found ($\frac{1}{2} < n < 1$), the system is classified as anomalous.

Drug release studies are a special case of solvent diffusion studies. In this case, solvent intake is an intermediate step in the release of a drug that has been mixed with the polymer. A detailed study of this release is important in the pharmaceutical science, as the release of drugs from tablets is a crucial step in the successful application of a drug. Drug release is closely related to the solvent intake studies, as it is caused by dissolution of the drug in the ingressed solvent. One advantage of drug release studies is that it introduces a new parameter that can easily be monitored, namely the amount of drug released from the tablet. In a typical dissolution test, a tablet is immersed in fluid and the drug concentration in this fluid is monitored, normally by high pressure liquid chromatography (HPLC) or ultraviolet (UV) spectrometry. Many parameters can influence the release profile observed in a dissolution test, e.g. stirring of the fluid, additives, temperature, etc. A number of organisations have therefore set detailed regulations to ensure reproducibility of the dissolution test. The widely accepted set of regulation is devised by the *United States Pharmacopeia* (USP). The theories and vocabulary used in drug release studies resemble those used for solvent intake. The division of glassy polymers into Case I and Case II, based on the relative rates of relaxation and solvent intake, is also used to classify release properties. In this case, the parameter entered as $M(t)$ into Equation (10.4) is normally the release profile, and accordingly the general interpretation of the classes has changed (Siepmann and Peppas, 2001). Case II now indicates that the drug release is constant in time, while Case I indicates that the release is proportional to the square root of time. This formulation of Case I and Case II does not contradict the earlier interpretation (based on relative velocities of solvent intake and polymer changes), but rather is an effect of it.

The dissolution test and the classification into Case I and Case II release have been used for many years, and have proven useful in many cases. However, there are some fundamental issues that arouse suspicion:

- Drug release profiles obtained by dissolution studies are often used to characterise processes inside a tablet. An evaluation of the validity of the applied theory is not possible.
- One obvious shortcoming is that the most calculations assume constant diffusion coefficients. This assumption is an easy aid in calculations, as it allows an analytical solution for Fick's second

law of diffusion. However, in most systems, D is directly related to the tablet matrix concentration. Polymer solutions will normally be viscous and the diffusion coefficient will thus be reduced.

In the last decade, dissolution studies have therefore been supplemented by a number of imaging studies.

A number of case studies will be presented below to highlight recent imaging studies and will describe the applied data processing methods. These cases are presented to provide a general overview rather than an extensive review:

- photographic and NMR studies;
- Raman line imaging;
- MIR transmission studies.

10.2.2   Optical and NMR Studies

NMR and optical techniques such as photography do not form an obvious group. The connecting factor for the studies described here is that they are imaging techniques used in a nonspectroscopic approach for the study of solvent intake and dissolution of pharmaceutical tablets. Accordingly, the spectrum dimension (see Figure 10.1) contains only one or a few variables. Spectroscopic imaging is possible for both techniques, but this would result in long acquisition times for NMR imaging and only a minor improvement of the chemical specificity for visual light spectroscopic imaging, as pharmaceutical tablets are usually all white. A short description of these nonspectroscopic approaches is included here, as it has been shown possible to make a basic distinction between polymer, solvent and (where present) drug using specific labels, and because of their general value for the field of diffusion studies.

Photography has been used extensively in solvent diffusion studies. Ueberreiter (Ueberreiter, 1968) provides a clear overview of microscopic studies in polymer dissolution. In these studies, a thin film is clamped between two windows, exposed to a solvent, and studied during dissolution of the polymer. Based on these studies, the dissolution process is normally classified as 'normal' or 'stress-cracking' (Ueberreiter, 1968; Miller-Chou and Keonig, 2003). Normal dissolution is characterised by the formation of a surface (gel) layer around a polymer upon solvent intake. The surface layer can be distinguished from the dry core of the polymer due to the changing optical properties of the polymer film upon interaction with the solvent. The boundary between the dry core and the surface layer is visible as a front that moves into the polymer. In the stress-cracking mechanism, the polymer crumbles upon solvent intake and small pieces leave the surface in a kind of eruption process.

Photography normally suffices to make these simple classifications, but it is very limited in cases where better quantification or a higher chemical specificity is needed. Some experiments have been designed to allow quantitative observations, such as the interferometry (Ueberreiter, 1968) and a calibration based on the scattering properties of glassy polymer (Gao and Meury, 1996). In these studies, the applied data processing methods aim at establishing distance–concentration curves, which can be used to calculated the diffusion coefficient. While useful for polymer dissolution studies, the value of these more quantitative photographic studies for drug release studies is limited, as the number of components in the system increases from two (solvent and polymer) to three (solvent, polymer and drug). The intrinsic low chemical specificity fails to distinguish between these materials, unless a special marker is used. Adler *et al.* (Adler *et al.*, 1999) investigated an HPMC tablet loaded with fluorescent particles. Their study shows a nice example of particle tracking to compress the image dimensions (shown in Figure 10.6). A range of studies (Colombo *et al.*, 1996, 1999; Bettini *et al.*, 2001) reported the investigation of tablets containing coloured drugs, to achieve a rough estimate on the distribution of the drug. In these tablets, three migrating fronts are observed during dissolution, labelled the swelling, dissolution, and erosion fronts, as shown in Figure 10.8. Data processing in these studies normally focuses on the time-dependent position of these fronts rather than the spatially resolved concentration, as the quantitative properties of the technique are poor.

NMR imaging (MRI) studies have been used for solvent intake and drug release studies. A large advantage of MRI is the flexible adjustment of the FOV: it is possible to acquire images in one, two or three dimensions. As a result, the inside of a tablet can be studied even if it is completely immersed in water. Interfacial effects, which might occur when the tablet is clamped between two windows or against a crystal, can thus completely be excluded. Unfortunately, the noise levels of MRI analyses are inherently high and limit the spatial and/or temporal resolutions that can be achieved. Most studies therefore restrict their observations to one dimension perpendicular to the polymer surface. Along the FOV, the amount of solvent in a polymer slab or pharmaceutical

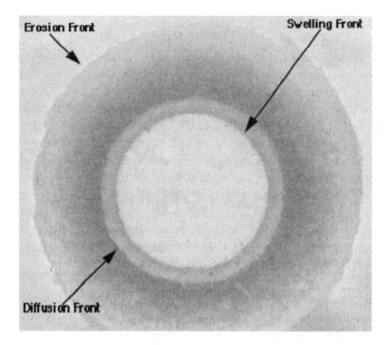

Figure 10.8 Picture of a pharmaceutical tablet containing a coloured drug during dissolution. Three fronts are visible in this sample (Colombo *et al.*, 1999). Reprinted from J. Control. Release, 61, P. Colombo, R. Bettini and N.A. Peppas, Observation of swelling process and diffusion front position during swelling in hydroxypropyl methyl cellulose (HPMC) matrices containing a soluble drug, 83–91, 1999 with permission from Elsevier (see Plate 33)

tablet can be quantified. Besides, additional NMR parameters such as T_1 and T_2 are acquired. The value of T_2, which is the most commonly used, is related to the mobility of the solvent. The polymer itself, being a solid, cannot be directly observed in these studies, but some elegant solutions have been proposed to circumvent this. Fyfe and Blazek (Fyfe and Blazek, 1997) prepared a number of polymer (HPMC) solutions in water and found that the T_2 was strongly dependent on the polymer concentration. This is to be expected as the increased viscosity of the polymer solution will reduce the mobility of the solvent, and thereby affect the T_2. The relationship between HPMC concentration and T_2 was subsequently used to calibrate the HPMC concentration of a tablet during dissolution. In another study (Snaar *et al.*, 1998; Narasimhan *et al.*, 1999), the possibility to directly measure self diffusion coefficients by MRI was used to study the inside of a poly(vinyl alcohol) (PVA) tablet.

A number of approaches have been proposed to improve the chemical specificity of MRI. The components in a binary solvent mixture, as well as their intake profile into a polymer, could be measured individually by application of ^{13}C-^{1}H cyclic cross-polarisation (Sackin et al., 2001). Fyfe and Blazek-Welsh found an elegant way to quantify the amount of dissolved drug by using different ^{19}F labelled drugs (Fyfe and Blazek-Welsch, 2000). A more general overview of the applications of MRI in the investigation of pharmaceutical tablets has been presented by Melia (Melia et al., 1998).

Both photography and MRI generally yield only a few variables. As a result, the possibilities for multivariate data processing are limited to processing in the image dimensions. Next to the nonspectroscopic nature of photography and MRI, there are a number of additional similarities. For both techniques there have been attempts to generalise the results using mathematical models: Narasimhan provides an extensive model and compares the results with experimental results obtained by MRI (Narasimhan et al.,1999), while Kiil reports an extensive mathematical model, based on the experimental work by the group of Colombo (Kiil and Dam-Johansen, 2003).

10.2.3 Line Imaging

Raman spectroscopy using a line focus laser enables the simultaneous study of several points along a line (Bowden et al., 1990). In the study on solvent diffusion by Gupper et al. (Gupper et al., 2004b), a polymer slab was clamped between two windows and exposed to solvent vapour. A line parallel to the dimension of solvent intake was imaged and the sample was assumed homogeneous in the perpendicular dimension. The resulting Raman spectra were used to quantify the amounts of solvent ingressed into the polymer at different positions and times. Additionally, conformational changes taking place in the polymer (solvent-induced crystallisation) could be analysed based on the Raman spectra. Good quality spectra of all components in the system could be acquired, and a CLS approach was therefore used to process the results. Figures presented in this study show a nice feature of the 1-D analyses: plots can be made in which the second dimension is occupied by the time axis. Thus, the whole time resolved experiment is shown in only a few plots (Figure 10.9). Based on these plots, the solvent diffusion coefficient in the polymer was calculated.

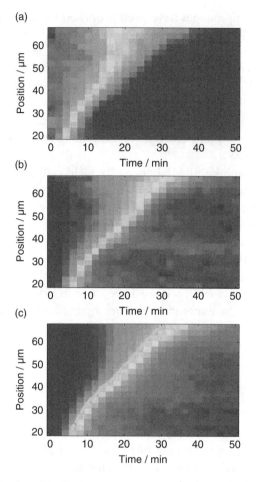

Figure 10.9 Line imaging Raman spectroscopy of toluene intake in syndiotactic polystyrene (sPS). The Raman spectra were processed using CLS using spectra of amorphous sPS, crystalline sPS and toluene. The plots show the spatially and temporally resolved amounts of (a) amorphous sPS, (b) crystalline sPS and (c) toluene (Gupper *et al.*, 2005) (see Plate 34)

10.2.4 Global MIR Imaging Studies of Solvent Intake

FTIR imaging studies have been used in a number of polymer dissolution studies since the technique was introduced almost a decade ago. Most of the early studies are presented by the group of Jack Koenig. In a series of articles, the intake of solvent and solvent mixtures into polymers has been shown. Snively and Koenig (Snively and Koenig, 1999) showed the intake of a liquid crystal (5CB) into poly (butyl methacrylate) (PBMA) at different temperatures. Polymer and solvent profiles are shown for different times, and diffusion distances are calculated from these concentration–distance curves. Calculated values for the diffusion exponents n [see Equation (10.4)] range from 0.4 to 0.54, and diffusion was surprisingly classified as anomalous. (Usually this classification is used for exponents with values of n between 0.5 and 1.) In a second way of data processing, the total mass of the solvent intake was calculated by integration of the concentration profile between the borders of the polymer and the diffusion front. It appeared that the mass of the ingressed solvent is proportional to the distance of solvent diffusion front. This proportionality is to be expected from the square root relations (Crank and Park, 1968) discussed above.

In a series of articles, the intake of solvent mixtures into a polymer, mostly poly(α-methyl styrene) (PAMS), was studied using a mixture of solvents (Ribar *et al.*, 2000; Miller-Chou and Koening, 2002), a mixture of a solvent and a nonsolvent (Ribar *et al.*, 2001), and a mixture of nonsolvents (Gonzalez-Benito and Koening, 2002). It was shown that normal dissolution and stress cracking (Ueberreiter, 1968) could well be distinguished by FTIR imaging (Ribar *et al.*, 2000; Miller-Chou and Koenig, 2002). Stress cracking was observed as a roughening of the polymer surface. The surface roughening was quantified by the 'root mean square roughness', basically defined as the standard deviation of the 64 values found for the position of the polymer front (the applied imaging instrument analyses 64 rows of spectra).

This series of articles shows a number of good examples of the possibilities of FTIR imaging in polymer dissolution studies. The emphasis is on the discrimination and quantification of individual solvents in the polymer–solvent mixtures. The concentration of different components is determined by single band integration. This simple way of data processing seems sufficient in these cases, as components with well separated absorption bands have been chosen. In this way, the concentrations of the different solvents were determined independently for every spectrum, and their ratio could be calculated. This ratio, shown in Figure 10.10, provides a good indication of the sequence of processes taking place and helps to answer the questions whether the solvents ingress simultaneously, or subsequently. The spatial dimension in these studies are normally compressed using the easy method shown in Figure 10.4(a), as the polymer samples are positioned parallel to one of the image dimensions.

Another study by Gupper *et al.* (Gupper *et al.*, 2004a) reports the intake of solvent vapour into a syndiotactic polystyrene. The temperature and concentration of the solvent were controlled in a special cell. The solvent front positions, found by integration of a single absorption band, was determined as a function of time to classify the type of diffusion as Fickian [$n = 0.49$ to 0.59, see Equation (10.4)], and to determine the diffusion coefficient D. Spectral changes of the polymer were used to monitor the conformation change of the syndiotactic polystyrene from amorphous to crystalline. It was found that the increase of the amount of solvent in polymer, and thus that of solvent induced crystallisation, is dependent on the distance from the original solvent surface, as is expected from the root mean square rule (Crank and Park, 1968). This was elegantly used to show that the rate of crystallisation, both inside and outside the polymer being dissolved, depends on the solvent concentration, as shown in Figure 10.11. This study utilised the power of FTIR imaging to study the dynamic process of solvent-induced crystallisation.

Kazarian and Chan (Kazarian and Chan, 2004) studied dynamic processes in polymer systems under high-pressure carbon dioxide by spectroscopic imaging. The experimental set-up, based on a diamond ATR accessory, allows the study of supercritical carbon dioxide ($scCO_2$), an inexpensive but valuable solvent for 'green' chemistry. In the cited study, intake of $scCO_2$ into different polymer blends was studied and the resulting phase separation was observed. Images were prepared from univariate peak integration, and selected spectra were shown for chemical analysis. In this way, swelling and melting of the polymer, as well as phase separation due to ingress of CO_2, were visualised.

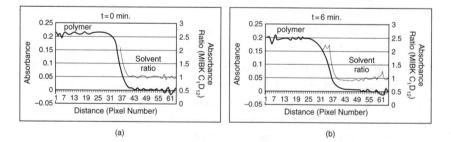

(a)   (b)

Figure 10.10 (a) Absorbance profile for the polymer and a profile of the ratio of the solvent mixture components' absorbance for 95:5 cyclohexane-d:methyl isobutyl ketone at the start of an experiment. (b) A similar plot for a time corresponding to 6 min of dissolution (Miller-Chou and Koenig, 2002). Reprinted with permission from Macromolecules 35 (2), FT-IR imaging of polymer dissolution by solvent mixtures. 3. Entangled polymer chains with solvents, 440–444, B. A. Miller-Chou and J. L. Koenig. © 2002 American Chemical Society

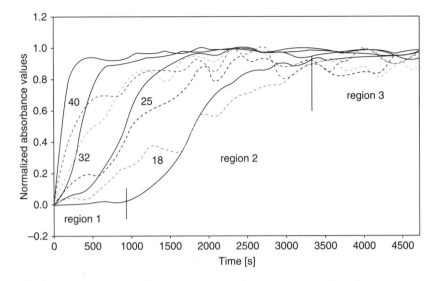

Figure 10.11 Profiles of crystalline sPS (solid line) and toluene (dotted line) as a function of solvent exposure time. The numbers next to the lines show the corresponding focal plane array (FPA) detector pixel number (a higher number is closer to the polymer/vapour interface). Reprinted with permission from Macromolecules, 37, FT-IR imaging of solvent-induced crystallization in polymers, 6498–6503, Gupper, A, Chan, KLA, Kazarian, SG. © 2004 American Chemical Society

10.3 DRUG RELEASE

The high chemical specificity of FTIR imaging has proven very useful in drug release studies. MIR as well as NIR imaging studies have been used to elucidate the distribution of different components inside a tablet (Chan *et al.*, 2003; Clarke, 2004).

A number of studies describe the diffusion of drugs through a membrane. The aim of these experiments is a simulation of transdermal drug delivery (i.e. through the skin). The proposed approaches include single point ATR measurements, where the front side of the membrane is exposed to a drug-containing solution or ointment, while the concentration of drug at the back-side of the membrane is monitored by ATR-FTIR (Pellett *et al.*, 1997a, Cantor, 1999; Hanh, *et al.*, 2000a,b,c, 2001; Dias *et al.*, 2004). Depth profiles through skin or a polymer model have been obtained using photo-acoustic spectroscopy (PAS) (Hanh *et al.*, 2000a, 2001), Raman spectroscopy (Puppels *et al.*, 2001) and FTIR imaging (Rafferty and Koenig, 2002). Additionally, transmission FTIR imaging has been used to study sorption of water into a pharmaceutical formulation comprising of griseofulvin and poly (ethylene glycol) kept under controlled humidity (Chan and Kazarian, 2004b). With this approach, it was possible to discriminate the sorption of water into the different constituents of the tablet.

The possibility to study the changing distribution of materials in a tablet during exposure to water was realised in two different groups almost simultaneously (Coutts-Lendon et al., 2003; Kazarian and Chan, 2003). In the group of Koenig, tablet dissolution was an obvious extension of the polymer dissolution work described above and the experimental set-up was identical (Coutts-Lendon et al., 2003). The experiment thus required the preparation of a thin custom-made tablet between two windows. Tablets were prepared by solvent casting from chloroform. After preparation, the tablet was analysed by transmission FTIR imaging. The distribution of the drug in the dry tablets was studied by single band integration of a drug-specific peak. Histograms of this absorbance value were presented to investigate the homogeneity of the dry tablets. Such a histogram would show a very narrow band if the distribution of the drug were finer than the spatial resolution of the imaging system, while separate bands would be found if the domains were much larger than the spatial resolution. The presented plots showed relatively wide bands, indicating that the particle sizes are in the same order of magnitude as the spatial resolution. Solvent was brought in contact with the tablet by the capillary force between these windows. D_2O was chosen as a solvent, presumably to prevent the strong water absorbances in important spectral regions. After wetting of the sample, several datasets were acquired as a function of time. These datasets were used to monitor the time development of the polymer/solvent boundary (inflection point), and classify them as anomalous using the power law [Equation (10.4)]. A series of absorption plots indicate that drug in tablets with a lower loading (10–20 %) did dissolve simultaneously with the polymer [poly (ethylene oxide)], while hardly any dissolution of drug could be seen at higher drug loadings (30–40 %). The amount of drug present in the tablets was calculated by integration of the drug absorbance over part of the FOV. The drug release can thus be calculated by subtracting this integrated amount from the original value. In theory, the resulting value is comparable with the drug release measured by a dissolution test, but the proposed experimental set-up does not provide a possibility to evaluate this.

10.3.1   ATR-FTIR Imaging

A series of publications from the group of Kazarian introduces the coupling of FTIR imaging with an attenuated total reflection (ATR) accessory. This set-up, shown in Figure 10.12, has a number of advantages over the transmission measurements introduced by the group of Koenig. The most important advantage is the redundancy of the requirement of very thin tablets. Transmission measurements of tablets thicker than about 10 μm lead to complete absorption of the IR beam, or at best a compromised quantitative accuracy. On the other hand, the effective path length of a light beam in a sample is only a few micrometers for an ATR analysis. The depth of penetration is determined by the refractive indices of the ATR crystal and the sample, as well as the wavelength and incident angle of the probing light, but it is independent of the thickness of the tablet. Consequently, tablets to be studied by ATR need not be very thin, and the use of D_2O is not necessary. Kazarian and Chan studied comparable tablets by transmission and ATR approaches (Kazarian and Chan, 2003). They found that the dry core of the polymer matrix disappeared in about 20 min in an ATR measurement, while a major part of the tablet was still intact after 90 min in a 6 μm transmission cell. It was inferred that the inhibition of the dissolution process was caused by crystallites of poorly soluble drug: crystallites, trapped between the two transmission windows around the tablet, hindered the water supply to the tablet. Such crystallites are not only formed by contact with liquid water (Kazarian and Chan, 2003), but even by contact with water vapour (Chan and Kazarian, 2004a). Their presence invalidates the transmission approach for poorly soluble drugs.

The easiest way to prepare tablets to be investigated by ATR imaging is by melting on top of a heated ATR crystal (Chan and Kazarian, 2004a; Kazarian and Chan, 2003). When this is impossible, *in situ* compaction of tablets seems the best alternative, as compaction is one of the most common ways to make pharmaceutical tablets. A compaction/flow-through cell based on the ATR set-up has been presented by van der Weerd *et al.* (van der Weerd *et al.*, 2004; van der Weerd and Kazarian, 2004a). This set-up is similar to the cell in Figure 10.12, but additionally enables the formulation of a tablet from powder on top of an ATR crystal by compaction. This *in situ* compaction and flow-through cell has a number of advantages:

- Good contact between the tablet and the ATR crystal. This contact is important for ATR analyses, as the short penetration depth

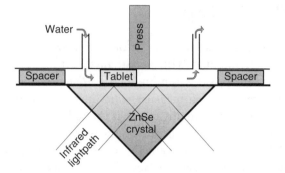

Figure 10.12 Experimental set-up to study tablet dissolution by ATR-FTIR imaging spectroscopy. Reproduced by permission of Society of Applied Spectroscopy, Appl. Spectrosc., 58, 2004, Validation of macroscopic attenuated total reflection-Fourier transform infrared imaging to study dissolution of swelling pharmaceutical tablets, J. van der Weerd, S. G. Kazarian, Society for Applied Spectroscopy

> renders a sample that is not in close contact with the ATR crystal invisible. Besides, imperfect contact may give rise to a water flow along the tablet/ATR crystal interface (leaking) rather than through the tablet.
> - The flow-through ability enables simultaneous analysis of a single tablet by both FTIR and dissolution test.

Obviously, the high pressures involved in this process require the ATR crystal to be very hard, making diamond the only appropriate choice. In a subsequent study (van der Weerd and Kazarian, 2004b), it was found that a reduced pressure (= reduced contact) does reduce the quality of the acquired IR spectra, but does not induce leaking for hydroxypropyl methylcellulose (HPMC) tablets.

The system was applied to analyse the study of a tablet containing HPMC and caffeine particles (van der Weerd et al., 2004) and showed the dislocation of a caffeine particle due to polymer swelling. Data processing was performed by single band integration. However, the good quantitative qualities attainable by ATR spectroscopy make the application of multivariate techniques promising. This was realised by van der Weerd and Kazarian in a subsequent study (van der Weerd and Kazarian, 2004a). The main focus of this study is the coupling of a flow-through cell with a UV spectrometer. This enables the quantification of the amount of dissolved drug in the solvent flow. This information is comparable to a conventional dissolution test, but a derivative signal of drug concentration is obtained as a result of the flow-through approach.

A quality check of the imaging spectra based on PCA of the calibration set was introduced. This method allows a comparison between original and reconstructed spectra, and forms an easy way to find erratic spectra, and the sources of these errors. Quantification of all materials in the system was acquired using PLS in an unfolding approach. Several mixtures of HPMC, niacinamide, and water were prepared in known ratios as a calibration set. This set was used to formulate three PLS-1 models, one for each component. The quantitative results were used to calculate the original amount of drug in the tablet, and the amount of released drug during tablet dissolution. The amount of released drug in this system is thus obtained by two independent measurements, namely the FTIR imaging and the dissolution test using a UV detector. It was shown that the obtained profiles did correspond fairly well (Figure 10.13), thereby validating the combination of techniques.

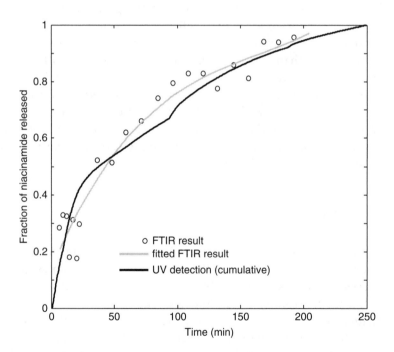

Figure 10.13 Comparison of drug release profiles obtained by FTIR imaging data and by dissolution test. Open circles show the release as calculated from the integrated amounts of niacinamide determined by FTIR imaging. The black solid line shows the cumulative results of the flow-through dissolution test. Reprinted from J Control release, Vol 98, J. van der Weerd, S. G. Kazarian, Combined approach of FTIR imaging and conventional dissolution tests applied to drug release, 295-305, 2004 with permission from Elsevier

The calibration set obtained in this study was also applied in a subsequent study (van der Weerd and Kazarian, 2004b), in which the influence of the compaction force on the water ingression speed was monitored. The main focus of this study is the determination of whether or not leakage of water between the ATR crystal and the tablet occurs. This would seriously affect ATR measurements, as the processes at the tablet surface which are analysed by ATR, would not form a valid representation of the core of the tablet. It was found that the water intake is virtually independent of the pressure with which the tablet is pressed on the ATR crystal. The leakage of water along the ATR crystal/polymer interface is thus ruled out for the investigated polymer, HPMC. After calibration by PLS, the spatial dimensions were reduced by assuming that the water intake was parallel with one of the image dimensions. This simplification was justified by the small FOV in comparison with the tablet size.

The most recent study on drug release from this group reports the investigation of a poorly soluble drug by FTIR imaging using a flow-through cell similar to the one in Figure 10.12 (van der Weerd and Kazarian, 2005). It was found that both the flow-through and the FTIR analyses offer considerable advantages over the conventional dissolution test. CLS was used in this study to compress the spectral dimension. Subsequently, the image dimensions were compressed using the method described in Figure 10.4(b), taking the curvature of the tablet into account. Reduction of the image dimensions to a single dimension (radius) allowed the summarising of a complete image series in a few plots.

10.4   CONCLUSIONS

The first section of this chapter introduced multivariate movies, and provided several ways to reduce the size of the obtained dataset. Data compression is necessary to be able to interpret and visualise results of the vast datasets in multivariate movies. The compression of different (spectrum, image, and time) axes are all well covered in the literature, but do traditionally fall in different scientific fields (chemometrics, image analysis, and time series analysis), which makes the task of a combined approach to a single dataset exciting, but laborious.

The application of the multivariate movies has been shown in the second part of this chapter. The applied data processing methods are in most cases straightforward as a result of the relative novelty of laboratory based imaging systems. It is to be expected that the exploitation of multivariate movies will improve with a wider application of the various powerful techniques available.

The application of optical photography, MRI, Raman line imaging, and FTIR imaging in the study of solvent diffusion in polymers, polymer dissolution, polymer crystallisation, and drug release from pharmaceutical tables have been reviewed. Specific applications of multivariate techniques, such as PLS and CLS, have been used to analyse imaging datasets obtained by ATR-FTIR imaging of tablet dissolution. These examples demonstrated applicability of multivariate technique to tablet dissolution and drug release and show that multivariate analysis of dynamic systems can help to reveal the key features of the mechanism that governs polymer dissolution and drug release in an aqueous environment. It is hoped that this chapter can stimulate further applications of chemometrics to imaging of multi-component systems changing with time. The combination of chemometrics, image processing, and time series analysis with time- and space-resolved spectroscopic techniques has a great potential which is yet to be fully realised.

ACKNOWLEDGEMENT

We would like to thank EPSRC for grant GR/S03942/01.

REFERENCES

Adler, J., Jayan, A. and Melia, C. D. (1999) A method for quantifying differential expansion within hydrating hydrophilic matrixes by tracking embedded fluorescent microspheres, *J. Pharm. Sci.* **88**, 371–377.

Andrew, J. J. and Hancewicz, T. M. (1998) Rapid analysis of Raman image data using two-way multivariate curve resolution, *Appl. Spectrosc.* **52**, 797–807.

Balik, C. M. and Simendinger, W. H. (1998) An attenuated total reflectance cell for analysis of small molecule diffusion in polymer thin films with Fourier-transform infrared spectroscopy, *Polymer* **39**, 4723–4728.

Barbillat, J., Dhamelincourt, P., Delhaye, M. and Dasilva, E. (1994) Raman confocal microprobing, imaging and fiberoptic remote-sensing – a further step in molecular analysis, *J. Raman Spectrosc.* **25**, 3–11.

Barrer, R. (1946) Measurement of diffusion and thermal conductivity 'constants' in nonhomogeneous media, and in media where these 'constants' depend respectively on concentration or temperature, *Proc. Phys. Soc.* **58**, 321.

Beebe, K. R., Pell, R. J. and Seasholtz, M. B. (1998) *Chemometrics a Practical Guide*, John Wiley & Sons, Ltd, New York.

Bettini, R., Catellani, P. L., Santi, P., Massimo, G., Peppas, N. A. and Colombo, P. (2001) Translocation of drug particles in HPMC matrix gel layer: effect of drug solubility and influence on release rate, *J. Controlled Release* **70**, 383–391.

Bovensmann, H., Burrows, J. P., Buchwitz, M., Frerick, J., Noel, S., Rozanov, V. V., Chance, K. V. and Goede, A. P. H. (1999) SCIAMACHY: mission objectives and measurement modes, *J. Atmos.Sci.* **56**, 127–150.

Bowden, M., Gardiner, D. J., Rice, G. and Gerrard, D. L. (1990) Line-scanned micro Raman-spectroscopy using a cooled CCD imaging detector, *J. Raman Spectrosc.* **21**, 37–41.

Brantley, N. H., Kazarian, S. G. and Eckert, C. A. (2000) In situ FTIR measurement of carbon dioxide sorption into poly(ethylene terephthalate) at elevated pressures, *J. Appl. Polym. Sci.* **77** , 764–775.

Bro, R. (1997) PARAFAC. Tutorial and applications, *Chemometrics Intell. Lab. Syst.* **38**, 149–171.

Burka, E. M. and Curbelo, R. (2000) Imaging ATR Spectrometer. US Patent 6 141 100.

Cantor, A. S. (1999) Drug and excipient diffusion and solubility in acrylate adhesives measured by infrared-attenuated total reflectance (IR-ATR) spectroscopy, *J. Controlled Release* **61**, 219–231.

Caprioli, R. M., Farmer, T. B. and Gile, J. (1997) Molecular imaging of biological samples: Localization of peptides and proteins using MALDI-TOF MS, *Anal. Chem.* **69**, 4751–4760.

Challa, S. R., Wang, S. Q. and Koenig, J. L. (1996) In situ diffusion studies using spatially resolved infrared microspectroscopy, *Appl. Spectrosc.* **50**, 1339–1344.

Challa, S. R., Wang, S. Q. and Koenig, J. L. (1997) In situ diffusion and miscibility studies of thermoplastic PDLC systems by FT-IR microspectroscopy, *Appl. Spectrosc.* **51**, 297–303.

Chan, K. L. A. and Kazarian, S. G. (2003) New opportunities in micro- and macro-attenuated total reflection infrared spectroscopic imaging: spatial resolution and sampling versatility, *Appl. Spectrosc.* **57**, 381–389.

Chan, K. L. A. and Kazarian, S. G. (2004a) FTIR spectroscopic imaging of dissolution of a solid dispersion of nifedipine in poly(ethylene glycol), *Mol. Pharmaceut.* **1** , 331–335.

Chan, K. L. A. and Kazarian, S. G. (2004b) Visualisation of the heterogeneous water sorption in a pharmaceutical formulation under controlled humidity via FT-IR imaging, *Vib. Spectrosc.* **35**, 45–49.

Chan, K. L. A., Hammond, S. V. and Kazarian, S. G. (2003) Applications of attenuated total reflection infrared spectroscopic imaging to pharmaceutical formulations, *Anal. Chem.* **75**, 2140–2146.

Chatfield, C. (1984) *The Analysis of Time Series: an Introduction*, Chapman and Hall, Boca Raton.

Chen, Y. H., Rangarajan, G., Feng, J. F. and Ding, M. Z. (2004) Analyzing multiple nonlinear time series with extended Granger causality, *Phys. Lett.* A **324**, 26–35.

Clarke, F. (2004) Extracting process-related information from pharmaceutical dosage forms using near infrared microscopy, *Vib. Spectrosc.* **34**, 25–35.

Colombo, P., Bettini, R., Santi, P., DeAscentiis, A. and Peppas, N. A. (1996) Analysis of the swelling and release mechanisms from drug delivery systems with emphasis on drug solubility and water transport, *J. Controlled Release* **39**, 231–237.

Colombo, P., Bettini, R., and Peppas, N. A. (1999) Observation of swelling process and diffusion front position during swelling in hydroxypropyl methyl cellulose (HPMC) matrices containing a soluble drug, *J. Controlled Release* **61**, 83–91.

Cotugno, S., Larobina, D., Mensitieri, G., Musto, P. and Ragosta, G. (2001) A novel spectroscopic approach to investigate transport processes in polymers: the case of water-epoxy system, *Polymer* **42**, 6431–6438.

Coutts-Lendon, C. A., Wright, N. A., Mieso, E. V. and Koenig, J. L. (2003) The use of FT-IR imaging as an analytical tool for the characterization of drug delivery systems, *J. Controlled Release* **93**, 223–248.

Crank, J. (1975), *The Mathematics of Diffusion*, Oxford University Press , Oxford.

Crank, J. and Park, G. (Eds) (1968) Methods of measurement. In *Diffusion in Polymers*, Academic Press , London.

de Juan, A., Tauler, R., Dyson, R., Marcolli, C., Rault, M. and Maeder, M. (2004) Spectroscopic imaging and chemometrics: a powerful combination for global and local sample analysis, *Trends Anal. Chem.* **23**, 70–79.

Dias, M., Hadgraft, J., Raghavan, S. L. and Tetteh, J. (2004) The effect of solvent on permeant diffusion through membranes studied using ATR-FTIR and chemometric data analysis, *J. Pharm. Sci.* **9**, 186–196.

Ebizuka, N., Wakaki, M., Kobayashi, Y. and Sato, S. (1995) Development of a multi-channel Fourier-transform spectrometer, *Appl. Optics* **34**, 7899–7906.

Esbensen, K. (2001) *Multivariate Data Analysis in Practice*, Camo, Trondheim.

Fyfe, C. A. and Blazek, A. I. (1997) Investigation of hydrogel formation from hydroxypropylmethylcellulose (HPMC) by NMR spectroscopy and NMR imaging techniques, *Macromolecules* **30**, 6230–6237.

Fyfe, C. A. and Blazek-Welsh, A. I. (2000) Quantitative NMR imaging study of the mechanism of drug release from swelling hydroxypropylmethylcellulose tablets, *J. Controlled Release* **68**, 313–333.

Gao, P. and Meury, R. H. (1996) Swelling of hydroxypropyl methylcellulose matrix tablets. 1. Characterization of swelling using a novel optical imaging method, *J. Pharm. Sci.* **85**, 725–731.

Geladi, P. (2003) Chemometrics in spectroscopy. Part 1. Classical chemometrics, *Spectrochim. Acta, Part B* **58**, 767–782.

Geladi, P. and Grahn, H. (1996) *Multivariate Image Analysis*, John Wiley & Sons, Ltd, Chichester.

Geladi, P., Sethson, B., Nyström, J., Lillhonga, T., Lestander, T. and Burger, J. (2004) Chemometrics in spectroscopy. Part 2. Examples, *Spectrochim. Acta, Part B* **59**, 1347–1357.

Goebel, R., Roebroeck, A., Kim, D. S. and Formisano, E. (2003) Investigating directed cortical interactions in time-resolved fMRI data using vector autoregressive modeling and Granger causality mapping, *Magn. Reson. Imaging* **21**, 1251–1261.

Gonzalez-Benito, J. and Koenig, J. L. (2002) FTIR imaging of the dissolution of polymers. 4. Poly(methyl methacrylate) using a cosolvent mixture (carbon tetrachloride/methanol), *Macromolecules* **35**, 7361–7367.

Granger, C. W. J. and Newbold, P. (1977) *Forecasting Economic Time Series*, Academic Press, London.

Gupper, A., Chan, K. L. A. and Kazarian, S. G. (2004a) FT-IR imaging of solvent-induced crystallization in polymers, *Macromolecules* **37**, 6498–6503.

Gupper, A., van der Weerd, J and Kazarian, S. G. (2004b) Micro-Raman spectroscopic investigation of solvent diffusion in polymers, *FACSS 2004*, Portland, OR.

Gurden, S. P., Lage, E. M., de Faria, C. G., Joekes, I. and Ferreira, M. M. C. (2003) Analysis of video images from a gas-liquid transfer experiment: a comparison of PCA and PARAFAC for multivariate image analysis, *J. Chemom.* **17**, 400–412.

Hanh, B. D., Neubert, R. H. H., Wartewig, S., Christ, A. and Hentzsch, C. (2000a) Drug penetration as studied by noninvasive methods: Fourier transform infrared-attenuated total reflection, Fourier transform infrared, and ultraviolet photoacoustic spectroscopy, *J. Pharm. Sci.* **89**, 1106–1113.

Hanh, B. D., Neubert, R. H. H. and Wartewig, S. (2000b) Investigation of drug release from suspension using FTIR-ATR technique: part II. Determination of dissolution coefficient of drugs, *Int. J. Pharm.* **204**, 151–158.

Hanh, B. D., Neubert, R. H. H. and Wartewig, S. (2000c) Investigation of drug release from suspension using FTIR-ATR technique: part I. Determination of effective diffusion coefficient of drugs, *Int. J. Pharm.* **204**, 145–150.

Hanh, B. D., Neubert, R. H. H., Wartewig, S. and Lasch, J. (2001) Penetration of compounds through human stratum corneum as studied by Fourier transform infrared photoacoustic spectroscopy, *J. Controlled Release* **70**, 393–398.

Harris, R. and Sollis, R. (2003) *Applied Time Series Modelling and Forecasting*, John Wiley & Sons, Ltd, Chichester.

Huang, J., Wium, H., Qvist, K. B. and Esbensen, K. H. (2003) Multi-way methods in image analysis: relationships and applications, *Chemom. Intell Lab. Syst.* **66**, 141–158.

Johansson, T. and Pettersson, A. (1997) Imaging spectrometer for ultraviolet–near-infrared microspectroscopy, *Rev. Sci. Instrum.* **68**, 1962–1971.

Kaiser, J. W., Eichmann, K. U., Noel, S., Wuttke, M. W., Skupin, J., von Savigny, C., Rozanov, A. V., Rozanov, V. V., Bovensmann, H. and Burrows, J. P. (2004) SCIA-MACHY limb spectra, *Adv. Space Res.* **34**, 715–720.

Kazarian, S. G. and Chan, K. L. A. (2003) 'Chemical photography' of drug release, *Macromolecules* **36**, 9866–9872.

Kazarian, S. G. and Chan, K. L. A. (2004) FTIR imaging of polymeric materials under high-pressure carbon dioxide, *Macromolecules* **37**, 579–584.

Kiil, S. and Dam-Johansen, K. (2003) Controlled drug delivery from swellable hydroxypropylmethylcellulose matrices: model-based analysis of observed radial front movements, *J. Controlled Release* **90**, 1–21.

Krueger, A. J. and Jaross, G. (1999) TOMS ADEOS instrument characterization, *IEEE Trans. Geosci. Remote Sensing* **37**, 1543–1549.

Lasch, P., Haensch, W., Lewis, E. N., Kidder, L. H. and Naumann, D. (2002) Characterization of colorectal adenocarcinoma sections by spatially resolved FT-IR microspectroscopy, *Appl. Spectrosc.* **56**, 1–9.

Lasch, P., Haensch, W., Naumann, D. and Diem, M. (2004) Imaging of colorectal adenocarcinoma using FT-IR microspectroscopy and cluster analysis, *Biochim. Biophys. Acta* **1688**, 176–186.

Lewis, E. N., Treado, P. J., Reeder, R. C., Story, G. M., Dowrey, A. E., Marcott, C. and Levin, I. W. (1995) Fourier-transform spectroscopic imaging using an infrared focal-plane array detector, *Anal. Chem.* **67**, 3377–3381.

Linossier, I., Gaillard, F., Romand, M. and Feller, J. F. (1997) Measuring water diffusion in polymer films on the substrate by internal reflection Fourier transform infrared spectroscopy, *J. Appl. Polym. Sci.* **66**, 2465–2473.

Massart, D., Vandeginste, B., Deming, S., Michotte, Y. and Kaufman, L. (1988) *Chemometrics: a Textbook*, Elsevier, Amsterdam.

Melia, C. D., Rajabi-Siahboomi, A. R. and Bowtell, R. W. (1998) Magnetic resonance imaging of controlled release pharmaceutical dosage forms, *Pharm. Sci. Technol. Today* **1**, 32–39.

Miller-Chou, B. A. and Koenig, J. L. (2002) FT-IR imaging of polymer dissolution by solvent mixtures. 3. Entangled polymer chains with solvents, *Macromolecules* **35**,

440–444.

Miller-Chou, B. A. and Koenig, J. L. (2003) A review of polymer dissolution, *Prog. Polym. Sci.* **28**, 1223–1270.

Musto, P., Mensitieri, G., Cotugno, S., Guerra, G. and Venditto, V. (2002) Probing by time-resolved FTIR spectroscopy mass transport, molecular interactions, and conformational ordering in the system chloroform-syndiotactic polystyrene, *Macromolecules* **35**, 2296–2304.

Narasimhan, B., Snaar, J. E. M., Bowtell, R. W., Morgan, S., Melia, C. D. and Peppas, N. A. (1999) Magnetic resonance imaging analysis of molecular mobility during dissolution of poly(vinyl alcohol) in water, *Macromolecules* **32**, 704–710.

Niblack, W. (1986) *An Introduction to Digital Image Processing*, Prentice-Hall International, London.

Noordam, J. C. and van den Broek, W. (2002) Multivariate image segmentation based on geometrically guided fuzzy C-means clustering, *J. Chemom.* **16**, 1–11.

Noordam, J. C., van den Broek, W. and Buydens, L. M. C. (2003) Unsupervised segmentation of predefined shapes in multivariate images, *J. Chemom.* **17**, 216–224.

Paatero, P. and Juntto, S. (2000) Determination of underlying components of a cyclical time series by means of two-way and three-way factor analytic techniques, *J. Chemom.* **14**, 241–259.

Pellett, M. A., Watkinson, A. C., Hadgraft, J. and Brain, K. R. (1997a) Comparison of permeability data from traditional diffusion cells and ATR-FTIR spectroscopy. 1. Synthetic membranes, *Int. J. Pharm.* **154**, 205–215.

Pellett, M. A., Watkinson, A. C., Hadgraft, J. and Brain, K. R. (1997b) Comparison of permeability data from traditional diffusion cells and ATR-FTIR spectroscopy. 2. Determination of diffusional pathlengths in synthetic membranes and human stratum corneum, *Int. J. Pharm.* **154**, 217–227.

Petrou, M., and Bosdogianni, P. (1999) *Image Processing, the Fundamentals*, John Wiley & Sons, Ltd, Chichester.

Pollock, H. M., Hammiche, A., Dupas, E., Price, D. M., Reading, M. and Bozec, L. (2000) Microthermal probing of polymers: dynamic localized thermomechanical analysis, localized IR spectroscopy, *Pap. Am. Chem. Soc.* **220**, 28.

Pravdova, V., Boucon, C., de Jong, S., Walczak, B. and Massart, D. L. (2002) Three-way principal component analysis applied to food analysis: an example, *Anal. Chim. Acta* **462**, 133–148.

Price, D. M., Reading, M., Hammiche, A. and Pollock, H. M. (1999) Micro-thermal analysis: scanning thermal microscopy and localised thermal analysis, *Int. J. Pharm.* **192**, 85–96.

Puppels, G., Bakker Schut, T., Caspers, P., Wolthuis, R., Van Aken, M., van der Laarse, A., Bruining, H., Buschman, H., Sim, M. and Wilson, B. (2001) In vivo Raman spectroscopy. In *Handbook of Raman Spectroscopy from the Research Laboratory to the Process Line* (Lewis, I. and Edwards, H., Eds), Marcel Dekker Inc., New York, vol. 28, pp. 549–574.

Rafferty, D. W. and Koenig, J. L. (2002) FTIR imaging for the characterization of controlled-release drug delivery applications, *J. Controlled Release* **83**, 29–39.

Ribar, T., Bhargava, R. and Koenig, J. L. (2000) FT-IR imaging of polymer dissolution by solvent mixtures. 1. Solvents, *Macromolecules* **33**, 8842–8849.

Ribar, T., Koenig, J. L. and Bhargava, R. (2001) FTIR imaging of polymer dissolution. 2. Solvent/nonsolvent mixtures, *Macromolecules* **34**, 8340–8346.

Rosenfeld, A. and Kak, A. C. (1982) *Digital Picture Processing, Volume 2,* Academic Press, New York.

Sackin, R., Ciampi, E., Godward, J., Keddie, J. L. and McDonald, P. J. (2001) Fickian ingress of binary solvent mixtures into glassy polymer, *Macromolecules* 34, 890–895.

Schoonover, J. R., Marx, R. and Zhang, S. L. L. (2003) Multivariate curve resolution in the analysis of vibrational spectroscopy data files, *Appl. Spectrosc.* 57, 154A–170A.

Schowengerdt, R. A. (1997) *Remote Sensing, Models and Methods for Image Processing,* Academic Press, San Diego.

Schueler, B. (2001) Time of flight mass analysers. In *ToF-SIMS: Surface Analysis by Mass Spectrometry* (Vickerman, J. C. and Briggs, D., Eds), IM Publications, Chichester, pp. 75–94.

Seul, M., O'Gorman, L. and Sammon, M. J. (2000) *Practical Algorithms for Image Analysis, Description, Examples, and Code,* Cambridge University Press, Cambridge.

Shaw, R. A., Mansfield, J. R., Rempel, S. P., Low-Ying, S., Kupriyanov, V. V. and Mantsch, H. H. (2000) Analysis of biomedical spectra and images: from data to diagnosis, *THEOCHEM* 500, 129–138.

Siepmann, J. and Peppas, N. A. (2001) Mathematical modeling of controlled drug delivery, *Adv. Drug Deliv. Rev.* 48, 37–138.

Snaar, J. E. M., Bowtell, R., Melia, C. D., Morgan, S., Narasimhan, B. and Peppas, N. A. (1998) Selfdiffusion and molecular mobility in PVA-based dissolution-controlled systems for drug delivery, *Magn. Reson. Imaging* 16, 691–694.

Snively, C. M. and Koenig, J. L. (1999) Studying anomalous diffusion in a liquid crystal/polymer system using fast FTIR imaging, *J. Polym. Sci., Polym. Phys. Ed.* 37, 2261–2268.

Tauler, R. and Barcelo, D. (1993) Multivariate curve resolution applied to liquid- chromatography diode-array detection, *Trends Anal. Chem.* 12, 319–327.

Ueberreiter, K. (1968) The solution process. In *Diffusion in Polymers* (Crank, J. and Park, G., Eds), Academic Press, London, pp. 219–257.

van der Weerd, J. and Kazarian, S. G. (2004a) Combined approach of FTIR imaging and conventional dissolution tests applied to drug release, *J. Controlled Release* 98, 295–305.

van der Weerd, J. and Kazarian, S. (2004b) Validation of macroscopic ATR-FTIR imaging to study dissolution of swelling pharmaceutical tablets, *Appl. Spectrosc.* 58, 1413–1419.

van der Weerd, J. and Kazarian, S. (2005) Release of poorly soluble drugs from HPMC tablets studied by FTIR imaging and flow-through dissolution tests, *J. Pharm. Sci.* 94, 2096–2109.

van der Weerd, J., Chan, K. L. A. and Kazarian, S. G. (2004) An innovative design of a compaction cell for in situ FT-IR imaging of tablet dissolution, *Vibr. Spectrosc.* 35, 9–13.

Williams, K. P. J., Pitt, G. D., Batchelder, D. N. and Kip, B. J. (1994) Confocal Raman microspectroscopy using a stigmatic spectrograph and ccd detector, *Appl. Spectrosc.* 48, 232–235.

Zhang, L., Small, G. W., Haka, A. S., Kidder, L. H. and Lewis, E. N. (2003) Classification of Fourier transform infrared microscopic imaging data of human breast cells by cluster analysis and artificial neural networks, *Appl. Spectrosc.* 57, 14–22.

Zuur, A. F., Fryer, R. J., Jolliffe, I. T., Dekker, R. and Beukema, J. J. (2003) Estimating common trends in multivariate time series using dynamic factor analysis, *Environmetrics* 14, 665–685.

11

Multivariate Image Analysis of Magnetic Resonance Images: Component Resolution with the Direct Exponential Curve Resolution Algorithm (DECRA)

Brian Antalek, Willem Windig and Joseph P. Hornak

11.1 INTRODUCTION

Magnetic resonance imaging (MRI) is routinely used in the clinical setting to characterize tissue damage or identify pathology. Typically, an optimized single image is adequate as a result of the excellent contrast provided by the distribution of hydrogen nuclear magnetic resonance (NMR) spin-spin and spin-lattice relaxation times, T_2 and T_1, respectively, within the tissues. However, obtaining further information using MRI is not so straightforward. For example, physicians may want to follow the progress of drug therapy by quantifying the size reduction of pathology or find evidence of pathology in its very early stages. To do so, one may introduce contrast agents or multivariate imaging schemes. The work presented in this chapter will focus on the latter.

Techniques and Applications of Hyperspectral Image Analysis   Edited by H. F. Grahn and P. Geladi
© 2007 John Wiley & Sons, Ltd

Because contrast in MRI depends on T_2 and T_1, they have been used to characterize living tissues (Peemoeller *et al.*, 1980; Kroeker and Henkelman, 1986; Just and Thelen, 1988; Armsach *et al.*, 1991; English *et al.*, 1991; Stark and Bradley, 1992; Fletcher *et al.*, 1993; Kao *et al.*, 1994; Ordidge *et al.*, 1994; Ma *et al.*, 1997). These biological samples are complex, composed of several different spin systems that might exhibit exchange behavior. As such, several models have been proposed to explain the spin-relaxation behavior of such heterogeneous systems (Kroeker and Henkelman, 1986). One model proposes that a single tissue may be composed of different nonexchanging spin types and, therefore, possess multiexponential relaxation behavior characteristics of the sum of the individual components. At the opposite extreme is a model that proposes that the different spin types are coupled by strong exchange, and hence, possess one characteristic T_2 and T_1. In the middle of these two extremes are tissues exhibiting weak exchange and possessing multiexponential relaxation characteristics that are different from the pure spin components. There is no reason to believe that all tissues behave as one of the three possibilities or that a given tissue will be spatially homogeneous. Because the practical size of the volume element (voxel) is large enough to contain several tissues, partial volume effects are typically seen (i.e., a single signal value representing a combination of a mixture of more than one component). It is only at the small voxel limit that the question of which of the three models is appropriate for a given tissue can be answered. Even if the voxel size approaches the size of the cellular level, we should expect multiexponential relaxation behavior as a result of the heterogeneity of the living cell anatomy. Several chemically different hydrogens exist in the cell, primarily including alkyl (from lipids and fat) and water (intra- and extracellular). Furthermore, as the voxel size decreases, translational diffusion of the water becomes an important consideration.

Models for such complex systems must be manageable, however. Technical accuracy and practical considerations must be balanced, and the parameters must be accessible by measurement. It is favorable to institute a method that will identify a small number of key components that describe the vast majority of the data.

Various methods for studying the relaxation behavior of tissues have been described. T_2 and T_1 weighted images of the human body are routinely acquired in clinical MRI (Stark and Bradley, 1992). These images are easy to obtain, avoid the need for exponential fitting, and provide some insight into the gross variations in the tissue T_2 and T_1 values. Calculated apparent monoexponential T_1 (Gong and

Hornak, 1992) and T_2 (Li and Hornak, 1994) images were used to study variations in T_1, T_2, and spin density (ρ) of brain tissues (Fletcher *et al.*, 1993). These images can take up to an hour to acquire using a spin-echo sequence, but clearly demonstrate the partial volume effect on the measured relaxation times. Although faster pulse sequences, such as echo-planar imaging (Mansfield, 1977) and Look–Locker (Look and Locker, 1970) exist, they also demonstrate the partial volume effect. Multiexponential and continuous distribution analyses have been employed and begin to reveal the complexity of the spin system in biological tissues (Kroeker and Henkelman, 1986; Armsach *et al.*, 1991; English *et al.*, 1991; Li and Hornak, 1994). These techniques require large amounts of data, which is due to the short Δt requirement that makes it time-consuming to collect with imaging techniques.

Another image analysis approach involves principal component analysis (PCA). Geladi and Grahn and colleagues (Geladi *et al.*, 1989; Grahn *et al.*, 1989a,b) and Grahn and Sääf (Grahn and Sääf, 1992) have utilized PCA for analyzing sets of correlated MRI images. This method is fast, does not require a large number of images, and results in a set of extracted images that are based upon principal components. However, the extracted images are abstract and not strictly related to any real physical parameters, such as T_2 or T_1.

In this chapter, we describe a data analysis method that overcomes the effects caused by spin environment heterogeneity and extracts pure T_2 and T_1 information. We have pursued an approach similar to PCA, which is based upon a procedure developed by Kubista (Kubista, 1990; Scarminio and Kubista, 1993) and expressed in terms of the generalized rank annihilation method (GRAM) (Booksh and Kowalski, 1994). Kubista's technique was utilized previously by Schulze and Stilbs (Schulze and Stilbs, 1993) on a pulsed-gradient spin-echo (PGSE) nuclear magnetic resonance (NMR) dataset for separating highly overlapped spectra. Antalek and Windig (Antalek and Windig, 1996; Windig and Antalek, 1997) utilized the GRAM method on PGSE NMR data using a somewhat improved scheme and denoted the method as direct exponential curve resolution algorithm (DECRA). A review (Antalek, 2002) describes DECRA more fully in the context of PGSE NMR. The algorithm that we use for the image analysis is based upon DECRA. The application of DECRA for the analysis of MRI images was first described with the use of a well-characterized phantom (Windig *et al.*, 1998) and then, in a companion article, with images of the human brain (Antalek *et al.*, 1998).

The exponential profiles necessary to apply DECRA are present in a series of registered magnetic resonance images whose signal depends upon T_2 or T_1 (for which a transformation is needed and will be introduced below). Using the standard single-slice, single-echo, spin-echo sequence, and assuming that the repetition time (TR) is much larger than the echo time (TE), the signal S of a component m is described by Equation (11.1):

$$S \propto \sum_m \rho_m \, e^{-TE/(T_2)_m} \left(1 - e^{-TR/(T_1)_m}\right) \tag{11.1}$$

where ρ is the spin density, and magnetization exchange among the components are either very fast or very slow on the NMR timescale. Varying TE and keeping TR constant or vice versa will result in a series that depends upon T_2 or T_1, respectively. Because of the chemical nature and physical environments of the ^{1}H nuclei (described by their relaxation behavior), a series of these T_2 (or T_1) images can provide a spatial variation in signal intensity. If we consider the images to be comprised of a finite number of components, each having its own characteristic and discrete T_2 (or T_1) profile, it is possible to extract the pure T_2 (or T_1) images. The problem may be defined as mixture analysis.

In the following section, we first describe the basic concept of DECRA as applied to data containing exponential relationships. Then we show two examples of DECRA applied to MRI: (1) a phantom constructed explicitly to show the utility of DECRA in resolving the partial volume problem; and (2) the human head. Because of the nature of the mathematical approach (which models exponential functions), DECRA resolves pure components that are directly related to physically real parameters that describe exponential functions (namely, T_2 and T_1). The pure components are few in number and are regarded as the primary components that describe the vast majority of the image signal distribution. Images that are reconstructed from the resolved components, multiplied by their respective proportionality constants, are virtually identical to the original images.

11.2 DECRA APPROACH

We start by first considering a PGSE NMR dataset that has the following structure:

$$D = CP^T \tag{11.2}$$

D is a matrix of n spectra containing m data points each (size nm; n rows and m columns); C is the matrix of size nr containing concentration profiles for each of the r pure components; P is the matrix of size mr containing the spectra for each of the r pure components; and T stands for the matrix transpose. Multivariate analysis is normally used in cases where every object (spectrum) is an array. A series of objects forms a matrix, which is a two-way array. In multivariate image analysis, the object is an image, and a series of images forms a three-way array. In order to be able to apply DECRA, each image is reorganized into an array by appending the rows or columns of pixels. After the analysis, the resolved pure components (spectra) are converted to images by reconstructing the row/column structure.

Solving Equation (11.2) is a classic problem in spectroscopy. Data representing mixtures are the superposition of the spectra of the individual (pure) components weighted by their composition. The solution is trivial if the pure spectra are known. However, if these are not known, the solution is not simple at all. As such, there is a great body of work devoted expressly to solving Equation (11.2) for C and P (Malinowski and Howery, 1991). The problem lies in the fact that there are an infinite number of solutions possible. Many methods involve the use of constraints such as pure variables, non-negativity, etc. A pure variable is a point in the spectrum (ppm value, pixel coordinate, etc.) that changes within the dataset only because of a single pure component. This often does not hold true, and even so, the results might vary widely.

Kubista and Scarminio (Kubista, 1990; Scarminio and Kubista, 1993) showed that it is possible to obtain only one solution (the correct solution) to Equation (11.2) if two datasets exist that are proportional, as is shown with Equations ((11.3)) and (11.4):

$$A = CP^T \tag{11.3}$$

$$B = C\beta P^T \tag{11.4}$$

The diagonal matrix β of size rr defines a scaling factor between datasets A and B. Booksh and Kowalski (Booksh and Kowalski, 1994) later showed that the Kubista method may be expressed in terms of GRAM, and the problem can be solved analytically (i.e., no iterative or 'best fit' method). For details about GRAM consult the literature (Sanchez and Kowalski, 1988; Wilson *et al.*, 1989). With the two proportional datasets in hand, one can resolve the pure spectra and the diffusion coefficients without *a priori* knowledge of the spectral bandshapes of the pure components. How does one obtain the two datasets required for the analysis? Schulze and Stilbs (Schulze and Stilbs, 1993) used the Kubista method on data acquired from a variation of the conventional PGSE NMR experiment that produced two correlated datasets, but at the expense of the spectral bandshape. Antalek and Windig (Antalek and Windig, 1996; Windig and Antalek, 1997) showed that one might obtain two datasets from a single, conventional PGSE NMR dataset by splitting the set into two parts (DECRA). By using data from a single PGSE NMR experiment, problems relating to spectral registration and spectral distortion are eliminated.

Typically, two datasets are constructed such that one contains spectra 1 to $(n - 1)$ and the other contains spectra 2 to n. In this context, 'spectra' refer to the actual spectra acquired in a PGSE NMR experiment or an image reorganized into a single row. The two datasets formed are proportional because of their exponential nature. This is described by Windig and Antalek (Windig and Antalek, 1999) and is illustrated in Figure 11.1. Let us start with an example dataset D, containing four spectra from a simulated PGSE NMR experiment and shown in Figure 11.1(a). Each of the four spectra in the dataset is a mixture spectrum comprising two pure spectra that are weighted by a concentration factor. Because it is a PGSE NMR experiment, each pure component spectrum has a concentration profile that is exponential. Therefore, the analysis assumes that the spectrum from each component in the mixture decays with a pure (discrete) exponential. The first step is to split the dataset into two parts, A and B [shown in Figure 11.1(b) and (c)]. There is an infinite number of pairs of pure spectra that, when combined, will reproduce each of the mixture spectra. Two such solutions are shown in Figure 11.1(d). The two datasets, subset A and subset B, are shown on the left. Note that the vertical scale is different for each. The two possible pairs of spectra (I and II) that may be resolved are at the top. The concentration of each of the resolved pairs within each mixture spectrum is shown in the boxes. For example, 27 (left spectrum in spectral pair I) + 8 (right spectrum in spectral pair I) = spectrum 1. Because the concentration profiles for each pure component within datasets A and B are exponential, the correct solution comprises two pure component spectra whose concentrations are proportional between the two datasets. Thus, the two datasets are proportional. For example, if we compare the concentrations of the pairs of spectra (spectrum 1 from A and spectrum 2 from B, spectrum 2 from A and spectrum 3 from B, and spectrum 3 from A and

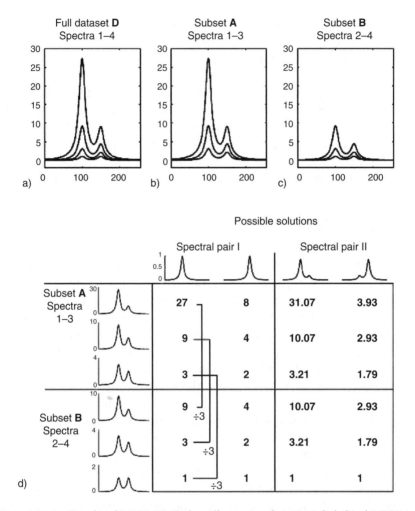

Figure 11.1 Simulated PGSE NMR data illustrating the principle behind DECRA. (a) Full dataset showing four mixture spectra comprising the superposition of two components having different diffusion coefficients. (b) Dataset 1; spectra 1–3 of dataset **D**. (c) Dataset 2; spectra 2–4 of dataset **D**. (d) A grid demonstrating proportionality. Note the change in intensity scale. Reproduced from *Concepts Magn. Reson.*, **14**, 225–258, © (2002) John Wiley & Sons Inc.

spectrum 4 from **B**), we can see that $27 \div 3 = 9$, $9 \div 3 = 3$, $3 \div 3 = 1$, and $8 \div 2 = 4$, $4 \div 2 = 2$, $2 \div 2 = 1$. The other solution, pair II, does not satisfy this behavior. The proportionality factors are contained within the matrix:

$$\beta = \begin{bmatrix} 3 & 0 \\ 0 & 2 \end{bmatrix}$$

The simple step of splitting the data enables us to satisfy all of the requirements for the GRAM analysis! An important aspect that should be clear is that the data must be collected with equal q^2 spacing (usually equal g^2 spacing) to satisfy the proportionality requirement (signal $\propto e^{q^2(\Delta-\delta/3)}$). Implicit in the algorithm is the discrete exponential behavior. Despite this seemingly strict requirement, DECRA may still be used for polymeric mixtures having distributions in decay rates, and it performs amazingly well (Windig and Antalek, 1999; Windig et al., 1999). In this case, the precision in the diffusion dimension is reduced because of the nondiscrete nature of the diffusion coefficients, and the resolved diffusion coefficient will be discrete as well as most closely related to a weight-average diffusion coefficient. DECRA has several advantages including the ability to resolve spectra with low signal to noise ratio (SNR), small differences in diffusion coefficients, and severe spectral overlap. It is a fast method that may be applied to all or part of a dataset. Typically, a set of 16 spectra with 32000 real points each may be fully processed in under 10 s using a typical UNIX-based workstation. A priori knowledge of the number of components is necessary. The only user input, in fact, is the number of pure components. Because a direct solution is found, the algorithm is very fast; therefore, many guesses can be made within a short period of time. Generally, if too many pure components are chosen, the extra component will look like noise only or have small peaks that have roughly equal amounts of positive and negative. There is a limit to the number of resolved components; five seems to be a practical maximum (Windig et al., 1999). There is also a practical limit to the resolution in the diffusion dimension that will depend on the polydispersity of the diffusion coefficient for each component, the SNR, and the general quality of the data. Differences as small as 20 % were reported (Windig and Antalek, 1997). Because exponential processes are prevalent in spectroscopy, DECRA is broadly applicable. In addition to PGSE NMR data of small molecules and polymers, DECRA was also applied to MRI (Antalek et al., 1998; Windig et al., 1998), kinetic (Windig et al., 1999; Bijlsma et al., 1998; Windig et al., 2000), and solid-state NMR data (Zumbulyadis et al., 1999).

The problem analysis for MRI data is identical to PGSE NMR. The PGSE NMR dataset comprises a series of mixture spectra of chemical components characterized by different exponential decays that depend on diffusivities. DECRA calculates the spectra of the pure chemical components and their contributions (concentrations) in the original spectra. The profiles formed by the contributions of a single chemical component over the whole series of spectra are exponential.

From the derived exponentials, the diffusivities are obtained. The MRI dataset comprises a series of mixture images of ^{1}H environments characterized by different exponential decays that depend on spin-spin (or spin-lattice) interactions. DECRA calculates the images of the pure ^{1}H environments and their contributions (concentrations) in the original images. The profiles formed by the contributions of a single ^{1}H environment over the whole series of images are exponential. From the derived exponentials the T_2 (or T_1) values are obtained.

The model that we adopt to describe the relaxation behavior is simple. The image comprises signal contributions from a finite number of components exhibiting discrete exponential behavior. Any exchange behavior in the system is considered to be very fast on the NMR timescale, resulting in a single exponential for the exchanging spin populations (and providing a single exponential), or the exchange behavior is very slow (i.e., nonexchanging and providing more than one discrete exponential). There is no provision in the DECRA algorithm for continuous exponential behavior having a broad distribution (which is analogous to the distributions of diffusion coefficients found in polymers).

11.3   DECRA ALGORITHM

This section will show the algorithm used for the analysis of the data presented in this chapter. It is based on the paper of Sanchez and Kowalski (Sanchez and Kowalski, 1988). Equations (11.3) and (11.4) can be rewritten as:

$$C = A(P^T)^+ \tag{11.5}$$

$$C\alpha = B(P^T)^+ \tag{11.6}$$

where $(P^T)^+$ represents the pseudoinverse of the matrix P^T. Postmultiplying the left-hand and right-hand side of Equation (11.5) by α results in:

$$C\alpha = A(P^T)^+\alpha \tag{11.7}$$

Combining Equations (11.6) and (11.7) results in:

$$A(P^T)^+\alpha = B(P^T)^+ \tag{11.8}$$

which can be rewritten as:

$$AZ\alpha = BZ \tag{11.9}$$

where

$$Z = (P^T)^+ \tag{11.10}$$

The expression in Equation (11.9) is known as the generalized eigenvector problem, where Z contains the eigenvectors and α contains the eigenvalues. The decay values can be calculated directly from the eigenvalues. In order to solve the generalized eigenvector problem, the matrices A and B need to be square. This can be achieved by projecting A and B into a common PCA space (Windig and Antalek, 1997).

In order to be able to apply the algorithm to the images, the images were reorganized into arrays, and the arrays into a data matrix. For example, a series of 10, 256×256 images results in a 10×65536 data matrix. On a typical personal computer, the calculation time is on the order of seconds.

11.4 ^{1}H RELAXATION

A standard spin-echo imaging sequence (Stark and Bradley, 1992) is often used to produce images for calculating T_2 and T_1. T_2 is calculated from a series of images with a constant TR and variable TE. T_1 is calculated from a series of images with constant TE and variable TR. Given that an exponential relationship exists between the signal and the intrinsic sample parameters T_2 and T_1, DECRA may be used to extract images of the unique T_2 and T_1 values. These images are a function of the spin densities and the T_2 and T_1 values. This relationship can best be seen by rewriting Equation (11.1) in terms of the i unique T_2 and j unique T_1 values found by DECRA:

$$S \propto \sum_i \sum_j \rho_{i,j} e^{-TE/T_{2i}} (1 - e^{-TR/T_{1j}}) \tag{11.11}$$

Each image resolved by DECRA is representative of the special distribution of a component possessing a discrete T_2 (or T_1) value. The actual spin density is not established in a straightforward way with the methods described here (Antalek et al., 1998).

11.5 T_1 TRANSFORMATION

For this section, Equation (11.1), with constant TE, is simplified as follows:

$$a_1(1 - e^{-bx}) \tag{11.12}$$

This function does not show the proportional behavior of the exponential as described in Figure 11.1(d), and the DECRA technique cannot be applied. If the term a_1 is known, it would be simple to subtract this constant. Because a_1 is not known, a subtraction procedure is not a viable option. However, a simple mathematical procedure can be applied to make it possible to use DECRA. Equation (11.12) can be rewritten as:

$$a_1(e^{0x} - e^{-bx}) \tag{11.13}$$

This shows that the expression basically is a linear combination of two exponential functions. This 'dataset' cannot be resolved using DECRA, because we have two exponentials in a dataset of rank 1 (i.e., the number of *linearly independent* components is 1). However, because one of the exponentials is known, e^{0x}, it is possible to add this as a new component to the dataset in the form of an extra variable with a constant value. This can be extended to complete datasets of T_1 character and basically consists of adding a column to the dataset where all the elements have a constant value. Simulated datasets were used to confirm this surprisingly simple procedure.

In practice, it appeared that the resolved data of T_1 images were relatively noisy. By applying singular value decomposition to the dataset prior to adding the column with constant values and reproducing the dataset using the same number of singular values as the number of components in the dataset (two for the phantom), the noise was reduced significantly.

11.6   IMAGING METHODS

A simple phantom was constructed to test DECRA and is shown in Figure 11.2. It consists of a plastic [poly(vinyl chloride) or PVC] box having the dimensions $18\,\text{cm} \times 10\,\text{cm} \times 1\,\text{cm}$ and containing two compartments separated by a thin plastic (PVC) sheet of thickness $0.18\,\text{cm}$. The compartments, therefore, formed two wedge-shaped

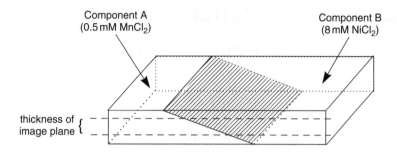

Figure 11.2 Schematic representation of the phantom used in the study. Reprinted from Journal Magnetic Reson., 132 with permission from Elsevier

spaces. The angle of the wedge is 6.2°. These spaces were filled with water that contained a specific amount of paramagnetic salt. One space contained 0.5 mM $MnCl_2$ and the other 8 mM $NiCl_2$. These concentrations were chosen to provide water T_2 values of nearly 30 and 150 ms for the $MnCl_2$ and $NiCl_2$ solutions, respectively. Consequently, the target T_1 values are 160 ms and 320 ms for the $MnCl_2$ and $NiCl_2$ solutions, respectively. The plastic separation sheet was glued to completely seal one compartment from the other. The solution containing the $MnCl_2$ salt will be designated component A, and that containing the $NiCl_2$ will be designated salt component B. Component A will be depicted toward the left side of the figures.

A General Electric Signa 1.5 T whole-body MRI imager, employing a standard spin-echo pulse sequence and standard birdcage radio frequency coil, was used for all of the image acquisitions. For the phantom 15, 256×256 pixels, images were acquired of a 5-mm-thick plane passing through the long axes of the object and parallel to the 18×10 cm surfaces. The image plane is represented in Figure 11.2 with a dashed line. The parameter, TE, in the spin-echo pulse sequence [see Equation (11.1)] was varied starting at 15 ms and incremented by 15 ms for each image. TR was a constant of 2000 ms, and the field of view (FOV) was 20 cm. Varying TE resulted in a dependence in signal intensity for each image based upon T_2 relaxation. The phantom was constructed so that one may obtain, using a 5-mm-thick slice, an image with three general regions, two regions where the signal represents the pure decay behavior of each component (close to the ends), and a middle region that represents a weighted sum of the two.

For the T_1 image series, the same experimental procedure and parameters were used, except that 10 images were acquired with a fixed TE of 15 ms, a TR starting at 200 ms, and incremented by 200 ms.

For the brain, the image plane passed through the head of the 42-year-old, healthy male volunteer at the level of the lateral ventricles. This slice contained six primary tissue types: cerebrospinal fluid (CSF), grey matter, white matter, meninges, adipose and muscle. Two sets of images of this slice were acquired: a set used to calculate T_2 in which TE was varied, and a set used to calculate T_1 in which TR was varied. The T_2 image set consisted of 14 images with a fixed TR = 1000 ms, and a TE that varied between 15 ms and 210 ms in 15 ms steps. The T_1 image set consisted of 15 images with a fixed TE = 15 ms, and a TR that varied between 200 ms and 3000 ms in 200 ms steps. Each 24 cm FOV, 5-mm-thick slice image was acquired with 256 phase-encoding steps to form a 256×256 pixel image. The motion of the volunteer was found to be minimal during the course of data collection, so no attempt was made to register the pixels within the series of correlated brain images. The image slice is treated as a mixture of pure T_2 and T_1 components. DECRA was applied to the T_2 and T_1 image sets in order to extract the pure T_2 and T_1 components present, respectively, as well as to obtain the representative images displaying the relative quantity of each component in a voxel.

11.7   PHANTOM IMAGES

11.7.1 T_2 Series

Figure 11.3(a) and (b) shows the first and last images of the T_2 image series. The tops of the images show the contribution of only the compartment with $MnCl_2$, and the bottoms of the images show the contribution of only the compartment with $NiCl_2$. In the middle region, the signal is a mixture of the two compartments. The intensity is lower in the middle because of the presence of the divider. Comparing Figure 11.3(a) and (b), it is clear that the $MnCl_2$ compartment has the fastest decay rate. The dataset was split into two parts for DECRA: images 1–14 for the first part and images 2–15 for the second part. The calculations take 40 s. The output of DECRA (after reorganizing the output into images) using two components is given in Figure 11.3(c) and (d). In order to verify that using two components describes the original dataset appropriately, the first image was reconstructed from the resolved results. The reconstructed image in Figure 11.3(e) is indistinguishable from the original image in Figure 11.3(a).

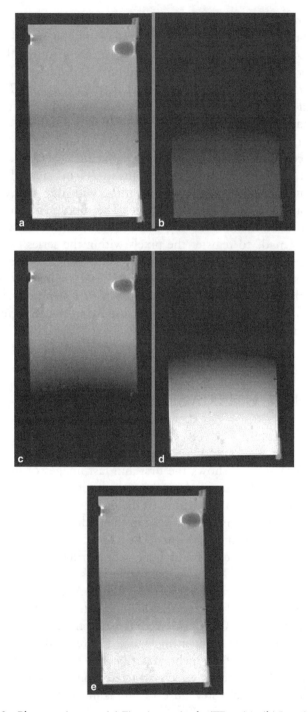

Figure 11.3 Phantom images. (a) First image in the TE series. (b) Last image in the TE series. (c) DECRA-resolved component 1. (d) DECRA-resolved component 2. (e) First image in the TE series reconstructed from the two DECRA-resolved images

When using three components for DECRA, the extra component extracted only represents noise, which is similar to the experience with PGSE NMR spectra (Windig and Antalek, 1997); the contribution profile was negative and the image, dominated by noise, had negative intensities that were in the same range as the positive intensities. This clearly indicates that the dataset has two components.

In order to judge the images more objectively, 'image profiles' are plotted. An image profile is obtained by averaging the rows of the images. The original data can now be plotted in a single figure [see Figure 11.4(a); the pixel positions are numbered along the bottom]. In order to check the quality of the dataset, the standard deviation image profile of the original dataset is given in Figure 11.4(b). As expected, the areas in the phantom where the pure components can be observed (approximately from pixel positions 20 to 100, and from pixel positions 200 to 240) are horizontal. The image profiles on Figure 11.4 show a dip in intensity between pixel positions 100 and 200, which is due to the thickness of the wall that separates the two compartments, and which gives no signal.

In Figure 11.4(c), the image profiles of the two resolved components are given. These image profiles are scaled to reproduce the first image. The dashed line that increases between pixel positions 125 and 200 represents the $NiCl_2$ compartment of the phantom, and the dashed line that decreases between pixel positions 100 and 175 represents the $MnCl_2$ compartment of the phantom. The sum of the two image profiles is represented by a solid line. In order to show how well the original image has been reproduced, the image profile of image 1 is also given with a dashed line. Because of the high overlap, it cannot be distinguished from the sum of the two resolved image profiles [Figure 11.4(c)].

The contribution profiles of the T_2 phantom are shown in Figure 11.4(d). The contributions from the first part of the split data (images 1–14) set are plotted using crosses. The contributions for the second part of the split dataset (images 2–15) are calculated by multiplying the contribution profile of the first dataset by its eigenvalue and are plotted using open squares. Because of the nature of the exponential function, the two profiles have an overlapping, proportional profile. It is clear from the plots in Figure 11.4(d) that the assumption of proportionality was correct, which implies an exponential relationship. The solid line in this figure is calculated from the eigenvalues and shows excellent fit. The T_2 values obtained from the eigenvalues are 28 ms for $MnCl_2$ (target value 30 ms) and 140 ms for $NiCl_2$ (target value 150 ms).

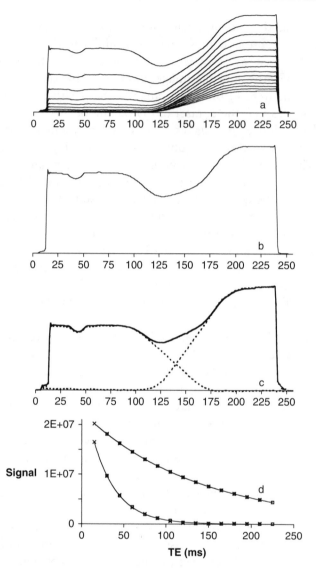

Figure 11.4 (a) Image profiles of each image in the TE series. (b) Standard deviation of the image profiles. (c) Resolved image profiles (dashed lines) for the first image. The sum of the resolved image profiles and the image profile from the first image are superimposed (solid lines). (d) The contribution profiles for the two resolved images. Reprinted from Journal Magnetic Reson., 132 with permission from Elsevier

It is possible to resolve the two components with only three images. DECRA was used on sets comprising images 1,5 and 5,9. Remarkably, the result was nearly identical to the results obtained from the entire dataset (Windig *et al.*, 1998). The calculated T_2 values are 29 ms and 140 ms for $MnCl_2$ and $NiCl_2$, respectively.

11.7.2 T_1 Series

The image profiles of the T_1 image series and its standard deviation image profile are shown in Figure 11.5(a) and (b). The standard deviation image profile shows that the pure regions for each of the

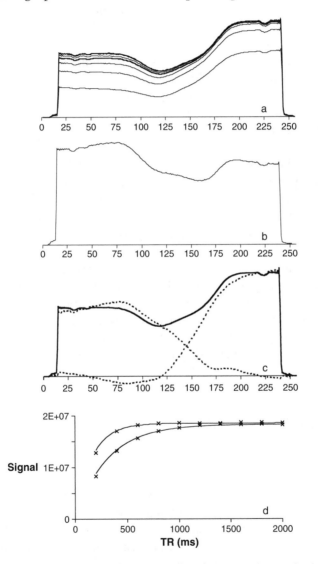

Figure 11.5   (a) Image profiles of each image in the TR series. (b) Standard deviation of the image profiles. (c) Resolved image profiles (dashed lines) for the first image. The sum of the resolved image profiles and the image profile from the first image are superimposed (solid lines). (d) The contribution profiles for the two resolved images. Reprinted from Journal Magnetic Reson., 132 with permission from Elsevier

compartments (pixel positions 15–77 and pixel positions 175–240) are not horizontal, as in the standard deviation image profile of the T_2 images in Figure 11.4(b), although the standard deviation values in each of the pure regions should be constant. The intensities are increasing toward the wall that separates the two compartments. Furthermore, the standard deviation image profile of the T_1 series is less smooth than that of the T_2 series in Figure 11.4(b), which indicates a higher noise level.

In order to resolve the T_1 dataset, a column with constant values was added to the dataset. This results in an extra image, which can be ignored. In order to obtain the contribution profiles from the images with the original dataset, the resolved images were regressed against the original dataset.

The DECRA analysis is shown in Figure 11.5(c) and (d). Deviations from the ideal behavior can be observed, which is not surprising, considering the deviations in the original dataset. Nevertheless, the resolved image profiles are close to the expected profiles. Despite the deviations, the sum of the resolved profiles, scaled to reconstruct the tenth image, again cannot be distinguished from the original image profile.

Selecting an extra component in DECRA resulted in an image and corresponding contribution profile that were dominated with noise. This result shows clearly that the proper number of components is two.

Figure 11.5(d) shows the resolved contribution profiles, which show the expected exponential behavior. The T_1 values derived from the eigenvalues were 160 ms for $NiCl_2$ (target value: 160 ms) and 300 ms for $MnCl_2$ (target value: 320 ms).

11.8 BRAIN IMAGES

11.8.1 T_2 Series

Three of the images used in analyzing the T_2 series are shown in Figure 11.6. Spatially dependent signal-decay variation is clearly seen. Selecting the number of components is a subjective aspect of the analysis at this time. The choice presents itself clearly, however. There are several criteria to examine when establishing the number of components present in a given dataset: (1) the pixel intensities and contribution profiles must be positive; (2) the noise level in the images and contribution profiles should be low; (3) the relative contribution of the component

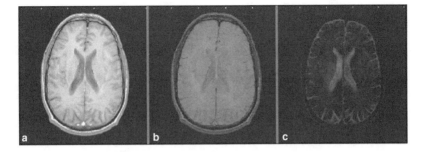

Figure 11.6 TE image series. (a) Image 1. (b) Image 4. (c) Image 14

is significant; and (4) the functional form of the contribution profile needs to fit the model. Generally, a combination of these criteria should be used.

DECRA was applied to the T_2 image set and three components were found. Three were chosen because a fourth comprised only noise. With each component there is associated both a T_2 value and an image. The three components, referred to as components C_1, C_2 and C_3, have T_2 values of 22, 64 and 290 ms, respectively. Images representing the amount of each component in the signal are presented in Figure 11.7. C_1, shown in Figure 11.7(a), is found predominantly in meninges, muscle, and adipose tissue. C_2, shown in Figure 11.7(b), is present largely in approximately equal amounts in grey and white matter, and to a lesser extent in meninges, muscle, and adipose tissue. C_3, shown in Figure 11.7(c), is found extensively in CSF. Although a biological classification is tempting, it is important to realize that the resolved components are defined by a pure exponential decay with its associated T_2 value. Tissue types are complicated mixtures that are largely comprising common building blocks such as intracellular and extracellular water,

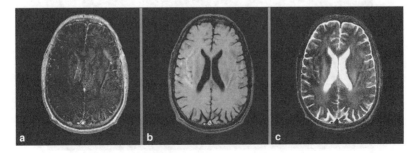

Figure 11.7 DECRA-resolved images from the TE image series. (a) Component 1. (b) Component 2. (c) Component 3

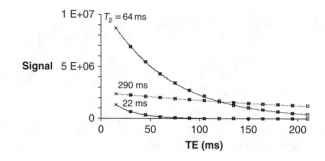

Figure 11.8 Contribution profiles for the three resolved images from the TE series. Reprinted from Journal Magnetic Reson., 132 with permission from Elsevier

lipid membranes, etc., which can describe the majority of the relaxation behavior. Further work is required to assign the resolved components to a specific material or materials.

The exponential contribution profiles for the three T_2 components are shown in Figure 11.8. The resolved contributions representative of the first (images 1–13) part of the split dataset overlap favorably with those from the second (images 2–14) part of the split dataset, calculated by multiplying the contribution profile of the first dataset by its associated eigenvalue. This clearly shows that the assumption of proportionality between the points is correct. The lines drawn through the points are calculated directly from the eigenvalues, and the normalization factor is used for the contribution data.

Finally, the three components may be combined in proportion to reconstruct image 1 from the original dataset. The two images are compared in Figure 11.9. DECRA describes the entire dataset very accurately with just three images.

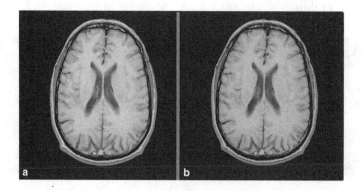

Figure 11.9 Side-by-side comparison of (a) image 1 of the TE series and (b) image 1. reconstructed from the three DECRA-resolved T_2 images

11.8.2 T_1 Series

DECRA was applied to the T_1 image set (with the added constant) and two components, referred to as C_4 and C_5, were found. These are shown in Figure 11.10. C_4, shown in Figure 11.10(a), is found in the grey and white brain matter, and in the adipose and muscle tissues. The largest concentration of C_4 appears to be in the adipose tissue. The concentration of C_4, immediately adjacent to the folds in the grey matter, is less than that in deeper grey and white matter of the brain. C_5, shown in Figure 11.10(b), is found predominantly in the CSF, occupying both the ventricles and the folds in the grey matter. This component also appears to be present in the grey matter adjacent to the folds, but to a lesser concentration than in pure CSF. DECRA did find some C_5 in the tissues outside of the skull.

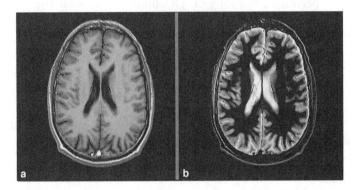

Figure 11.10 DECRA-resolved images from the TR image series. (a) Component 4. (b) Component 5

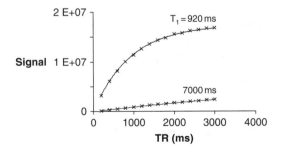

Figure 11.11 Contribution profiles for the two resolved images from the TR series. Reprinted from Journal Magnetic Reson., 132 with permission from Elsevier

The exponential contribution profiles for the two T_1 components are shown in Figure 11.11(a). For the resolution of the T_1 image series, a transformation of the dataset is required (Antalek and Windig, 1996). This results in correct (for the original data) images but transformed contribution profiles. The exponential contribution profiles of the entire original dataset are calculated as the contributions of the resolved images in this dataset. As a consequence, we have a single exponential contribution profile for each component instead of two overlapping profiles. The lines drawn through the points are calculated directly from the eigenvalues, and the normalization factor is used for the contribution data. C_4 had a T_1 of 0.92 s, while C_5 had a T_1 of 7.0 s. It is noted that a T_1 value of 7.0 s appears to be unrealistic (e.g. we should expect a value near 3 s for water at room temperature). Based only on the experimental values used, namely TR, we cannot claim this resolved T_1 value to be accurate. In order to obtain an accurate T_1 value in a saturation recovery experiment, the longest TR should be at least four times the longest T_1 found. Clearly, this requirement was not met. Additionally, others (Mansfield, 1977; Fletcher et al., 1993) have reported long T_1 values in excess of 5 s for CSF using a saturation recovery pulse sequence at a field strength of 1.5 T. They attributed this to a saturation effect arising from the flowing CSF. We make no further attempt to clarify the issue.

The resultant images based upon the DECRA analysis are significantly different from the T_2 or T_1 images obtained through single-exponential fitting routines of registered pixels within image datasets. The single-exponential fit results in a value representative of a linear combination of exponential time constants for all of the various ^{1}H environments represented in the volume element. DECRA extracts independent exponential values and renders images based upon these pure values. [1]H nuclear environments are segmented in this fashion, but tissues are not. For example, the component with the shortest T_2, Figure 11.7(a), most likely represents predominantly the methylene signal from the long alkyl chains in the fat and lipid tissues. The fat and muscle tissue are highlighted. T_2 values are primarily shorter for these environments as a result of restricted molecular mobility.

11.9   REGRESSION ANALYSIS

After resolving a multivariate image dataset, there might still be parts of the data that are not expressed clearly. Such a part might be collinear with several of the components, in which case it will be 'divided' over these collinear components. In order to enhance such a part, image regression can be applied by a method called discriminant regression (Geladi and Grahn, 1996) as follows:

(a) The part of interest in the image is indicated by a polygon.
(b) A binary image is created with ones within the polygon and with zeroes elsewhere.
(c) A linear combination of the unfolded resolved images that reproduces the unfolded logical image is calculated.
(d) The obtained image is refolded and displayed.

If the resolved unfolded images are in matrix C, and the binary image is in array b, we can calculate an array x with the following relation to C and b:

$$Cx = b \tag{11.14}$$

C has the size nr, x has r elements, and b has n elements. Because we have unfolded images, n is the product of the number of pixels in the x-direction of the image and the y-direction of the image. The least-squares solution for x is:

$$\hat{x} = (C^T C)^{-1} C^T b \tag{11.15}$$

The least-squares approximation of the resolved images for b is then:

$$\hat{b} = C\hat{x} \tag{11.16}$$

As discussed above, the T_2 and T_1 resolved images reflect certain tissue properties. However, there might be areas of interest in the brain that are not resolved properly, for example, if such an area is a mixture of the resolved tissues. We will use one of the resolved images, the second resolved image of the T_1 series, as an example (see Figure 11.12). The region of interest is contained in the boxed region shown in Figure 11.12(a) and expanded fully in Figure 11.12(b). We have an area of interest that is not clearly represented in any of the images. This area is represented by a polygon in Figure 11.12(c). We can see some of it in the first and second resolved T_2 images. In order to enhance this area of interest, we zoom in before we indicate the area of interest with a

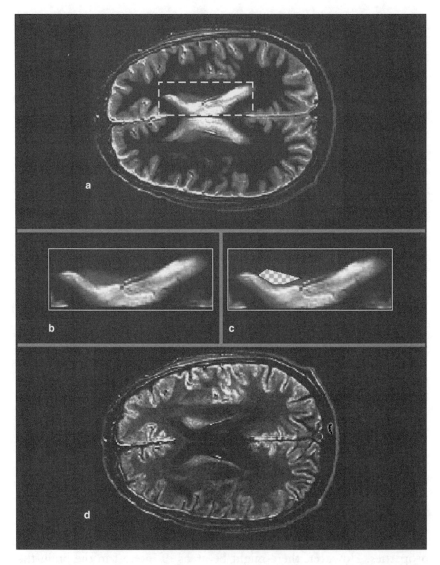

Figure 11.12 Image regression. (a) A small section of the first resolved image of the TR series is shown in the boxed region. (b) Expansion of the boxed region in (a). (c) The area of interest is represented by a polygon. (d) Image produced after the regression of the five resolved images to reproduce the area represented by the polygon in (c)

polygon. This must be done in order to avoid labeling areas related to the area of interest, unknown at this point, with zeroes.

In order to reproduce the binary image represented by the polygon in Figure 11.12(c) with the (expanded) resolved images, the array **x** is calculated using Equations (11.14) and (11.15) and applied to the complete images using Equation (11.16). The result is shown in Figure 11.12(d). The area of interest is clearly enhanced. It is also clear that other areas are related to this area, as indicated by the highlight, which is the reason why a zoomed area had to be used.

An image constructed by a purely mathematical approach does not necessarily provide physically meaningful information and have only positive pixel contributions. The binary image has positive pixel contributions; therefore, we only show the positive part of this regression image. Despite the less desirable properties of negative intensities, the image enhancement thus obtained makes this regression a worthwhile tool.

11.10   CONCLUSIONS

We have described a method to process a series of MRI images obtained in such a way as to produce a variation in signal intensity that is exponential in nature. The method, called DECRA, resolves images from the dataset that are representative of pure T_1 or T_2 components. Because the variation in the data is natural, the information provided in the resolved images is physically relevant. The method is fast, as it involves no iteration.

We have shown its ability to solve the problem of partial volume in imaging with the use of a phantom constructed in a way to include more than one component in the same voxel across a large part of the image. In this case, the two components were cleanly separated in each of the two series of images designed to create a signal variation based on both T_2 and T_1.

In applying the method to images of the human brain, we adopt a simple model where tissues comprise a finite number of components possessing discrete T_2 and T_1 values. DECRA then resolves images representative of the distribution of these components across the image. The process does not result in tissue segmentation but, rather, the differentiation of hydrogen chemical environments.

Image regression has been applied using all of the resolved images in the brain study to produce an image containing the properties found in a small region of interest.

REFERENCES

Antalek, B. (2002) Using pulsed gradient spin echo NMR for chemical mixture analysis: how to obtain optimum results. *Concepts Magn. Reson.* **14**, 225–258.

Antalek, B. and Windig, W. (1996) Generalized rank annihilation method applied to a single multicomponent pulsed gradient spin echo NMR data set. *J. Am. Chem. Soc.* **118**, 10331–10332.

Antalek, B., Hornak, J. P. and Windig, W. (1998) Multivariate image analysis of magnetic resonance images with the direct exponential curve resolution algorithm (DECRA). Part 2: Application to human brain images. *J. Magn. Reson.* **132**, 307–315.

Armsach, J.-P., Gounot, D., Rumbach, L. and Chambron, J. (1991) In vivo determination of multiexponential T2 relaxation in the brain of patients with multiple sclerosis. *Magn. Reson. Imag.* **9**, 107–113.

Bijlsma, S., Louwerse, D. J., Windig, W. and Smilde, A. K. (1998) Rapid estimation of rate constants using on-line SW-NIR and trilinear models. *Anal. Chim. Acta* **376**, 339–355.

Booksh, K. S. and Kowalski, B. R. (1994) Comments on data analysis (datan) algorithm, rank annihilation factor analysis for the analysis of correlated spectral data. *J. Chemom.* **8**, 287–292.

English, A. E., Joy, M. L. G. and Henkelman, R. M. (1991) Pulsed NMR relaxometry of striated muscle fibers. *Magn. Reson. Med.* **21**, 264–281.

Fletcher, L. M., Barsotti, J. B. and Hornak, J. P. (1993) A multispectral analysis of brain tissues. *Magn. Reson. Med.* **29**, 623–630.

Geladi, P. and Grahn, H. (1996) *Multivariate Image Analysis.* John Wiley & Sons, Ltd, Chichester.

Geladi, P., Isaksson, H., Lindqvist, L., Wold, S. and Esbensen, K. (1989) Principal component analysis of multivariate images. *Chemom. Intell. Lab. Syst.* **5**, 209–220.

Gong, J. and Hornak, J. P. (1992) A fast T1 algorithm. *Magn. Reson. Imag.* **10**, 623–626.

Grahn, H. and Sääf, J. (1992) Multivariate image regression and analysis. Useful techniques for the evaluation of clinical magnetic resonance images. *Chemom. Intell. Lab. Syst.* **14**, 391–396.

Grahn, H., Szeverenyi, N. M., Roggenbuck, N. W., Delaglio, F. and Geladi, P. (1989a) Data analysis of multivariate magnetic resonance images. I. A principal component analysis approach. *Chemom. Intell. Lab. Syst.* **5**, 311–322.

Grahn, H., Szeverenyi, N. M., Roggenbuck, M. W. and Geladi, P. (1989b) Tissue discrimination in magnetic resonance imaging: a predictive multivariate approach. *Chemom. Intell. Lab. Syst.* **7**, 87–93.

Just, M. and Thelen, M. (1988) Tissue characterization with T1, T2, and proton density values: results in 160 patients with brain tumors. *Radiology* **169**, 779–785.

Kao, Y.-H., Sorenson, J. A., Bahn, M. M. and Winkler, S. S. (1994) Dual-echo MRI segmentation using vector decomposition and probability techniques: a two-tissue model. *Magn. Reson. Med.* **32**, 342–357.

Kroeker, R. M. and Henkelman, R. M. (1986) Analysis of biological NMR relaxation data with continuous distributions of relaxation times. *J. Magn. Reson.* **69**, 218–238.

Kubista, M. (1990) A new method for the analysis of correlated data using Procrustes rotation which is suitable for spectral analysis. *Chemom. Intell. Lab. Syst.* **7**, 273–279.

Li, X. and Hornak, J. P. (1994) Accurate determination of T2 images in MRI. *J. Imag. Sci. Tech.* **38**, 154–157.

Look, D. C. and Locker, D. R. (1970) Time saving in measurement of NMR and EPR relaxation times. *Rev. Sci. Instrum.* **41**, 250–251.

Ma, J., Wehrli, F. W., Song, H. K. and Hwang, S. N. (1997) A single-scan imaging technique for quantitation of the relative contents of fat and water protons and their transverse relaxation times. *J. Magn Reson.* **125**, 92–101.

Malinowski, E. R. and Howery, D. G. (1991) *Factor Analysis in Chemistry.* Wiley-Interscience, New York.

Mansfield, P. (1977) Multi-planar image formation using NMR spin echoes. *J. Phys. C* **10**, L55–L58.

Ordidge, R. J., Gorell, J., Deniau, J., Knight, R. A. and Helpern, J. A. (1994) Assessment of relative brain iron concentrations using T2-weighted and T2*-weighted MRI at 3 Tesla. *Magn. Reson. Med.* **32**, 335.

Peemoeller, H., Pintar, M. M. and Kydon, D. W. (1980) Nuclear magnetic resonance analysis of water in natural and deuterated mouse muscle above and below freezing. *Biophys. J.* **29**, 427.

Sanchez, E. and Kowalski, B. R. (1988) Generalized rank annihilation factor analysis. *Anal. Chem.* **58**, 496–499.

Scarminio, I. and Kubista, M. (1993) Analysis of correlated spectral data. *Anal. Chem.* **65**, 409–416.

Schulze, D. and Stilbs, P. (1993) Analysis of multicomponent FT-PGSE experiments by multivariate statistical methods applied to the complete bandshapes. *J. Magn. Reson. Ser. A* **105**, 54–58.

Stark, D. D. and Bradley, W. G. (1992) *Magnetic Resonance Imaging,* 2nd Edn. Mosby-Year Book, St Louis.

Wilson, B., Sanchez, E. and Kowalski, B. R. (1989) An improved algorithm for the generalized rank annihilation method. *J. Chemom.* **3**, 493–398.

Windig, W. and Antalek, B. (1997) Direct exponential curve resolution algorithm (DECRA): a novel application of the generalized rank annihilation method for a single spectral mixture dataset with exponentially decaying contribution profiles. *Chemom. Intell. Lab. Syst.* **37**, 241–254.

Windig, W. and Antalek, B. (1999) Resolving nuclear magnetic resonance data of complex mixtures by three-way methods: examples of chemical solutions, the human brain. *Chemom. Intell. Lab. Syst.* **46**, 207–219.

Windig, W., Hornak, J. P. and Antalek, B. (1998) Multivariate image analysis of magnetic resonance images with the direct exponential curve resolution algorithm (DECRA). Part 1: Algorithm, model study. *J. Magn. Reson.* **132**, 298–306.

Windig, W., Antalek, B., Sorriero, L., Bijlsma, S., Louwerse, D. J. and Smilde, A. (1999) Applications, new developments of the direct exponential curve resolution algorithm (DECRA). Examples of spectra, magnetic resonance images. *J. Chemom.* **13**, 95–110.

Windig, W., Antalek, B., Robbins, M. J., Zumbulyadis, N. and Heckler, C. E. (2000) Applications of the direct exponential curve resolution algorithm (DECRA) to solid state nuclear magnetic resonance, mid-infrared spectra. *J. Chemom.* **14**, 213–227.

Zumbulyadis, N., Antalek, B., Windig, W., Scaringe, R. P., Lanzafame, A. M., Blanton, T. and Helber, M. (1999) Elucidation of polymorph mixtures using solid-state 13C CP/MAS NMR spectroscopy, direct exponential curve resolution algorithm. *J. Am. Chem. Soc.* **121**, 11554–11557.

12

Hyperspectral Imaging Techniques: an Attractive Solution for the Analysis of Biological and Agricultural Materials

Vincent Baeten, Juan Antonio Fernández Pierna and Pierre Dardenne

12.1 INTRODUCTION

At the time of writing this chapter, the European probe Huygens was sending data from Titan, the largest of Saturn's moons. A huge amount of the data was coming from an imaging system installed on board the spacecraft. This instrument was set up several years ago and has travelled fixed to the Casini-Huygens spacecraft during its 7-year journey from Earth to Titan. During this period, hyperspectral imaging systems have been applied to biological and agricultural materials. An important evolution in this field is apparent when looking at the ABES (agricultural, biology and environmental sciences) current contents, which includes most of the scientific papers already published. As an example, for the keywords 'hyperspectral' and 'imaging' nine

Techniques and Applications of Hyperspectral Image Analysis   Edited by H. F. Grahn and P. Geladi
© 2007 John Wiley & Sons, Ltd

papers were referenced for the 1998–2001 period, but 40 papers were referenced for the 2002–2004 period. This change reflects a growing awareness of the potential of hyperspectral imaging systems to solve analytical challenges (Baeten and Dardenne, 2002).

'Hyperspectral imaging history is at the early stage of its development and will occupy the analytical scene in the 21st century' (McClure, 2003).

For the first time in the scientific world, an analytical tool is available that allows researchers to acquire spatially resolved chemical information about biological and agricultural materials. With imaging systems working in the ultraviolet, visible, near infrared (NIR), infrared and Raman spectral range of the electromagnetic spectrum, researchers are able to obtain information not only about composition, but also about its distribution. They are also able to obtain chemical information at the microscopic level as well as at the macroscopic level. For the research community, this is an important revolution. Spectral imaging techniques offer the potential to collect thousands of spectra of one sample or process. This can be done nondestructively and without interfering with the composition of the sample or with the process.

Several studies have shown the potential of spectral imaging systems for studying biological and agricultural materials. The studied applications include satellite and aircraft remote-sensing, and macroscopic and microscopic imaging. Agricultural applications relate, for instance, to the use of imaging systems in the remote-sensing of crop production problems and to the use of this technology in the laboratory.

In the following sections, the perspectives offered by hyperspectral imaging for studying the composition and the distribution of chemical components as well as detecting defects and contamination of biological and agricultural materials are illustrated and discussed. In the final section of the chapter, other potential applications are reviewed briefly. Special emphasis is placed on instruments working in the NIR region, as they appear to be very interesting for studying heterogeneous biological and agricultural samples. NIR radiation penetrates these samples rather well and gives diffuse reflectance information from both the outside and inside of the samples (Geladi et al., 2004). Most of the studies considered here concern hyperspectral imaging techniques, but there are also brief references to multispectral imaging studies performed on agro-food products. The images shown in the following sections were collected at the Walloon Agricultural Research Centre (CRA-W, Gembloux, Belgium) with a

Matrix NIR™ instrument (Spectral Dimensions Inc., Olney, USA). This instrument includes an InGas array detector (240 × 320 pixels) active in the 900–1700 nm region of the electromagnetic spectrum.

12.2 SAMPLE CHARACTERIZATION AND CHEMICAL SPECIES DISTRIBUTION

Hyperspectral imaging spectroscopy enables analysts to characterize the chemical species present in a sample and to study their distribution in the sample. It provides the analyst with a nondestructive and rapid technique that can be used to acquire simultaneously spatial and spectral information of heterogeneous biological and agricultural products. Hyperspectral imaging techniques, mainly NIR imaging, have the flexibility to tackle all samples, whatever their size, i.e. they are able to deal with microscopic particles, a single kernel or the whole sample (i.e. fruit) in order to study the distribution of a wide range of chemical compounds. For analysing biological and agricultural products in the field, imaging technology is rapidly replacing the time-consuming mapping techniques (Koehler *et al.*, 2002). Figure 12.1 shows score images of the transversal section of a banana.

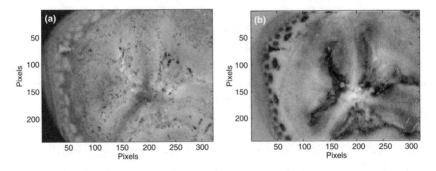

Figure 12.1 NIR score images of the transversal section of a banana: (a) image reconstructed using the first principal component (PC1); (b) image reconstructed using the third principal component (PC3)

12.2.1 Analysis of Fruit

One of the first studies in regarding plant was conducted by Taylor and McClure (Taylor and McClure, 1989). They used NIR imaging technology to undertake a plant physiology study. They worked with a charge coupled device (CCD) video camera with an effective sensitivity in the range 400–1100 nm and narrow-band interference filters. They showed the potential of this technique for studying the distribution of chlorophyll in plant tissues and water on different surfaces. They also used this technique to study the damaged parts of tobacco leaves. Okumura *et al.* (Okumura *et al.*, 1992) used the NIR imaging technology in the analysis of pears in a study of the correlation between water content and relative brightness. More recently, Tran and Grishko (Tran and Grishko, 2004) used this technology to determine the water content of olive leaves.

Martinsen and Schaare (Martinsen and Schaare, 1998) and Martinsen *et al.* (Martinsen *et al.*, 1999) used an NIR hyperspectral imaging spectrometer to measure the spatial distribution of soluble solids across cut surfaces of kiwifruit. The spectral volumes were recorded as reflectance spectra from the illuminated surfaces of fruit cut near the longitudinal centre. This first study on fruit highlighted the potential of using imaging technology to study the spatial distribution of the samples' constituents. The hyperspectral imaging instruments used in this study included a diffraction grating system (i.e. a system involving scanning, step by step, the reflected light of a line of the sample) and a CCD array detector. The spectral volumes measured contained 150×242 pixels over a sample about 50 mm^2 with a spatial resolution of 1.2×2.0 line pairs per mm, corresponding to a horizontal spatial resolution of 0.5 mm and a vertical spatial resolution of 0.7. The spectra were measured between 650 nm and 1100 nm, with a resolution of 5 nm. In order to reduce the noise, the spectral volumes were smoothed. In addition, the spectra were corrected in order to compensate for the nonunity reflectance of the reference used, which was obtained from scans of photographic gray card collected at the beginning of each session. The partial least squares (PLS) and inverse least squares (ILS) methods applying different pretreatments were used for data modelling, i.e. to build the calibration models. The best calibration model for determining soluble solids had a standard error of 1.2° Brix over a range of 4.7–14.1° Brix. The calibration was used to study the spatial distribution of the soluble solids in cut pieces of kiwifruit. This study would hardly have been possible using a traditional refractometer. It showed that the major limitation of this technique was the significant specular reflection due to the high water concentration (80–90 %) of the samples. Similar work has been done by Sugiyama (Sugiyama, 1999) using multispectral imaging systems for the analysis and visualization of the sugar content in the flesh of a melon (Tsuta *et al.*, 2002) and by Peng and Lu (Peng and Lu,

2004) for predicting the firmness and soluble solids content of apples.

Hyperspectral spectroscopic techniques have been also used to study the ripeness of tomatoes (Polder *et al.*, 2000, 2002, 2003). They used a hyperspectral imaging system which includes a prism–grating–prism (PGP) dispersive element, adapted transmission optics and a CCD detector that enabled them to collect spectra in the 430–900 nm spectral range. The spectral volumes were constructed step by step, one axis of the array detector being used as the spatial axis and the other as the spectral axis. In order to obtain the bi-dimensional image, the sample had to be presented step by step to the camera. The spectral resolution was 1.3 nm and the spatial resolution was about 30 μm. Principal component analysis (PCA) and Fisher's linear discriminant analysis (LDA; a classical classification method) were used, after normalization, to visualize the data and calibrate the instrument, respectively. LDA was used to establish classification rules, i.e. to define optimal boundaries between the various ripeness classes by maximizing the difference between them. An unknown object was assigned to the class it most resembled. This study showed that hyperspectral images offer more discriminating power than RGB images for measuring the ripening stages in tomatoes. The classification error of individual pixels was reduced from 51 to 19 %. This work also demonstrated the potential of the methodology for studying the total concentration of lycopene and chlorophyll, and the spatial distribution of this concentration.

Peirs *et al.* (Peirs *et al.*, 2003) proposed a new technique based on hyperspectral NIR imaging to determine the maturity stage of pre-climacteric apples. They set up an instrument based on line-by-line scanning technology. The system consisted of a spectrograph, mounted on a camera, that dispersed the captured light of one spatial line into the individual spectral components. To construct the entire spectral volume, the sample was placed on a table allowing it to be moved step by step under the camera. The instrument was active in the 900–1700 nm region, with a spectral resolution of 2.2 nm and a spatial resolution of 1.5 mm. The data were first analysed by applying a Savitsky–Golay smoothing algorithm in order to reduce the noise level of the obtained images. Then a study using PCA was conducted to distinguish the starch concentrations within one apple and among several apples during maturation. The soluble solids content was predicted using PLS models involving all the pixels from the image. All the analyses were performed on cut apples. The results showed that hyperspectral imaging systems could be used to measure fruit quality parameters (starch index) during maturation and to study spatial starch degradation.

Hyperspectral imaging systems are also useful for assessing the firmness of peaches, avoiding the need to use the conventional destructive techniques (Lu, 2004; Lu and Peng, 2004). The authors used a CCD camera that enabled them to collect hyperspectral scattering images from peaches in the 500 and 1040 nm region of the electromagnetic spectrum, with a spectral resolution of 1.65 nm. In order to predict peach firmness, a procedure comprising outlier detection methods, space reduction techniques (PCA) and the neural network approach was applied. This procedure proved to be useful for assessing peach firmness with reasonable standard errors of prediction.

12.2.2   Analysis of Kernels

The advantages of hyperspectral imaging techniques are also apparent in the studies undertaken to increase the understanding of plant physiology at the cellular level. Indeed, the ability of these imaging techniques to combine chemical and spatial information in a single analysis enables researchers to reach a new level of understanding of the analytical information and to conduct in-depth studies of the morphological structure of biological materials. In the case of wheat or corn kernels, for instance, qualitative and quantitative analytical results can be linked to morphological information (Budevska, 2002). One of the first microspectroscopic studies of agricultural products involved the use of the infrared microspectroscopic technique to analyse and characterize biological materials. The mapping strategy involving the successive acquisition of spectra sections of wheat was first used to generate functional group contour maps and surface plots. Generally, the maps and plots created were based on single wavelength absorbance. Infrared microspectroscopic mapping instruments were also used to study cellular and subcellular chemical heterogeneities (Wetzel et al., 1998). The use of microscopic mapping techniques was limited because of the complexity and heterogeneity of biological materials and the difficulty of performing a complete analysis. The introduction of focal plane array (FPA) detectors helped improve the microspectroscopic technique by facilitating the collection of images of plant materials. The first study involving an FPA detector concerned the analysis of a wheat section in order to show the structure of different parts of the kernel (Marcott et al., 1999). False colour images were constructed using the information at dedicated frequencies. The hyperspectral microspectroscopic imaging techniques were used, for instance, to study the endosperm/aleurone/pericarp area of a mature corn kernel (Budevska, 2002). In this study, the authors used the absorbance at different frequencies to study the distribution of the various constituents, such as proteins, carbohydrates and lipids, of the kernel section. This and other studies have shown that hyperspectral imaging in combination with sophisticated chemometric tools present an important opportunity for rapid chemical imaging of biological and agricultural products. At the cellular level, the ability to fingerprint the chemical composition in a space-resolved manner makes this a valuable tool for characterizing genetically altered materials (Budevska, 2002).

Hyperspectral imaging systems have also been used to develop single kernel methods to determine quality parameters. Cogdill *et al.* (Cogdill *et al.*, 2002, 2004) proposed an NIR hyperspectral imaging technique for the quality analysis of a single corn kernel (Stevermer *et al.*, 2003). They focused on calibrating the hyperspectral imaging instrument to predict the constituent concentrations of single kernels from NIR hyperspectral images. The instrument worked in transmission in the 750–1090 nm region. PLS and principal component regression (PCR) algorithms were applied to develop mathematical calibrations for determining moisture and oil content. The moisture and oil calibrations had a standard error of cross-validation (SECV) of 1.20 % and 1.738 %, respectively. They showed that reference chemistry contributed more than 50 % of the total variance of the oil calibration. The authors also developed an automated kernel analysis and sorting system to single out maize kernels and place them in the field of view (FOV) of the NIR imaging system.

Hyperspectral imaging techniques have also proved to be a valuable tool for analysing the wheat fraction. Robert *et al.* (Robert *et al.*, 1991) constructed one of the first NIR imaging systems and demonstrated its potential for wheat fraction analysis. Discriminant analysis methods were used to construct discriminant models in order to classify ingredients according to their sources. The instrument enabled the authors to collect spectra in the range of 900–1900 nm, with a resolution of 50 nm. In order to test the instrument, the authors studied its ability to discriminate isolated wheat fractions (pellets made by brand, gluten and carbohydrate) (Bertrand and Dufour, 2000; Bertrand *et al.*, 1996). Using the absorbances at six wavelengths (900, 950, 1000, 1450, 1500 and 1600 nm), they were able to discriminate the different fractions with a level of error between 0 % and 10 %, depending on the fraction considered.

12.2.3   Analysis of Food and Feed Mixtures

Baeten *et al.* (Baeten and Dardenne, 2001; Baeten *et al.*, 2001, 2004) proposed the use of NIR microscopic techniques for the complete screening of animal feed. Their research was in the support of the Proposal for a European Parliament and Council Directive amending Directive 79/373/EC on the marketing of compound feed. In this document, the EC highlights the advantages of the labelling requirements for compound feed for livestock in order to facilitate the tracing of compound feed. FPA NIR imaging spectroscopy was proposed as a more efficient method than those currently available, which tend to be time-consuming and to require significant analytical expertise. Fernández Pierna *et al.* (Fernández Pierna *et al.*, 2004a,c) applied chemometric classification strategies to automate the method and to reduce the need for constant expert analysis of the data. They compared the performance of various discrimination methods for multivariate data. Support Vector Machines (SVM) were shown to be the best technique for classifying feed particles as either meat and bone meal (MBM) or vegetal, using the spectra from NIR images.

These imaging techniques could also be very useful for studying the distribution of products in a mixture (Fernández Pierna *et al.*, 2004b). An example is the development of techniques for rapid, precise and reliable screening of compound feed. One possible way of conducting this study is to construct a classification tree by arranging the various products in a dichotomist way, where each node of the tree constitutes a discriminating chemometric model. In the case of food products, imaging techniques could be also useful for discriminating between two products (i.e. salt and sugar, as shown in Figure 12.2). Similarly, various spices could be discriminated using data obtained with the hyperspectral imaging technique.

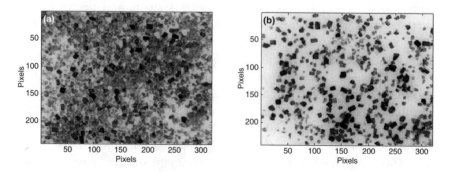

Figure 12.2   Images reconstructed from the first principal component for (a) a pure sample of sugar and (b) a mixture of 50/50 sugar and salt

12.3 DETECTING CONTAMINATION AND DEFECTS IN AGRO-FOOD PRODUCTS

In efforts to improve the quality and safety of food, rapid and noninvasive methods, which can be implemented to assess hazardous conditions in agricultural and food production, are essential. Vibrational spectroscopy is one of the techniques chosen by the agro-food sector to evaluate rapidly and at a reasonable cost the quality and safety of food products. Conventional vibrational spectroscopic methods have the disadvantage that the analysis focuses on only a relatively small part of the material analysed, while the imaging spectroscopic approach allows researchers to characterize the whole spatial variability. This approach is particularly suitable for detecting localized defects in agro-food products, as well as detecting contamination of feed.

12.3.1 Detecting Contamination in Meat Products

A multispectral imaging system has been proposed for the quality control of many food products. It has been used for the inspection of poultry carcasses during processing in order to identify cadaver diseases and tumours and to detect surface contaminants. The contamination of meat products is an important concern because bacterial pathogens are a major cause of illness and death. The contamination of poultry carcasses with bacterial food-borne pathogens can occur through exposure to ingesta or faecal material during the processing of the food product. Recently, hyperspectral imaging instruments have been used to identify faecal and ingesta contamination of poultry carcasses (Park et al., 2002a, 2002b). In this study, the authors used a visible NIR hyperspectral imaging system working in the range of 400–900 nm. The spectral resolution of the hyperspectral images was approximately 0.9 nm. The instrument was a line-scan imaging instrument and the time required to scan a whole carcass was about 34 s. They used hyperspectral images at key wavelengths, as well as wavelength ratio images to highlight carcass contamination. Background noise was eliminated by using a masking process and contamination was separated from the carcass using a stretching procedure. In another study by Lawrence et al. (Lawrence et al., 2003), the authors demonstrated the usefulness of PCA for distinguishing the contaminants inside poultry carcasses. From their analysis of 80 carcasses they showed, with this limited sample set, the potential of imaging systems for detecting faecal and ingesta contamination. More than 96 % of the contaminants were detected. The authors felt that more work was needed on vibrational spectroscopy using hyperspectral data. The drawback of the hyperspectral imaging approach in this application was that it was too slow for real-time on-line processing. But hyperspectral imaging systems can be more widely used to select the most adapted frequencies for a multispectral system used for routine analysis.

The same hyperspectral imaging technique has been also used to detect poultry skin tumours (Chao *et al.*, 2002). Vibrational spectroscopic imaging showed potential for detecting localized poultry diseases/defects, including include skin tumours and inflammatory process. Poultry skin tumours are ulcerous lesions surrounded by a rim of thickened skin and dermis. The authors used a hyperspectral imaging system to select wavelengths for designing a multispectral imaging system. The imaging system worked in the range of 420–850 nm and included a CCD camera equipped with an imaging spectrograph. The instrument enabled them to capture the spectral information from one of the spatial dimensions. The second spatial dimension was achieved through successive scans of samples moved along on a conveyor belt. For this study, they analysed eight poultry carcasses with tumours. The image processing included masking in order to separate the chicken image from the background. The spatial region of interest, containing skin defects, was then selected and PCA was applied to select useful bands for detecting tumorous regions. This information was used to set up a multispectral instrument that was able to collect the images simultaneously at the three narrow-band wavelength regions. Feature selection was then performed in order to apply the fuzzy logic classifiers technique. The features included mean, standard deviation, skewness, coefficient of variation and kurtosis. As the results showed, they were able to separate normal from tumorous skin areas with increasing accuracy as more features were included. In the best result, 91 and 86 % of normal and tumorous tissues, respectively, were detected.

12.3.2 Detecting Contamination and Defects in Fruit

Kim *et al.* (Kim *et al.*, 2001) proposed a hyperspectral reflectance and fluorescence imaging system for assessing the quality and safety of food commodities. The instrument used a line-by-line scanning technique for a wide range of sample sizes and included a CCD detector and various lighting peripherals. It was active in the 430–930 nm range and had a spectral resolution of 10 nm and a spatial resolution of about 1 mm. This imaging resolution was considered adequate to detect the features or small anomalies that may be encountered in many food commodities. Calibrations and image correction procedures were proposed and discussed. The authors demonstrated the versatility of hyperspectral imaging systems using sample fluorescence and reflectance images of apples to assess their safety and quality (Kim et al., 2002; Chen and Kim, 2004). Apples are an important agricultural commodity and special attention needs to be paid to bacteria contamination. They can be contaminated with bacteria from animal faecal material or ingesta of an animal's gastrointestinal tracts. Material with diseased or fungal-contaminated surfaces or with open skin cuts and bruises may become sites of decay and bacterial growth. There is therefore a need to develop technologies and methodologies to detect fruit defects and contamination in the post-harvesting pre-processing stage.

For this reason, Kim et al. (Kim et al., 2002) proposed using a set of reflectance and fluorescence image data. One of their studies involved discriminating between a healthy apple and an apple with fungal contamination and bruising. In a second study, they tried to evaluate spatial and spectral responses of hyperspectral reflectance images of faecal-contaminated apples. In the first study, a complete examination of the spectra was done, providing a detailed elucidation of spectral features of apples and their variations caused by contamination. From this study, they concluded that multispectral imaging could become an integral part of food production industries for automated on-line application. In the second study, apple surfaces exhibited flat reflectance responses, while the thick faeces treatment showed a monotonic increase in intensities in the NIR region from approximately 730 to 850 nm. In the spectra of thin faecal treatment on shaded and sun-exposed sides of certain apple varieties, a slight slope change was observed in the same region. The spectral changes were less apparent when the faecal contamination was thinly applied; in these cases, the apple skin was visible through the thin smear. The authors paid special attention to the inherent surface morphology and skin coloration that can significantly affect the performance of the methodology if they are not properly addressed (Chen and Kim, 2004).

Spectral imaging techniques have been also proposed by other authors for detecting contamination and defects in apples. Mehl et al. (Mehl et al., 2002) proposed the use of the hyperspectral analysis to detect defects on selected apple cultivars. They used hyperspectral images to develop a multispectral technique. They worked with three apple cultivars selected for their shape and spectral differences, and used normal apples (i.e. without defects) as well as abnormal apples affected by bruising, diseases and contamination. The hyperspectral imaging system allowed them to determine scabs, fungal and soil contamination and bruising using either PCA or the absorption intensities at specific frequencies. It also allowed them to select three spatial bands capable of separating normal from contaminated apples. These spectral bands were used to develop a multispectral imaging system with specific band pass filters. The performance of the multispectral imaging system was tested on a set of 153 normal, and contaminated and/or damaged apples. The correct classification of the apples was found to vary from 76 to 95 %, depending on the cultivars analysed.

Also with the objective of detecting bruising, Lu (Lu, 2003) developed an automated system to help the fruit industry provide better fruit for consumers and reduce potential economic losses. This study used an NIR hyperspectral imaging system active in the range of 900–1700 nm, with a spectral resolution of 4.3 nm. The system had a spatial resolution of 3 mm and 2 mm, depending on the axis. A complete 3-D hyperspectral cube was obtained by scanning, step by step, the entire surface of the fruit. The author's objective was to develop an imaging system for bruising detection and to identify and segregate new and old bruises from the normal tissues of apples. For this, hyperspectral images of individual apples were acquired over a period of 47 days, at various intervals, to evaluate the changes in bruising over time. The bruise-free apple images were acquired before the study started and used as a reference to identify bruise evolution during storage. A bruise detection algorithm was developed. Using this algorithm, the original samples were first corrected by removing the background, and then normalization was performed to reduce the variations of reflectance caused by an illumination source in the image. A PCA [the author also proposed a minimum noise fraction (MNF)] was then applied to enhance bruising features and reduce data dimensionality. The study showed that the spectral region between 1000 nm and 1340 nm was the most appropriate for detecting apple bruising. It was also observed that bruising features changed over time and that the rate of the change varied with the variety and fruit sample.

More recently, Mehl et al. (Mehl et al., 2004) proposed using an hyperspectral imaging approach to detect apple surface defects and contamination. They used an instrument working in the range of 424–899 nm which scanned the samples line by line. Monochromatic images and second difference analysis methods were used to sort wholesome and contaminated apples. The authors worked with samples of four apple cultivars collected after harvesting and before processing. The results of direct monochromatic observation indicated that there were no single waveband images that allowed normal apples to be differentiated from all apples that are defective or contaminated. Using the second difference method, the contaminated and damaged portions of the apples were more distinguishable from the normal portions. The results showed that the bruised parts and soil contamination could be easily determined. An important output was that there were no differences in the observations of the various apple cultivars using the appropriate data treatment procedure.

12.3.3 Detecting Contamination and Defects in Cereals

Hyperspectral imaging techniques also facilitate the development of methods for assessing the level of contamination during cereal production, storage and processing. Cereal contamination includes adulteration by other cereal species or seeds from other crops, decayed and damaged grains (e.g. mouldy grains), animal faeces (e.g. from birds and rodents), seed contaminated with mycotoxins and insect infestations. Cereal contamination reduces the quality of the product, which results in financial loss if it is not detected at the early stage. Ridgway and Chambers (Ridgway and Chambers, 1998) demonstrated the potential of the multispectral NIR imaging technique for detecting insects inside wheat kernels (see also Ridgway *et al.*, 2001). Figure 12.3 shows the score images reconstructed with the third principal component of four

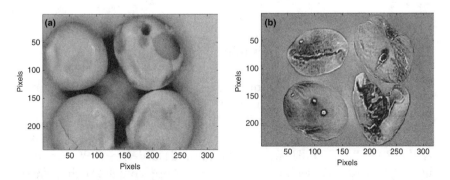

Figure 12.3 Images reconstructed from the third and the sixth principal component for four intact and infested (a) lupin seeds and (b) coffee beans, respectively

lupin seeds and the sixth principal component of four coffee beans. The image was collected using a hyperspectral imaging system active in the 900–1700 nm range. PCA allowed the insect infestation of the seed in the upper right part of the image to be enhanced. The same instrument was also used to acquire an image of wheat kernels in order to detect the presence of insects. Figure 12.4 is the reconstruction image based on the three first principal components, highlighting the contamination of the sample by an adult weevil.

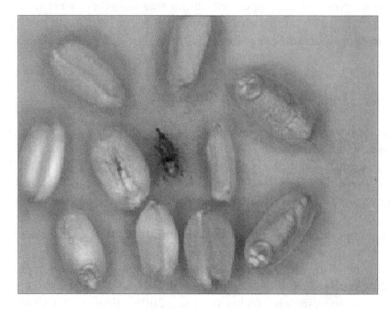

Figure 12.4 Reconstruction image based on the three first principal components of an image of a wheat kernel, including an adult weevil (see Plate 35)

12.3.4 Detecting Contamination in Compound Feed

Hyperspectral imaging techniques have been used for developing a new method to detect meat and bone meal (MBM) in compound feed. This is crucial in order to enforce legislation banning the use of animal proteins in compound feed enacted after the European 'mad cow' crisis (Gizzi *et al.*, 2003). Piraux and Dardenne (Piraux and Dardenne, 1999) proposed using FT-NIR microscopy, i.e. optic microscopy coupled with a classic Fourier-transformed spectrometry in the NIR spectral region. Baeten *et al.* (Baeten *et al.*, 2001) applied stepwise linear discriminant analysis

(SLDA), PLS and artificial neural networks (ANNs) on a large spectral library of more than 10 000 spectra from different animal and vegetal meals obtained with a NIR microscope. These studies demonstrated the relevance of the NIR spectra of particles for detecting the presence of low levels of MBM. However, the microscopic method proposed was time consuming. This led Baeten and Dardenne (Baeten and Dardenne, 2002) and Fernández Pierna et al. (Fernández Pierna et al., 2004c) to propose the use of NIR imaging spectroscopy to accelerate the process. The method proposed consisted of using a hyperspectral NIR camera that allowed the simultaneous detection of a larger number of ingredients in a single analysis, and was faster than classical and NIR microscopy because it collected thousands of spatially resolved spectra in a highly parallel fashion.

In their study, reflectance images were collected in the 900–1700 nm range, with a spectral resolution of 10 nm. The imaging spectrometer included an InGaAs FPA with 240×320 pixels (76 800 spectra per scan), along with a liquid crystal tunable filter (LCTF) for wavelength selection. The effective field of view (FOV) covered approximately $5 \, cm^2$, allowing simultaneous analyses of 300–400 particles. The spectral images obtained were background-corrected and converted to absorbance units prior to further analysis. Chemometric techniques were applied in order to discriminate according to the origin of the samples. Michotte Renier et al. (Michotte Renier et al., 2004) applied PLS and ANNs as discrimination methods. They demonstrated that the combination of a NIR camera and chemometrics could be a powerful tool for detecting animal particles in feed. They also showed that the detection limit can reach a value of 0.1 %, depending on the number of particles analysed. Fernández Pierna et al. (Fernández Pierna et al., 2004c) compared the performance of different chemometric techniques for discrimination applied to data obtained with the NIR camera. In this work, the performance of a new method for multivariate classification (SVMs) was compared with classical chemometric methods such as PLS and ANNs for classifying feed particles as either MBM or vegetal using the spectra from NIR images. They showed that while all the classification algorithms tested performed well, the ANN and SVM models showed superiority over PLS. In their paper, SVM is preferred to ANN because of the reduced amount of data that contribute to the solution, the much lower rate of false-positive detection and the fact that SVM leads to one global solution. Figure 12.5 presents the results of the best SVM model for detecting MBM in a sample that has been artificially spiked.

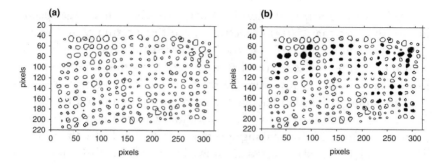

Figure 12.5 (a) Image of a sample artificially spiked with MBM and vegetal particles. (b) In grey, MBM particles detected in (a) using an SVM model

12.4 OTHER AGRONOMIC AND BIOLOGICAL APPLICATIONS

The application of hyperspectral imaging techniques to microscopic and macroscopic studies of biological and agricultural materials have been outlined and discussed. The story of the use of imaging systems started with a demonstration of the potential of these technologies for satellite and aircraft remote-sensing. Satellite-based images generally cover large areas over fixed-time intervals and with a relative low spatial resolution, while aircraft-based images cover smaller areas, are more flexible in acquiring the images and have a high spatial resolution. Various instruments have been developed for observations of Earth. These include the well-known Landsat 1, SPOT (Satellite pour l'observation de la terre) and AVHRR (Advanced very high resolution radiometer) satellites; and the AIS (Airborne imaging system), AVIRIS (Airborne visible and infrared imaging spectrometer) and CASI (Compact airborne spectrographic imager) aircraft equipment (Curran, 1989; Yao *et al.*, 2001)

The different equipments used for satellite and aircraft remote-sensing are active in the visible and infrared range and can be classified according to their use of multispectral or hyperspectral approach. The multispectral instruments (e.g. Landsat, SPOT) allow the spectral information of a limited number of bands to be collected and are generally dedicated instruments. In contrast, the hyperspectral instruments (e.g. AVIRIS, CASIS) allow the spectral information from narrow bands from a wide spectral range to be collected. These instruments seem to be able to identify most biological materials (Vane and Goetz, 1993; Jacobsen, 2000). It is generally accepted that the reflectance data extracted from hyperspectral instruments are more sensitive and have extended the application of the spectroscopic remote-sensing approach. Remote-sensing has proved to be a very valuable tool for obtaining essential data for a wide range of problems. One of the most well-known applications of remote-sensing is its use in environmental and ecological research, such as its use for the remote observation of ecosystems, the management of natural resources and the detection of pollution (Santos, 2000; Curran, 2001; Kerr and Ostrovsky, 2003).

Other applications concern the study of vegetation indices, the monitoring of crop and its use in precision agriculture. Vegetation indices are calculated using spectroscopic remote-sensing imaging data in order to study crop characteristics. The vegetation indices include leaf area index, wet biomass, plant height, grain yield and red-edges index that are affected by climate, soils, cultural practices, management and technological inputs. Traditionally, multispectral techniques have been applied to these problems, and combinations of reflectance at different wavelength bands have been used to minimize the effect of the differences in soil background and atmospheric conditions. Recent developments have shown that hyperspectral imaging data could improve the calculation of the vegetation indices (Wessman *et al.*, 1997; Clevers, 1999; Thenkabail *et al.*, 2000; Clevers and Jongschaap, 2001).

The airborne imaging system (AIS) has also demonstrated the use of hyperspectral imaging data for characterizing grassland canopy, which is an important component of the agricultural landscape. Buffet and Oger (Buffet and Oger, 2003a,b) tried to demonstrate this performance based on spectral remote-sensing and field observations of representative meadows in south-east Belgium. An interesting aspect of their study was that it showed the potential of combining information from sensors in different regions of the electromagnetic spectrum. Two sensors were used to acquire the hyperspectral data used in their study: a CASI sensor working in the 400–950 nm region, with a ground resolution of 2.5×2.5 m and a spectral resolution of 6 nm; and a SASI sensor working in the 850–2500 nm region, with a ground resolution of 2×2 m and a spectral resolution of 10 nm. They also showed the different relationships existing between physico-chemical parameters and hyperspectral data in order to assess grassland canopy and the possibility of discriminating between different types of meadows.

Precision agriculture is an important area for satellite and aircraft remote-sensing. Spectral information collected using remote-sensing systems provides information about the use of water, pesticides or fertilizers to meet crop needs. Seelan *et al.* (Seelan *et al.*, 2003) showed that these technologies used with a geographical information system (GIS) and global positioning system (GPS) may provide tools that will enable farmers to maximize the economic and environmental benefits of precision agriculture The spectral information obtained provides most of the information needed on plant stress and health. This information is combined with the spatial resolution, which is valuable for monitoring crop appearance. An important application is the possibility of the early detection of crop infestations. In this regard, Hamid and Larsolle (Hamid and Larsolle, 2003) proposed employing agricultural remote-sensing systems using imaging spectroscopy for the characterization and determination of fungal diseases in wheat. Their aim was to discriminate between healthy and diseased areas in a spring wheat crop suffering from fungal infestation. The plant-cover damage level in the affected areas was also studied in real-time. They used hyperspectral crop reflectance data consisting of 164 bands in the 360–900 nm region.

Another application in this area is the work by Yang *et al.* (Yang *et al.*, 2001, 2004). They have been using airborne hyperspectral imagery to estimate grain sorghum yield variability and also to map variable growing conditions and yields of cotton and corn.

12.5 CONCLUSION

We have shown in this chapter that hyperspectral imaging techniques provide an attractive solution for the microscopic and macroscopic analysis of biological, agricultural and environmental materials. The various applications outlined show the benefits of these techniques for sample characterization and chemical species distribution, for fruit and kernels and for food and feed mixtures. We also reviewed the advantages of NIR imaging methods for detecting contamination and defects in agro-food products such as meat, fruit, cereals and compound feed.

In the final part of the chapter we looked at the most recent findings from studies on hyperspectral imaging techniques, mainly those based on remote-sensing. These have proved to be a very valuable tool for a wide range of applications in environmental and ecological research. The use of hyperspectral imaging in the biological and agricultural sciences has increased in recent years because it provides an attractive analytical solution to meeting the justified consumer demand for quality and safe.

REFERENCES

Baeten, V. and Dardenne, P. (2001) The contribution of near infrared spectroscopy to the fight against the mad cow epidemic, *NIRS news*, **12**, 12–13.

Baeten, V. and Dardenne, P. (2002). Spectroscopy: developments in instrumentation and analysis, *Grasas y Aceites*, **53**, 45–63.

Baeten, V., Michotte Renier, A., Sinnaeve, G. and Dardenne, P. (2001) Analyses of feedingstuffs by near-infrared microscopy (NIRM): detection and quantification of meat and bone meal (MBM), *Proceedings of the Sixth Food Authenticity and Safety International Symposium (FACIS)*, 28–30 November, FACIS Organization Committee, Nantes, pp. 1–11.

Baeten, V., Michotte Renier, A., Fernández Pierna, J. A., Fissiaux, I., Fumière, O., Berben, G. and Dardenne, P. (2004) Review of the possibilities offered by the near infrared microscope (NIRM) and near infrared camera (NIR Camera) for the detection of MBM, *Proceedings of the International Symposium Food and Feed Safety in the Context of Prion Diseases*, 16–18 June, CRA-W, Namur, p. 26.

Bertrand, D. and Dufour, E. (2000) *La spectroscopie infrarouge et ses applications analytiques*, Editions Tec & Doc, Paris.

Bertrand, D., Robert, P., Novales, B. and Devaux, M. (1996) Chemometrics of multichannel imaging, in *Near Infrared Spectroscopy: the Future Waves* (Eds, A. M. C. Davies and P. Williams), NIR publications, Chichester, pp. 72–90.

Budevska, B. O. (2002) *Vibrational Spectroscopy Imaging of Agricultural Products*, John Wiley & Sons, Ltd, New York.

Buffet, D. and Oger, R. (2003a) Hyperspectral imagery for environmental mapping and monitoring: case study of grassland in Belgium, Geographical Information Systems and Remote Sensing: Environmental Applications, *Proceedings of the International Symposium*, 7–9 November, Volos, Greece.

Buffet, D. and Oger, R. (2003b) Characterisation of grassland canopy using CASI-SASI hyperspectral imagery, Presented at the *CASI-SWIR2002 Workshop*, Bruges, 4 September.

Chao, K., Mehl, P. M. and Chen, Y. R. (2002) Use of hyper- and multi-spectral imaging for detection of chicken skin tumours, *Applied Engineering in Agriculture*, **18** (1), 113–119.

Chen, Y. R. and Kim, M. S. (2004) Visible/NIR imaging spectroscopy for assessing quality and safety of agrofoods, *Proceedings of the 11th International Conference, on Near-infrared Spectroscopy* (Eds A. M. C. Davies and A. Garrido), NIR publications, Chichester, pp. 67–75.

Clevers, J. G. P. W. (1999) The use of imaging spectrometry for agricultural applications, *ISPRS Journal of Photogrammetry and Remote Sensing*, **54**, 299–304.

Clevers, J. P. G. W. and Jongschaap, R. (2001) Imaging spectrometry for agricultural applications, in *Imaging Spectrometry* (Eds F. D. van der Meer and S. M. de Jong), Kluwer Academic Publishers, pp. 157–199.

Codgill, R. P., Hurburgh, Jr, C. R., Jensen, T. C. and Jones, R. W. (2002) Single-kernel maize analysis by near-infrared hyperspectral imaging, *Proceedings of the 10th International Conference on Near-infrared Spectroscopy* (Eds A. M. C. Davies and R. K. Cho), NIR publications, Chichester, pp. 243–247.

Codgill, R. P., Hurburgh Jr, C. R. and Rippke, G. R. (2004) Single kernel maize analysis by near-infrared hyperspectral imaging, *Transactions of the ASABE*, **47**, 311–320.

Curran, P. J. (1989) Remote sensing of foliar chemistry, *Remote Sensing of Environment*, **29**, 271–278.

Curran, P. J. (2001) Imaging spectrometry for ecological applications, *International Journal of Applied Earth Observation and Geoinformation*, **3**, 305–312.

Fernández Pierna, J. A., Baeten, V., Michotte Renier, A. and Dardenne, P. (2004a) NIR camera and chemometrics: the winner combination for the detection of MBM, in *Book of Abstracts of the International Symposium on Food and Feed Safety in the Context of Prion Diseases*, 6–18 June, CRA-W, Namur, pp. 119–120.

Fernández Pierna, J. A., Didelez, G., Baeten, V., Michotte Renier, A. and Dardenne, P. (2004b) Chemometrics for multispectral and hyperspectral data: application to screening of compound feeds, *Proceedings of the International Conference of Chemometrics in Analytical Chemistry (CAC-2004)*, 20–23 September (Ed. J. Sampaio Cabral), Technical University of Lisbon, Lisbon, p. 176.

Fernández Pierna, J. A., Baeten, V., Michotte Renier, A., Cogdill, R. P. and Dardenne, P. (2004c) Combination of Support Vector Machines (SVM) and near infrared (NIR) imaging spectroscopy for the detection of meat and bone meat (MBM) in compound feeds, *Journal of Chemometrics*, **18**, 341–349.

Geladi, P., Burger, J. and Lestander, T. (2004). Hyperspectral imaging: calibration problems and solutions, *Chemometrics and Intelligent Laboratory Systems*, **72**, 209–217.

Gizzi, G., van Raamsdonck, L. W. D., Baeten, V., Murray, I., Berben, G., Brambilla, G. and von Holst, C. (2003) An overview of tests for animal tissues in animal feeds used in the public health response against BSE, *Scientific and Technical Review*, **22**, 311–331.

Hamid, M. H. and Larsolle, A. (2003) Feature-vector based analysis of hyperspectral crop reflectance data for discrimination quantification of fungal disease severity in wheat, *Biosystems Engineering*, **86**, 125–134.

Jacobsen, A. (2000) Analysing airborne optical remote sensing data from a hyperspectral scanner and implications for environmental mapping and monitoring – results from a study of *casi* data and Danish semi-natural, dry grasslands, *PhD thesis*, National Environmental Research Institute, Denmark.

Kerr, J. T. and Ostrovsky, M. (2003) From space to species: ecological applications for remote sensing,*Trends in Ecology and Evolution*, **18**, 299–305.

Kim, M. S., Chen, Y. R. and Mehl, P. M. (2001) Hyperspectral reflectance and fluorescence imaging system for food quality and safety, *Transactions of the ASABE*, **44**, 721–729.

Kim, M. S., Lefcourt, A. M., Chao, K., Chen, Y. R., Kim, I. and Chan, D. E. (2002) Multispectral detection of fecal contamination on apples based on hyperspectral imagery: Part I. Application of visible and near-infrared reflectance imaging, *Transactions of the ASABE*, **45**, 2027–2037.

Koehler, I. V. F. W., Lee, E., Kidder, L. H. and Lewis, E. N. (2002). Near infrared spectroscopy: the practical chemical imaging solution, *Spectroscopy Europe*, **14**, 12–19.

Lawrence, K. C., Windham, W. R., Park, B. and Buhr, R. J. (2003) A hyperspectral imaging system for identification of faecal and ingesta contamination on poultry carcasses, *Journal of Near Infrared Spectroscopy*, **11**, 269–281.

Lu, R. (2003) Detection of bruises on apples using near-infrared hyperspectral imaging, *Transactions of the ASABE*, **46**, 523–530.

Lu, R. (2004) Multispectral imaging for predicting firmness and soluble solids content of apple fruit, *Post Harvest Biology and Technology*, **31**, 147–157.

Lu, R. and Peng, Y. (2004) Hyperspectral scattering for assessing peach fruit firmness. *Proceedings of the ASAE/CSAE Annual International Meeting*, 1–4 August, Ottawa, Ontario, American Society of Agricultural and Biological Engineers, St Joseph, MI, p. 105.

Marcott, C., Reeder, R. C., Sweat, J. A., Panzer, D. D. and Wetzel, D. L. (1999) FT-IR spectroscopic imaging microscopy of wheat kernels using a mercury-cadmium-telluride focal-plane array detector, *Vibrational Spectroscopy*, 19, 123–129.

Martinsen, P. and Schaare, P. (1998) Measuring soluble solids distribution in kiwifruit using near-infrared imaging spectroscopy, *Postharvest Biology and Technology*, 14, 271–281.

Martinsen, P., Schaare, P. and Andrews, M. (1999) A versatile near infrared imaging spectrometer, *Journal of Near Infrared Spectroscopy*, 7, 17–25.

McClure, W. F. (2003) 204 years of near infrared technology: 1800–2003, *Journal of Near Infrared Spectroscopy*, 11, 487–518.

Mehl, P. M., Chao, K., Kim, M. and Chen, Y. R. (2002) Detection of defects on selected apple cultivars using hyperspectral and multispectral image analysis, *Applied Engineering in Agriculture*, 18, 219–226.

Mehl, P. M., Chen, Y. R., Kim, M. S. and Chan, D. E. (2004) Development of hyperspectral imaging technique for the detection of apple surface defects and contaminations, *Journal of Food Engineering*, 61, 67–81.

Michotte Renier, A., Baeten, V., Sinnaeve, G., Fernández Pierna, A. and Dardenne, P. (2004) The NIR camera: a new perspective for meat and bone meal detection in feedingstuffs, *Proceedings of the 11th International Conference on Near-infrared Spectroscopy* (Eds A. M. C. Davies and A. Garrido), NIR publications, Chichester, pp. 1061–1065.

Okumura, K., Sato, H., Muraki, H., Hiramatsu, M. and Shimomoto, Y. (1992). Imaging by using near infrared absorption, *Proceedings on the 7th Symposium on the Imaging and Sensing Technology in Industrial Fields*, 7–8 July, Tokyo, Japan.

Park, B., Windham, W. R., Lawrence, K. C., Buhr, R. J. and Smith, D. P. (2002a) Imaging spectrometry for detecting faeces and ingesta contamination on poultry carcasses, *Proceedings of the 10th International Conference on Near-Infrared Spectroscopy* (Eds A. M. C. Davies and R. K. Cho), NIR publications, Chichester, pp. 457–461.

Park, B., Lawrence, K. C., Windham, W. R., Chen, Y. R. and Chao, K. (2002b) Discriminant analysis of dual-wavelength spectral images for classifying poultry carcasses, *Computers and Electronics in Agriculture*, 33, 219–231.

Peirs, A., Scheerlinck, N., De Baerdemaeker, J. and Nicolaï, B. M. (2003) Starch index determination of apple fruit by means of a hyperspectral near infrared reflectance imaging system, *Journal of Near Infrared Spectroscopy*, 11, 379–389.

Peng, Y. and Lu, R. (2004). Predicting apple fruit firmness by multispectral scattering profiles, *Proceedings of the ASAE/CSAE Annual International Meeting*, 1–4 August, Ottawa, Ontario, American Society of Agricultural and Biological Engineers, St Joseph, MI, pp. 1–9.

Piraux, F. and Dardenne, P. (1999) Feed authentication by near-infrared microscopy, *Proceedings of the 9th International Conference on Near-infrared Spectroscopy* (Eds R. Giangiacomo and A. C. M. Davies), NIR publications, Chichester, pp. 535–541.

Polder, G., van der Heijden, G. W. A. M. and Young, I. T. (2000), Hyperspectral image analysis for measuring ripeness of tomatoes, *Presented at the 2000 ASAE International Meeting*, Paper number 003089, 9–12 July, Milwaukee, WI, American Society of

Agricultural and Biological Engineers, St Joseph, MI.

Polder, G., van der Heijden, G. W. A. M. and Young, I. T. (2002), Spectral image analysis for measuring ripeness of tomatoes, *Transactions of the ASABE*, 45, 1155–1161.

Polder, G., van der Heijden, G. W. A. M. and Young, I. T. (2003) Tomato sorting using independent component analysis on spectral images, *Real-Time Imaging*, 9, 253–259.

Ridgway, C. and Chambers, J. (1998) Detection of insects inside wheat kernels by NIR imaging, *Journal of Near Infrared Spectroscopy*, 6, 115–119.

Ridgway, C., Davies, R. and Chambers, J. (2001) Imaging for the high-speed detection of pest insects and other contaminants in cereal grain in transit, *Presented at the 2001 ASAE International Meeting*, Paper number 013056, 30 July – 1 August, Sacramento, CA. American Society of Agricultural and Biological Engineers, St Joseph, MI.

Robert, P., Bertrand, D., Devaux, M. F. and Sire, A. (1991). Identification of chemical constituents by multivariate near infrared spectral imaging, *Analytical Chemistry*, 64, 664–667.

Santos, A. M. P. (2000) Fisheries oceanography using satellite and airborne remote sensing methods: a review, *Fisheries Research*, 49, 1–20.

Seelan, S. K., Laguette, S., Casady, G. M. and Seielstad, G. A. (2003) Remote sensing applications for precision agriculture: a learning community approach, *Remote Sensing of Environment*, 88, 157–169.

Stevermer, S. W., Steward, B. L., Codgill, R. P. and Hurburgh, C. R. (2003) Automated sorting and single kernel analysis by near-infrared hyperspectral imaging, *Presented at the 2003 ASAE International Meeting*, Paper number 036159, 27–30 July, Las Vegas, NV, American Society of Agricultural and Biological Engineers, St Joseph, MI.

Sugiyama, J. (1999) Visualization of sugar content in the flesh of a melon by near infrared imaging, *Journal of Agricultural and Food Chemistry*, 47, 2715–2718.

Taylor, S. K. and McClure, W. F. (1989) NIR imaging spectroscopy: measuring the distribution of chemical components, *Proceeding of the 2nd International NIRS Conference* (Eds M. Iwamoto and S. Kawano), Korun, Tokyo, pp. 393–404.

Thenkabail, P. S., Smith, R. B. and De Pauw, E. (2000) Hyperspectral vegetation indices for determining agricultural crop characteristics, *Remote Sensing of Environment*, 71, 158–182.

Tran, C. D. and Grishko, V. I. (2004) Determination of water contents in leaves by a near-infrared multispectral imaging technique, *Microchemical Journal*, 76, 91–94.

Tsuta, M., Sugiyama, J. and Sagara, Y. (2002) Near-infrared imaging spectroscopy based on sugarabsorption band for melons, *Journal of Agricultural and Food Chemistry*, 50 (1), 48–52.

Vane, G. and Goetz, A. F. H. (1993) Terrestrial imaging spectrometry: current status, future trends, *Remote Sensing of Environment*, 44, 117–126.

Wessman, C. A., Bateson, C. A. and Benning, T. L. (1997) Detecting fire and grazing patterns in a tallgrass prairie using spectral mixture analysis, *Ecological Applications*, 7, 493–511.

Wetzel, D. L., Eilert, A. J., Pietrzak, L. N., Miller, S. S. and Sweat, J. A. (1998) Ultraspatially-resolved synchrotron infrared microspectroscopy of plant tissue in situ, *Cellular and Molecular Biology*, 44, 145.

Yang, C., Bradford, J. M. and Wiegand, C. L. (2001) Airbone multispectral imagery for mapping variable growing conditions and yields of cotton, grain sorghum, and corn, *Transactions of the ASABE*, 44, 1983–1994.

Yang, C., Everitt, J. H. and Bradford, J. M. (2004) Airbone hyperspectral imagery and yield monitor data for estimating grain sorghum yield variability, *Transactions of the ASABE*, **47**, 915–924.

Yao, H., Tian, L. and Noguchi, N. (2001) Hyperspectral imaging system optimization and image processing, *Presented at the 2001 ASAE International Meeting*, Paper number 01–1105, 30 July – 1 August, Sacramento, CA. American Society of Agricultural and Biological Engineers, St Joseph, MI.

13

Application of Multivariate Image Analysis in Nuclear Medicine: Principal Component Analysis (PCA) on Dynamic Human Brain Studies with Positron Emission Tomography (PET) for Discrimination of Areas of Disease at High Noise Levels

Pasha Razifar and Mats Bergström

13.1 INTRODUCTION

Positron emission tomography (PET) studies on human brain are usually performed dynamically where the brain is sequentially imaged at different time points after administration of a radiolabelled tracer.

Techniques and Applications of Hyperspectral Image Analysis Edited by H. F. Grahn and P. Geladi
© 2007 John Wiley & Sons, Ltd

In other words in these types of acquisition, datasets are generated in the form of images obtained from the same sector where each structure has a variable signal in-between images. This type of acquisition is used for providing quantitative data on physiological, biochemical and functional aspects of the human brain by analysing the kinetic behaviour of administered tracers in brain tissue in comparison with that in plasma. Individual images illustrate part of the kinetic behaviour of administered tracer in different structures in the brain within the time sequences. However, these images are inherently noisy due to limited amount of administered radioactivity, a usually short half-life of the labelled radionuclide, technical limitations, applied corrections and the reconstruction algorithm used.

Several factors make analysis of PET images difficult, such as a high noise magnitude and correlation between image elements in conjunction with a high level of nonspecific binding to the target and a sometimes small difference in target expression between pathological and healthy regions. This implies that the individual images are not optimal for the analysis and visualization of anatomy and pathology. Instead different types of averages or weighted average images are generated via kinetic models in order to accentuate the desired property in the images.

Principal component analysis (PCA) is one of the most commonly used multivariate image analysis tools for analysing dynamic PET data. PCA is mostly used in order to find variance-covariance structures of the input data aiming for reduction of dimensionality of these multivariate images. Since PCA is a data-driven technique, it has difficulty in discriminating between variance due to noise contra variance due to signal representing variation in kinetic behaviour between the regions, when input data are not accurately pre-normalized. In PET sequences with high and variable noise in-between images, PCA may be guided by the undesired variation and may emphasize noise instead of regions with different kinetic behaviours.

This chapter focuses on new approaches for application of PCA on dynamic PET images on human brain to overcome these problems. In these methodologies, PCA is used as a multivariate analysis technique that without modelling assumptions is able to generate images with improved quality, contrast, precision, visualization and discrimination between pathological and healthy regions, leading to optimized signal to noise ratio (SNR) and improvement of clinical interpretation and diagnosis.

13.2   PET

13.2.1   History

PET is a noninvasive tomographic and imaging technique based on the principle of annihilation coincidence detection (ACD) (Phelps *et al.*, 1975). The potential of this type of detection was recognized in the early 1950s (Wrenn Jr *et al.*, 1951; Brownell and Sweet, 1953). Its outstanding potential as a noninvasive tomographic tool used for generating images, which reveal functional or biochemical expression, was recognized on introducing this modality as positron emission transaxial tomography (PETT) by Ter-Pogossian and his colleagues in the mid 1970s (Ter-Pogossian *et al.*, 1975).

13.2.2   Principles

PET is based on detection of simultaneous emission of two back to back directed photons (photons in opposite directions) by detectors positioned on opposite sides of the detector rings. Administered radionuclide (either by injection or inhalation) decays by positron emission implying that a proton, ρ^+, from the unstable nucleus is transformed into a neutron, n, a positron, β^+, and a neutrino, v, with the release of energy:

$$\rho^+ \longrightarrow n + v + \beta^+ + \text{energy} \tag{13.1}$$

The emitted positron travels through surrounding tissues in a range up to a few millimetres depending on its energy up to 1.7 MeV (Lundqvist *et al.*, 1998), and material of the surrounding matter, until it has lost its energy in collisions with electrons of other atoms. After loss of the energy it is annihilated by an electron, yielding two photons with energy of 511 keV each (called annihilation photons) that leave the annihilation site in an antiparallel direction (Figure 13.1).

When annihilation photons are detected within a predefined coincidence timing window of ~10–12 ns, by a pair of detectors positioned along a line called the line of response (LOR), a 'true' coincidence is registered (Ollinger and Fessler, 1997) (Figure 13.2). Interaction between the photon and the crystal generates light flashes, which are converted to electric pulses that are recorded by the electronic and recording devices.

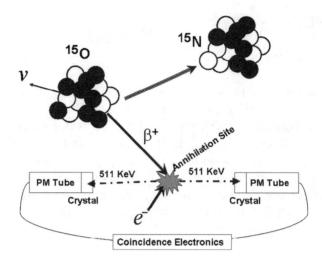

Figure 13.1 Schematic view of a coincidence channel. ^{15}O decays to a stable nuclide ^{15}N by emitting a positron with a range of 2 mm or less and a neutrino

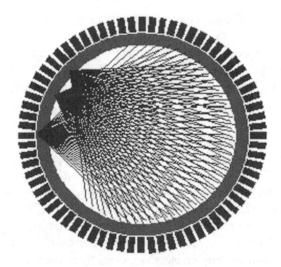

Figure 13.2 Detection procedure of the true coincidences by a pair of detectors positioned along the LOR

All LORs, which are parallel at a definite angle through the object, form a single projection view stored as a single row in a so-called sinogram. Projections obtained from all parallel LORs from different angles through the object are stored similarly but in different rows in the sinogram (sinogram domain). Generated sinograms are referred to as raw data in PET, and are used for further processing and reconstruction.

Different types of corrections need to be performed on the acquired PET data prior to the reconstruction procedure, including correction for differential detector efficiencies (normalization), random coincidence correction, dead-time correction, scatter correction and attenuation correction (Razifar, 2005). Acquired PET data after applying different types of corrections are reconstructed either analytically by filtered back projection (FBP) (Cho et al., 1974; Brooks and Di Chiro, 1975; Barrett and Swindell, 1977) or iteratively by ordered subset expectation maximization (OSEM) (Hudson and Larkin, 1994) (image domain). Independent of used reconstruction methodologies, images contain effects and errors caused by the reconstruction algorithm including different corrections in the image domain compared with the sinogram domain where statistical properties of the measurements are more well-defined.

Today, PET is established as a sophisticated technique based on tracing of molecules labelled with positron-emitting radionuclides to image metabolism, physiology and functionality in the body. With its wide range of tracer molecules, PET is an outstanding diagnostic imaging technique that has been shown to have a significant role in basic medical research as well as in clinical research and drug development (Ter-Pogossian et al., 1980).

13.2.3 Scanning Modes in PET

Conventional PET studies in humans are usually performed in either of two modes, providing different sets of data: static acquisition in which different parts of the human body or brain are imaged once during a predefined time point or frame, generating static datasets and dynamic acquisition in which the same sector is imaged repeatedly at different and specific time points, generating dynamic datasets.

Dynamic PET images can either be used directly or after kinetic modelling to extract quantitative values of a desired physiological, biochemical or pharmacological entity. However, due to limitations of the amount of the administered tracers that are usually short-lived labelled radionuclides, technical limitations, applied corrections and reconstruction algorithm used, generated dynamic images are noisy. This implies that the individual images are not optimal to be used as input dataset for the extraction of quantitative values using kinetic modelling methods and even for the analysis and visualization of anatomy and pathology. Figure 13.3 illustrates a conventional PET study of a human brain.

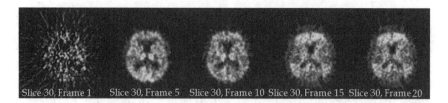

Figure 13.3 Dynamic PET images. The same image but from different frames has been illustrated, showing the uptake of the administered tracer in the human brain

13.2.4   Analysis of PET Data/Images

One of the common methods used for analysis of the dynamic PET images is summation, which generates 'sum images'. This method is based on pixel wise (PW) summation of the slices through a specific number of frames, which depend on the tracer used and its kinetic behaviour in the subject under study. Summation is performed in order to reduce noise and allowing quantitative measurements in dynamic PET images; however, this method has a tendency to reduce the contrast in the image, which causes poor discrimination between pathological and healthy regions. Summation over the whole sequence tends to emphasize the images in early frames that represent kinetic behaviour, which usually depend more on flow than binding of the administered tracer. Summation of the images through late frames of the study, tend to create images with high magnitude of the noise because of the limited number of counts due to radioactive decay. Therefore, it is essential to have a proper knowledge of the kinetic behaviour of the administered tracer in different regions in the human body or brain. This implies that this knowledge is necessary in order to sum the images through appropriate frames in the sequence, in which the specific signal is proportionally large enough to be able to generate images with high qualitative and quantitative value when using this technique.

Kinetic modelling methods, which generate 'parametric images', such as Patlak (Patlak *et al.*, 1983; Gjedde *et al.*, 1991), Logan plots (Logan, 2000), compartment modelling or extraction of components, such as in factor analysis or spectral analysis (Cunningham and Jones, 1993) and population approaches (Bertoldo *et al.*, 2004) are other methods, which are utilized to analyse dynamic PET images. These techniques are adequate but the generated parametric images usually suffer from poor quality and a nonoptimized SNR. This is because most of these techniques do not consider any SNR optimization during the extraction of physiological parameters from input images. Furthermore, there are often too many parameters and variables that need to be correctly considered to obtain proper results when applying these techniques on images generated using new tracers or when using existing tracers in different clinical applications.

Dynamic PET images can also be analysed utilizing different multivariate, statistical techniques such as PCA, which is one of the most commonly used multivariate image analysis tools, and has several applications, e.g. in nuclear medical imaging such as PET (Friston *et al.*, 1993, 1996; Pedersen *et al.*, 1993, 1994, 1995; Geladi and Grahn, 1996; Thireou *et al.*, 2001, 2003; Joliffe, 2002), computed tomography (CT) (Kalukin *et al.*, 2000) and in functional magnetic resonance imaging (fMRI) (Andersen *et al.*, 1999); Friston *et al.*, 1999; Hansen *et al.*, 1999). A majority of the results from these studies indicate that PCA has difficulty separating the signal from noise when the magnitude of the noise is relatively high, thus with a low magnitude SNR, since this a data-driven technique. Some studies have also indicated that variable noise levels and nonisotropic noise correlation (Razifar *et al.*, 2005a, b) in PET images in a dynamic sequence might dramatically affect the subsequent multivariate analysis unless properly handled. A few approaches have been proposed for this purpose (Pedersen *et al.*, 1993).

13.3   PCA

13.3.1   History

PCA was originally introduced as a method based on finding 'lines and planes of closest fit to systems of points in the space' by Pearson in the early 1900s (Pearson, 1901). It is known under several names in different fields, such as Hotelling transform (Gonzalez and Woods, 1992), the discrete Karhunen–Loéve expansion in the electrical engineering field, eigenvector transform in physical science and singular value decomposition (SVD) in numerical analysis (Eriksson *et al.*, 2001).

13.3.2   Definition

PCA is a multivariate and maximum variance projection method used to study and explain the variance-covariance or correlation structure of a multivariate dataset to simplify and reduce dimensionality. It implies that PCA is capable of transforming the input dataset into a more compact dataset without any loss of the main information. This type of transformation is performed by calculating transformation vectors (Principal Components, PCs), which characterize the directions of maximum (largest) variance of the dataset in the multidimensional feature space (Joliffe, 2002; Geladi and Grahn, 1996). Each PC is orthogonal to all the others in p-dimensional space; thus, the first PC (PC1) represents the linear transformation of the original variables which contains the largest variance and approximates the data in the least squares sense (Eriksson *et al.*, 2001). The second PC (PC2) is the combination, which explains the remaining variance as much as possible and is orthogonal to the previous one, and has no correlation with the previous one and so on.

If input matrix $X' = [X_1, X_2, X_3, \ldots, X_p]$ has variance-covariance matrix S with eigenvalues $\lambda = [\lambda_1, \lambda_2, \lambda_3, \ldots, \lambda_p]$ and corresponding eigenvectors $e = [e_1, e_2, e_3, \ldots, e_p]$, where $\lambda_1 \geq \lambda_2 \geq \lambda_3 \geq \ldots \geq \lambda_p \geq 0$ and p corresponds to the number of input columns in the matrix X. If $q = p$ then the qth PC is generated by:

$$Y_q = e'_q X = e_{q1} X_1 + e_{q2} X_2 + e_{q3} X_3 + \ldots + e_{qp} X_p \tag{13.2}$$

The condition of $\mathrm{cov}(Y_q, Y_i) = 0$ where $i \neq q$ is required, means that the components are uncorrelated to each other. PCs define and explain the magnitude of variance in decreasing order. Practically if 80–90 % of the total variance in a multivariate dataset can be accounted for by the first few PCs, corresponding to the largest eigenvalues, then the remaining components can often be rejected without much loss of information (Everitt and Dunn, 2001).

Here, each element in each eigenvector is utilized as a weight-factor used for creating 'PC images'. The term PC images refers to the 'score images' or 'classification images' introduced by Esbensen and Geladi (Esbensen and Geladi, 1989). PC images are generated by projecting all observations onto PCs in p-dimensional space, to get new values along each PC or simply visualization of the PC vectors as images.

13.3.3 Pre-processing and Scaling

Prior to application of PCA, input data are usually pre-processed aiming to transform the dataset into a more suitable form for analysis. There are two types of pre-processing procedures which need to be performed prior to PCA; Scaling and Mean-centring.

There are different types of scaling procedures that can be used to rescale the input data but one of the most common techniques is unit variance. This method is performed to rescale or compensate the size of the variance of the input variables aiming to equalize between the length of the vector and its standard deviation(Sd_i). The reason is that input variables might have substantially different ranges and then PCA is guided by the variable that has the largest range. Scaling is performed by measuring the standard deviation for each of the input variables X_i and dividing each element X_{ij} of the variables by the corresponding (Sd_i) using Equation (13.3), where X_i' refers to rescaled variables:

$$X_i' = \frac{X_{ij}}{Sd_i} \qquad (13.3)$$

After rescaling, the variance of each input variable is equal to one. It implies that the variables with large variance are 'down-weighted' and those with small variance are 'up-weighted' (Eriksson *et al.*, 2001). Rescaling transforms the dataset into a new set of data in which no variable dominates over the others in the dataset. PCA is sensitive to scaling implying that by performing different types of modification or transformation of the variance of the input variable, it is possible to gain different important information about them.

Mean-centring is the second pre-processing method applied on the rescaled dataset to transform the rescaled data into a new dataset in which the mean value of each variable is set equal to zero. Mean-centring is performed by measuring the mean $\left(\overline{X_i}\right)$ for each one of the rescaled input variables X_i' and subtracting each element X_{ij}' of the variables by the corresponding mean value using Equation (13.4), where X_i'' refers to rescaled and mean centred variables.

$$X_i'' = \frac{X_{ij}'}{\overline{X_i}} \qquad (13.4)$$

After applying these two types of pre-processing methods, the pretreated dataset is now used as the input dataset for further application of PCA, which can be applied using different approaches (Razifar, 2005).

13.3.4 Noise Pre-normalization

As mentioned earlier, independent of the correction methods and reconstruction algorithm used, PET images contain errors. The impact of noise is further complicated by its variations over time and its variability within the image, both with respect to magnitude and correlation of noise between the pixels. These factors affect the performance of the PCA if input images/data are not handled or pre-processed accurately (Pedersen *et al.*, 1994; Razifar *et al.*, 2005a, b). This means that if images are not pre-normalized for noise, PCA might emphasize the fluctuation of the noise instead of the important signals, which represent the kinetic behaviour and information. If PCA is performed on dynamic PET images without applied noise-normalization, the results might be almost identical (however, lower magnitude of noise because of scaling and mean-centring) to summation images over all frames but with poor discrimination between pathological and healthy regions.

Several approaches of pre-processing of input data have been proposed and implemented on dynamic PET data (Pedersen *et al.*, 1995; Sˇaʹmal *et al.*, 1999; Thireou *et al.*, 2003). These approaches are performed in the image domain to produce a new sequence of images in which the variance of the noise becomes as stable as possible in-between the frames and at the same time, to preserve the signal strength as much as possible. This allows PCA, which is sensitive to the type of scaling or transformation method used, to detect fluctuations of the signals through the time sequence and not be dominated by the noise variation.

13.4 APPLICATION OF PCA IN PET

PCA is used as an image analysis tool for transforming dynamic PET data into a new set of data in which the most relevant information is explained in a limited number of PCs. This is performed with the aim of dimension reduction, signal extraction and recently, improvement of image quality and noise removal (Esbensen and Geladi, 1989; Friston *et al.*, 1993, 1996; Pedersen *et al.*, 1993, 1994, 1995; Geladi and Grahn, 1996; Jung *et al.*, 1998; Sˇaʹmal *et al.*, 1999; Thireou *et al.*, 2001, 2003; Joliffe, 2002).

PCA is mostly applied on dynamic PET images in the image domain where dynamic PET data have been reconstructed either analytically by FBP or iteratively by OSEM. Another approach is to apply PCA on dynamic PET data, sinograms, which represent the set of projections generated by the tomograph, sinogram domain, to reduce computational cost and pre-processing (Wernick *et al.*, 1999) or for quantification and temporal compression (Chen *et al.*, 2004).

The following sections describe studies performed by Razifar *et al.* (Razifar *et al.*, 2005a, b) where PCA is used as a technique that without model assumptions can generate images with a low level of noise while enhancing contrast, thereby optimizing the SNR. These studies introduce and offer descriptions of three different approaches for application of PCA on dynamic PET images/data. These approaches are named slice or sinogram wise application of PCA (SWPCA), volume wise application of PCA (VWPCA) and masked volume wise application of PCA (MVWPCA) on either dynamic PET images or sinogram data. Furthermore, advantages and limitations of these approaches are discussed.

13.4.1 SWPCA

PCA is applied either slice wise on dynamic PET images in the image domain or sinogramwise on dynamic PET data in the sinogram domain. Both approaches are based on application of PCA on input matrices where each column vector contains pixel values of either images or sinograms from the same slice or sinogram but from different frames. This procedure is performed sequentially on each one of the slices or sinograms within the scanned volume. One of the reasons for performing the slice or sinogram wise approach is technical limitations of the computer to handle large amounts of data. The number of output components is equal to the number of input variables (number of frames), but usually only the first two to three are explored.

It is essential to emphasize that the input dynamic images/data have to be correctly pre-normalized based on the noise properties in different domains. SWPCA on dynamic PET images in the image domain is performed using the following steps:

(1) *Background noise pre-normalization.* The first pre-normalization is named 'background noise pre-normalization' and is performed according to Equation (13.5) on input images:

$$X'_{ij} = \frac{X_{ij}}{sd_i} \qquad (13.5)$$

where X'_{ij} refers to a new value of the pixel j of image i , X_{ij} refers to the original value of the corresponding pixel and sd_i refers to the standard deviation of pixel values within the predefined outlined mask. Figure 13.4 shows a mask used for applying

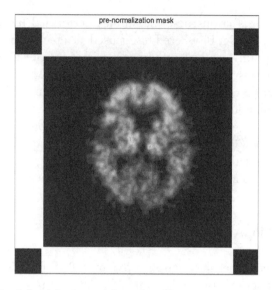

Figure 13.4 Predefined background mask, white areas, covering pixels used for determining background noise for pre-normalization

background noise pre-normalization. This mask is utilized aiming to improve the estimation of the standard deviation by covering pixels with noise from different positions in the background within the image. This is assumed to normalize for different levels of noise when the noise magnitude was the same all over the image field.

The reason for not including pixels at the corners of the images is that in PET studies on the human brain where data are reconstructed by FBP without applied zooming function, background pixels cover large portions of the images but the pixels in these corners contain zero values, which should be excluded. However, if zooming function is used then these pixels can be included for calculation of the standard deviation.

(2) *Outlining of reference region.* The next step after performing data treatment and background noise pre-normalization, is to calculate the time–activity curve (TAC) data for a reference region. From a kinetic point of view, a reference region is an area in the brain where there is no specific binding of administered tracer. However, from an image analysis point of view, a reference region is an area from which the deviation of the other structures in the brain is in focus.

Mostly, cerebellum is used as the reference region in human brain studies using different tracers. For better visualization of the cerebellum, SWPCA can be applied on images in the late or early frames depending on tracer used without any treatment or pre-normalization of the images. The result will be a single frame containing 63 planes, which have improved contrast specifically between grey and white matter. This improves the precision for the outlining of the reference region. The drawn region of interest (ROI) is imported into the corresponding slice in all frames in the treated and pre-normalized dynamic PET images and the mean value of the pixels within the ROI is calculated and an EXCEL file is created containing time–activity data of the reference region.

(3) *Contrast or kinetic pre-normalization.* This pre-normalization is based on dividing the noise pre-normalized value of each pixel j in a single image i, X'_{ij}, by the corresponding mean value $\overline{X}_{i_{ref}}$ of the pixels within a drawn ROI, masking the chosen reference region in each frame [Equation (13.6)], where $(X_{ij})_{new}$ is the new value of the pixel after applied kinetic pre-normalization:

$$(X_{ij})_{new} = \frac{X'_{ij}}{\overline{X}_{i_{ref}}} \qquad (13.6)$$

Performing the kinetic pre-normalization will reduce the values of the pixels (intensity) for regions that have similar kinetic behaviour as the reference region and enhance the contrast within the images. The pre-normalized, dynamic PET images are then used as the input dataset (matrix) for applying SWPCA. Figure 13.5 illustrates the qualitative comparison between images obtained using reference Patlak, summation and SWPCA in a human brain study, where the quality of the image and contrast and discrimination between affected and unaffected regions of the brain is improved in images obtained using SWPCA. This improves the interpretation, which leads to improvement of clinical diagnosis (Razifar, 2005).

One limitation of this approach is that PCA is performed on each slice/sinogram separately and independently from the other slices/sinograms implying separate treatment of structures in the brain that appear in different slices when the thickness of the structure is thicker than slice thickness. This might cause inconsistency in the resulting images within the volume.

(a)

(b)

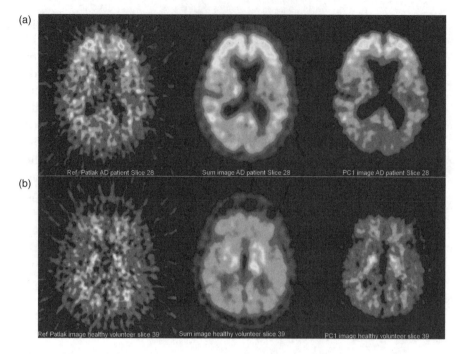

Ref. Patlak AD patient Slice 28   Sum image AD patient Slice 28   PC1 image AD patient Slice 28

Ref Patlak image healthy volunteer slice 39   Sum image healthy volunteer slice 39   PC1 image healthy volunteer slice 39

Figure 13.5 Qualitative comparison between images obtained by applying reference Patlak, summation and SWPCA on images. (a) An Alzheimer diseased (AD) patient and (b) a healthy volunteer

13.4.2 VWPCA

Another approach for application of PCA on either PET images or PET data is VWPCA. In contrast to the previous approach, PCA can be applied volume wise on dynamic PET volumes either in the image domain or in the sinogram domain. Here, the input matrix consists of variables, which are represented by each column vector that contains pixel values of the full image volume or sinogram volume but from different frames.

One of the advantages of applying VWPCA compared with MVWPCA is that PCA avoids separate treatment of structures in the brain that appear in different slices. Another advantage is that PCA is forced to find the largest variance using the dataset from the whole scanned volume. One limitation of applying this approach is that pixels representing the background of the images, which contain only noise, are included in the input dataset prior to application of VWPCA.

13.4.3   MVWPCA

Application in image domain

In this section, the third novel approach for application of PCA on either
PET images or PET data is introduced, which is called MVWPCA. In
contrast to the previous approach, PCA is applied on the whole masked
volumes, either in the image domain or in the sinogram domain. Here,
each input variable or column vector contains pixel values from either
the masked image volume or masked sinogram volume from different
frames. The masking procedure and MVWPCA includes the following
steps:

- Applying SWPCA on the PET image without any pre-
  normalization.
- Create 2-D binary images from the PC1 images using the Otsu
  method (Otsu, 1979).
- If there are holes, perform morphological operations to fill them.
- Select the main object from the background (masking out the back-
  ground pixels).
- Save the coordinates of the object pixels.
- Perform background noise pre-normalization.
- Draw ROI/ROIs representing the reference region.
- Perform kinetic pre-normalization.
- Perform PCA on the dataset representing the masked volume of the
  brain.
- Re-project the new pixel values back to empty matrices.

The reason for masking is to exclude background pixel values from
the image volume or the sinogram volume (Figure 13.6). This approach
forces the PCA to find the largest variance using the dataset from the
masked part of the whole scanned volume.

The reason for introducing this approach was that by masking the
brain (Figure 13.5), background data are excluded from input data
thereby avoiding the clusters of data with values around zero, which
are pixels containing only noise (Figure 13.7).

By applying MVWPCA, PCA avoids separate treatment of structures
in the brain that appear in different slices. In this approach PCA is
forced to find the largest variance using the dataset within the whole
volume of the brain (including all structures) but from different frames.

(a)   (b)   (c)

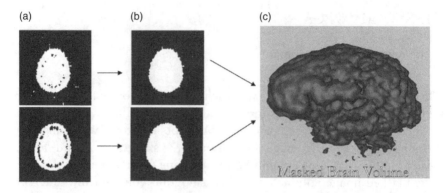

Figure 13.6 Masking procedure. (a) Results of applying iterative threshold technique on PCI images without any application of pre-normalization, using Otsu method as two, 2-D binary slices. (b) Same slices but the holes hae been filled using different morphological methods such as image closing followed by selecting the main object. (c) 3-D visualization of masked brain. Note: The extracranial tissues have been removed by applying a threshold for better visualization of the brain but the data from extracranial tissues are included in the masked input data for PCA

This improves the performance of the PCA by focusing on the main and essential part of the images.

In one of the studies by Razifar (Razifar, 2005), the performance of this method was studied by applying this approach to images obtained from PET studies on human brain of patients suffering from different neurological disorders and healthy volunteers, using several well-known tracers in clinical and research practice. Figure 13.8 illustrates the qualitative comparison between images obtained using MVWPCA, reference Patlak and summation on human brain study where the quality of the image has significantly been improved using MVWPCA compared with the other methods. This improves clinical interpretation and leads to improvement of clinical diagnosis.

Application in sinogram domain

This section describes MVWPCA on dynamic PET data in the sinogram domain. In contrast to the previous method, a new, PW noise pre-normalization method was applied on dynamic emission data to perform a more accurate transformation of the input data. The new transformed data were further used as input data in the reconstruction procedure utilizing both FBP and OSEM.

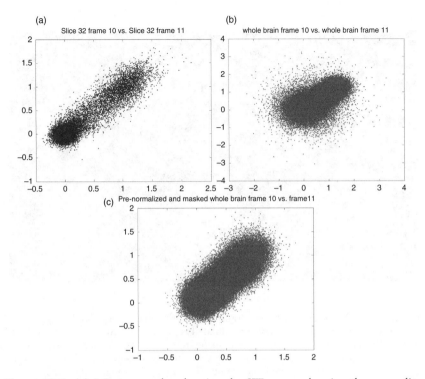

Figure 13.7 (a) 2-D scatter plot showing the SW approach using the same slice but from different frames. The large cluster of pixels with values around (0,0) represents predominantly pixels outside the brain. (b) 2-D scatter plot of data representing the whole volume of the brain without exclusion of the background data (masking). (c) Whole, masked volume of the brain with applied noise pre-normalization is plotted to show data distribution of input data compared with the other two approaches. Reprinted from Neuroimage, 33, issue 2, Razifar *et al.*, A new application of pre-normalized PCA...., 588–598, © 2006, with permission from Elsevier

Since PET is based on annihilation coincidences and photon detection, and radioactive decay obeys Poisson statistics it is assumed that the value of each of the pixels with coordinate (i, j) of the sinogram X_{ij} represents the number of detected coincidence counts for a given detector pair. The value of each pixel is assumed to have an uncertainty of $\sqrt{X_{ij}}$; therefore, the PW noise pre-normalization is applied on all detected counts (pixels) independently using the following equation:

$$X'_{ij} = \frac{X_{ij}}{\sqrt{|X_{ij}|}} \qquad (13.7)$$

(a)   (b)   (c)

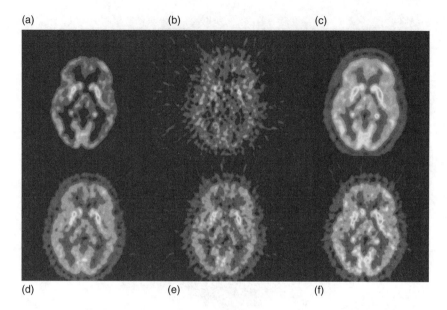

(d)   (e)   (f)

Figure 13.8 MVWPCA images compared with images obtained using other methods on dynamic PET study of human brain in a healthy volunteer. (a) PCI image, (b) same image but generated using reference Patlak and (c) sum image (sum of images from frames 3–17). (d) Corresponding sum image (from frames 15–17), (e) same single image from frame 17 and (f) same single image from frame 12. Reprinted from Neuroimage, 33, issue 2, Razifar *et al.*, A new application of pre-normalized PCA . . . ., 588–598, © 2006, with permission from Elsevier

where X'_{ij} refers to the new value of the pixel with coordinate (i, j). Figure 13.9 illustrates the principles of applying SWPCA and MVWPCA on dynamic PET data in the sinogram domain.

13.5   CONCLUSIONS

This chapter shows another feature of accurate application of the multivariate image analysis technique, such as PCA, in the medical imaging field compared with previous studies. PCA has mostly been used and applied to perform different statistical measurements or to perform quantitative analysis in combination with other studies (Barber, 1980; Rossion *et al.*, 2001; Zuendorf *et al.*, 2003; Weder *et al.*, 2005). However, here, the possibility of extracting different kinetic information from input images/data and representing it as images with high quality and improved signal extraction, in different components compared with

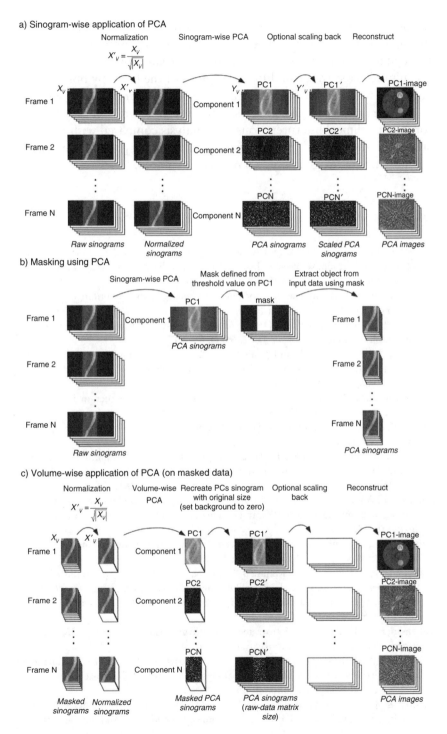

Figure 13.9 Schematic illustration of SWPCA and MVWPCA on dynamic PET data in the sinogram domain. Reproduced by permission of IEEE

the other methods is described. This shows the potential and possibilities of using PCA as a method that without kinetic assumptions can analyse the dynamic PET images/data aiming to explain kinetic behaviour of the administered tracer in different regions of the brain by observing its variation within the time sequence when it is accurately applied. This is another angle of extracting essential information that can be used for further clinical interpretations and analysis compared with the other studies, in the medical imaging field, where other aspects of application of multivariate image analysis have been explored.

REFERENCES

Andersen, A. H., Gash, D. M. and Avison, M. J. (1999) Principal component analysis of the dynamic response measured by fMRI: a generalized linear systems framework. *Magnetic Resonance Imaging* 17: 795–815.

Barber, D. C. (1980) The use of principal components in the quantitative analysis of gamma camera dynamic studies. *Physics in Medicine and Biology* 25(2): 283–292.

Barrett, H. H. and Swindell, W. (1977) Analog reconstruction methods for transaxial tomography. *Proceedings of the IEEE* 65(1): 89–107.

Bertoldo, A., Sparacino, G. and Cobelli, C. (2004) 'Population' approach improves parameter estimation of kinetic models from dynamic PET data. *IEEE Transaction on Medical Imaging* 23(3): 297–306.

Brooks, R. A. and Di Chiro, G. (1975) Theory of image reconstruction in computed tomography. *Radiology* 117: 561.

Brownell, G. L. and Sweet, W. H. (1953) Localization of brain tumors with positron emitters. *Nucleonics* 11: 40–45.

Chen, Z., Parker, B. J., Feng, D. D. and Fulton, R. (2004) Temporal processing of dynamic positron emission tomography via principal component analysis in the sinogram domain. *IEEE Transaction on Nuclear Science* 51(5): 2612–2619.

Cho, Z. H., Ahn, I., Bohm, C. and Huth, G. (1974) Computerized image reconstruction methods with multiple photon/X-ray transmission scanning. *Physics in Medicine and Biology* 19: 511–522.

Cunningham, V. J. and Jones, T. (1993) Spectral analysis of dynamic PET studies. *Journal of Cerebral Blood Flow Metabolism* 13: 15–23.

Eriksson, L., Johansson, E., Kettaneh-Wold, N. and Wold, S. (2001) *Multi- and Megavariate Data Analysis*. Umetrics Publishing Company, Umeå, pp. 43–69.

Everitt, B. S. and Dunn, G. (2001) *Applied Multivariate Data Analysis*, 2nd Ed. Arnold, New York, pp. 458–497.

Esbensen, K. and Geladi, P. (1989) Strategy of multivariate image analysis (MIA). *Chemometrics and Intelligent Laboratory Systems* 7: 67–86.

Friston, K. J., Frith, C. D., Liddle, P. F. and Frackowiak, R. S. J. (1993) Functional connectivity: the principal-component analysis of large (PET) datasets. *Journal of Cerebral Blood Flow Metabolism* 13: 5–14.

Friston, K. J., Poline, J.-B., Holmes, A. P., Frith, C. D. and Frackowiak, R. S. J. (1996) A multivariate analysis of PET activation studies. *Human Brain Mapping* 4: 140–151.

Friston, K., Phillips, J., Chawla, D. and Buchel, C. (1999) Revealing interactions among brain systems with nonlinear PCA. *Human Brain Mapping* 8: 92–97.

Geladi, P. and Grahn, H. (1996) *Multivariate Image Analysis*. John Wiley & Sons, Ltd, Chichester.

Gjedde, A., Reith, J., Dyve, S., Léger, G., Guttman, M., Diksic, M. *et al.* (1991) Dopa decarboxylase activity of the living human brain. *Proceedings of the National Academy of Sciences, Neurobiology, USA* 88: 2721–2725.

Gonzalez, R. C. and Woods, R. E. (1992) *Digital Image Processing*. Addison Wesley, New York, pp. 148–159.

Hansen, L. K., Larsen, J., Nielsen, F. A., Strother, S. C., Rostrup, E., Savoy, R. *et al.* (1999) Generalizable patterns in neuroimaging: how many principal components? *NeuroImage* 9: 534–544.

Hudson, H. M. and Larkin, R. S. (1994) Accelerated image reconstruction using Ordered Subsets of Projection Data. *IEEE Transactions on Medical Imaging* 13: 601–609.

Joliffe, I. T. (2002) *Principal Component Analysis*, 2nd Edn. Springer, New York.

Jung, T., Humphries, C., Lee, M., Iragui, V., Makeig, S. and Sejnowski, T. (1998) Removing electroencephalographic artifacts: Comparison between ICA and PCA. *IEEE International Workshop on Neural Networks for Signal Processing* 63–72.

Kalukin, A. R., Van Geet, M. and Swennen, R. (2000) Principal components analysis of multienergy X-ray computed tomography of mineral samples. *IEEE Transaction on Nuclear Science* 47: 1729–1736.

Logan, J. (2000) Graphical analysis of PET data applied to reversible and irreversible tracers. *Nuclear Medicine and Biology* 27: 661–670.

Lundqvist, H., Lubberink, M. and Tolmachev, V. (1998) Positron emission tomography. *European Journal of Physics* 19: 537–552.

Ollinger, J. M. and Fessler, J. A. (1997) Positron emission tomography. *IEEE Signal Processing Magazine* 14: 43–55.

Otsu, N. (1979) A threshold selection method from gray-level histograms. *IEEE Transaction on System, Man and Cybernetics* SMC-9(1): 62–66.

Patlak, C. S., Blasberg, R. G. and Fenstermacher, J. D. (1983) Graphical evaluation of blood-to-brain transfer constants from multiple-time uptake data. *Journal of Cerebral Blood Flow Metabolism* 3: 1–7.

Pearson, K. (1901) On lines and planes of closest fit to systems of points in space. *Philosophical Magazine* 6: 559–572.

Pedersen, F., Bergström, M., Bengtsson, E. and Maripuu, E. (1993) Principal component analysis of dynamic PET and gamma camera images: a methodology to visualize the signals in the presence of large noise. *IEEE Conference Record of Nuclear Science Symposium and Medical Imaging Conference* 3: 1734–1738.

Pedersen, F., Bergström, M., Bengtsson, E. and Långström, B. (1994) Principal component analysis of dynamic positron emission tomography images. *European Journal of Nuclear Medicine and Molecular Imaging* 21: 1285–1292.

Pedersen, F., Bengtsson, E. and Nordin, B. (1995) An extended strategy for exploratory multivariate image analysis including noise considerations. *Journal of Chemometrics* 9: 389–409.

Phelps, M. E., Hoffman, E. J., Huang, S. and Ter-Pogossian, M. M. (1975) Application of annihilation coincidence detection to transaxial reconstruction tomography. *Journal of Nuclear Medicine and Molecular Imaging* 16: 210–214.

Razifar, P. (2005) Novel Approaches for Application of Principal Component Analysis on Dynamic PET Images for Improvement of Image Quality and Clinical Diagnosis, PhD thesis, Uppsala University.

Razifar, P., Lubberink, M., Schneider, H., Långström, B., Bengtsson, E. and Bergström, M. (2005a) Non-isotropic noise correlation in PET data reconstructed by FBP but not by OSEM demonstrated using auto-correlation function. *BMC Med Imaging* 5:1.

Razifar, P., Sandström, M., Schneider, H., Långström, B., Maripuu, E., Bengtsson, E. and Bergström, M. (2005b) Noise correlation in PET, CT, SPECT and PET/CT data evaluated using autocorrelation function: a phantom study on data, reconstructed using FBP and OSEM. *BMC Med Imaging* 5: 3.

Rossion, B., Schiltz, C., Robaye, L., Pirenne, D. and Commerlinck, M. (2001) How does the brain discriminate familiar and unfamiliar faces? A PET study of face categorical perception (abstract). *Journal of Cognition Neuroscience* 13(7): 1019–1034.

S˘a′mal, M., K′arny′, M., Benali, H., Backfrieder, W., Todd-Pokropek, A. and Bergmann, H. (1999) Experimental comparison of data transformation procedures for analysis of principal components. *Physics in Medicine and Biology* 44: 2821–2834.

Ter-Pogossian, M. M., Phelps, M. E., Hoffman, E. J. and Mullani, N. A. (1975) A positron-emission transaxial tomograph for nuclear imaging (PETT). *Radiology* 114: 89–98.

Ter-Pogossian, M. M., Raichle, M. E. and Sobel, B. E. (1980) Positron Emission Tomography. *Scientific American* 243: 170–181.

Thireou, T., Strauss, L., Kontaxakis, G., Pavlopoulos, S. and Santos, A. (2001) Principal component analysis in dynamic PET. *Conference Record, CASEIB* 241–243.

Thireou, T., Strauss, L. G., Dimitrakopoulou-Strauss, A., Kontaxakis, G., Pavlopoulos, S. and Santos, A. (2003) Performance evaluation of principal component analysis in dynamic FDG-PET studies of recurrent colorectal cancer. *Computerized Medical Imaging and Graphics* 27(1): 43–51.

Weder, B. J., Shindler, K., Loher, T. J., Wiest, R., Wissmeyer, M., Ritter, P., Lovblad, K., Donati, F. and Missimer, J. (2005) Brain areas involved in medial temporal lobe seizures: a principal component analysis of ictal SPECT data. *Human Brain Mapping* (published on-line). ISSN 1097-0193. 22 September.

Wernick, M. N., Infusino, E. J. and Milosevic, M. (1999) Fast spatio-temporal image reconstruction for dynamic PET. *IEEE Transaction on Medical Imaging* 18: 185–195.

Wrenn, J.R. F R, Good, M. L. and Handler, P. (1951) The use of positron emitting radioisotopes for localization of brain tumors. *Science* 113: 525–527.

Zuendorf, G., Kerrouche, N., Herholz, K. and Baron, J.-C. (2003) Efficient Principal Analysis for multivariate 3D voxel-based mapping of brain functional imaging data sets as applied to FDG-PET and normal aging. *Human Brain Mapping* 18: 13–21.

14

Near Infrared Chemical Imaging: Beyond the Pictures

E. Neil Lewis, Janie Dubois, Linda H. Kidder and Kenneth S. Haber

14.1 INTRODUCTION

Many samples are both spatially and chemically heterogeneous, either as a design element or a structural flaw. In either case, a thorough understanding of both spatial and chemical heterogeneity is of great value in assessing sample functionality, such as product performance of an engineered sample, or to establish a diagnosis of pathology in biological tissue. Historically, manufactured products such as pharmaceuticals were analyzed for their chemical composition, whereas the emphasis for biological samples was on identifying differences in spatial morphology. The availability of near infrared chemical imaging (NIRCI) however has enabled these two types of information to be gathered simultaneously without the need for exogenous stains commonly used to enhance morphological contrast in biological tissue. The ability to rapidly acquire high quality, spatially resolved spectra, and to simultaneously derive intrinsic image contrast at high spatial resolution has significant implications in a variety of fields. These data can be analyzed quantitatively to produce objective and reproducible chemical and spatial metrics allowing easy comparisons between large numbers of samples.

Techniques and Applications of Hyperspectral Image Analysis   Edited by H. F. Grahn and P. Geladi
© 2007 John Wiley & Sons, Ltd

As the technique has become widely applied in a variety of application areas, the questions being addressed have become much more sophisticated, placing greater demands on appropriate sample presentation and the development of advanced analyses.

The purpose of this chapter is to impart an understanding of issues associated with data collection, both in terms of acquisition modalities and the selection of relevant sampling conditions, and also to present strategies that enable the extraction of a maximum amount of information from chemical images. This information can arise from both spatially bound quantitative information, where the spatial aspect of the data is critical for sample characterization, and also from spatially unmixed spectra acquired in a high throughput fashion, where the image is actually irrelevant. Examples will focus on pharmaceutical applications, but the principles clearly apply to any chemically heterogeneous system with a dimension scale compatible with optical imaging instrumentation. The approaches are equally applicable for a single sample or multiple samples in a high throughput configuration.

NIRCI microscopy is a relatively young technique that has rapidly gained popularity and acceptance for the analysis of complex samples. The technique is based on well-understood and established NIR spectroscopic theory that has been applied broadly to the analysis of homogeneous mixtures and pure components. The imaging approach, termed NIRCI, adds a spatial dimension to the chemical information, thereby dramatically broadening the scope of applications to include complex heterogeneous samples and other applications where chemical 'uniformity' is being probed (Koehler et al., 2002). While chemical heterogeneity is usually not considered (or ignored) in a traditional single-point NIR spectroscopic measurement, it is a fundamental prerequisite that drives the applications of NIRCI. It is important to understand that sample heterogeneity is a relative concept, and depending on the magnification utilized for measurement, many solid-state samples can be considered either chemically heterogeneous or homogeneous.

The effect of the magnification on the measurement can be explored with the following example: a product composed of an active pharmaceutical ingredient (API) and excipients that comprise the matrix that binds the tablet together, which gives it the bulk necessary to make it easily handled. When viewed at high magnification, it is possible to visualize micro spatial locations within a tablet that contain either predominantly API or predominantly excipient. If 20 tablets are imaged with high resolution, the pattern of chemical heterogeneity for each is 'unique', although the statistical summary is the same for all 20 tablets. At low magnification, multiple tablets may be imaged simultaneously, with the trade off being that the chemical heterogeneity is averaged over a greater spatial extent, and each tablet may now appear chemically homogeneous (Lewis *et al.*, 2005). In fact, single point NIR spectrometers (and wet chemistry techniques such as HPLC) are intentionally designed to average this information so that individual measurements of single tablets produce results that characterize their bulk chemical composition only. Chemical imaging provides the ability to tailor magnification, and therefore the size scale of chemical heterogeneity, which makes the technique suitable for questions that are simply not accessible with conventional NIR spectroscopy. However, just like its parent NIR spectroscopy, NIRCI requires little or no sample preparation, is readily commercially available, and is supported by state of the art sample presentation and data processing tools. The information accessed provides insight into both the chemical and morphological characteristics of a single sample, or to the comparative features of multiple samples analyzed simultaneously.

Processing and interpretation of NIRCI data benefit from both the spectroscopic and digital imaging origins of the technique; the hypercube may be analyzed with both the data processing and interpretation tools developed for single-point spectroscopy as well as the sophisticated image plane analysis routines developed for diagnostic and industrial imaging (Colarusso *et al.*, 1998). In addition, beyond 'traditional' data analysis of chemical images, where the chemical information is related to spatial location, methods exist that rely solely on spectral distribution statistics, regardless of spatial origin or arrangement.

In this chapter, we describe some of the data collection and processing strategies that enable NIRCI to be a valuable tool for the analysis of complex manufactured products. We also wish to emphasize aspects that impact proper interpretation of the data, thereby maximizing the amount of information available. This leads into a discussion of the concepts of spectral and spatial unmixing that enable NIRCI to provide quantitative and robust analyses that go well beyond the creation of pictures. While starting with some basic data analysis concepts and methods, this chapter will extend the discussion to describe powerful yet simple approaches. These analyses emphasize quantitative descriptions of chemical images that are not currently commonly applied. Specifically, we will discuss objective methods that do not rely upon subjective interpretation of the image, but that directly translate quantitative and statistical results into robust and reproducible analytical measurements. Such metrics are fundamental to the deployment of the technology in process, quality assurance (QA) and quality control (QC) environments.

14.2   DATA MEASUREMENT

The most common NIR spectral mapping/imaging modalities are: (1) single or multi-point (linear) mapping based on Fourier transform (FT) spectrometers; and (2) area imaging using tunable filter devices. FT instruments have higher spectral resolution than tunable filter instruments although high spectral resolution is generally unnecessary in the NIR spectral region because of the relatively broad intrinsic line widths of the absorption bands that occur in this spectral region. One of the advantages of the tunable filter approach over FT based instruments is that a tunable filter can acquire images at a user definable and computer-controlled subset of the available wavelengths, in effect 'tuning' only to spectral regions that contain analytically useful absorption bands (Kidder *et al.*, 2002). This arrangement results in very fast data collection speeds, and compact, feature rich datasets relative to the FT approach.

Single point FT mapping instruments suffer from a multiplex disadvantage compared with the multi-point FT mapping instruments. One advantage of the mapping FT schemes compared with area imaging is that only a single pixel or small line of pixels is directly illuminated. This has the potential of reducing inter-pixel mixing effects at the highest magnifications, although depth of penetration effects can still degrade the effective resolution of both area imaging and mapping methods. In general a tunable filter based instrument compared with a mapping implementation will provide a much greater degree of flexibility in the size scales and shapes of objects that can be measured. This allows a variety of large, irregularly shaped objects to be easily chemically imaged.

Ultimately, the selection of a particular experimental implementation will likely be dictated by the nature of specific application, or range of applications, to which the instrument is to be applied. In general an area imaging instrument will likely provide a greater degree of flexibility and fewer constraints on sample type, sample size and shape than a mapping instrument. This in turn results in the ability to measure samples with little or no preparation. It also should be noted that this flexibility allows an area imaging instrument to be quickly adapted to measure the spatial and chemical complexity of a sample over size scales that encompass statistically representative areas. There may be a tendency in

CI microscopy to believe that higher magnification and higher spatial resolution will result in superior data. However, it is much more critical to be able to adapt and match the spatial resolution and sampling area of the measurement to the problem at hand. This capability provides a significant advantage when a multi-purpose research tool is desirable.

Tailoring the experiment to the scale of the spatial and spectral complexity of the sample is of paramount importance in a CI experiment. On the one hand, the spatial resolution should be sufficient to resolve the spatial heterogeneity of the samples, and on the other hand, the extent of the image should be large enough to produce robust sample statistics. For instance, to characterize the distribution uniformity of an API within a 10 mm diameter tablet, imaging a 1×1 mm area with a magnification of 5 μm per pixel might yield a statistically meaningless result with respect to the distribution of components within the entire tablet. This area would contain 40 000 spectra but still only measure 1.25 % of the sample surface. Depending on the heterogeneity of the API distribution, this may not provide a robust statistical sample of the tablet. An area scan system has the ability to 'compromise' in terms of magnification, and can easily scan larger areas with lower magnification in the same amount of time. In the example above, the entire area of the tablet could be analyzed at once with a magnification of 40 μm per pixel. Ultimately, it must be determined if high resolution (5 μm per pixel) is needed to characterize the distribution of the API, or whether the sample heterogeneity can be characterized with 40 μm per pixel magnification. If both high resolution and large sample extent are necessary to address a specific problem, then even an area scan instrument must be raster-scanned, with the resulting series of images quilted together to form the larger image. Clearly an instrument with the ability to trade magnification and sampling area is an important attribute, even for samples as small as a single pharmaceutical tablet.

The third spatial dimension of the sampling volume should not be ignored in designing and interpreting NIR imaging experiments (Clarke et al., 2002). Depth of penetration is an important consideration in NIR imaging and working at high lateral magnification will not necessarily achieve the anticipated result. Absorption lengths may be hundreds to thousands of times the scattering lengths over portions of the NIR spectrum, and light may penetrate more than 100 μm into a sample. As a result, when imaging at higher magnifications it is often desirable to restrict the wavelength range of the dataset to the combination band region of the spectrum (1950 – 2500 nm) where the absorption lengths are short enough to reduce inter-pixel mixing effects.

Finally, optimizing the spectral resolution for the problem to be addressed can also be important for minimizing acquisition time. In the case of area imaging, the scanned wavelength range can be adapted to address the chemical differentiation desired. Collection of images at wavelengths that contain no analytically useful data wastes time and can degrade the performance of a spectral classifier designed to convert the image to a chemical map.

14.3   SELECTION OF SAMPLES AND ACQUISITION SCHEMES

The concept of spatial heterogeneity is often interpreted with a very limited scope, and it is an artificial limitation to restrict the definition of heterogeneity to the chemical composition of one physically independent sample. Indeed, heterogeneity can also arise from the differences seen in a number of individually homogeneous samples analyzed at once (Figure 14.1). Taking full advantage of the 2-D detector, the area imaging approach allows multiple entities to be imaged simultaneously, i.e. in a single field of view, and greatly accelerates the measurement process, and positions the technology as a 'high throughput' spectroscopic tool (Lewis, 2002).

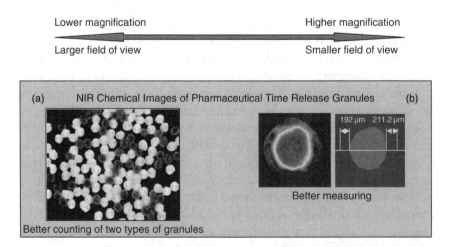

Figure 14.1   The scales of the heterogeneity can vary as a function to the information of interest. (a) The magnification allows counting of granules of a particular chemical makeup. (b) A high magnification image is better suited to measure structural elements such a a coating thickness

Having introduced the concept of inter-sample heterogeneity, it is appropriate to bring up the utility of NIRCI for liquids, suspensions and emulsions. There is typically no value in acquiring images of liquids due to their intrinsic chemical homogeneity. Measurements requiring very short acquisition times may be suitable to characterize certain types of emulsions or suspensions, but the heterogeneity must be stable enough to enable chemical 'snapshots' to be recorded at multiple wavelengths within timescales short enough to freeze sample motion or flow. One may consider acquiring a two or three wavelength NIR chemical image dataset in <1 s from a relatively stable emulsion. However, the high throughput approach (described previously) expands the use of NIR imaging to liquids. For example, an array of solutions of potentially different composition and/or concentration contained in isolated wells could be chemically analyzed simultaneously within a single field of view. In this case each individual well in the array would be treated as a separate sample, greatly increasing the analysis speed over that achievable using a sequential NIR spectral measurement. An additional advantage of this approach can be obtained if reference or calibration samples are placed in the same field of view and measured simultaneously with the 'unknown' samples. In this manner a single measurement may also contain 'calibration' data, which is free of artifacts arising from different experimental conditions or instrumental changes over time. High throughput applications are generally the only situations where NIRCI is suitable for liquids. A section of this chapter will discuss the high throughput approach in greater detail.

Finally, sample type will impact the suitability of one CI technique or approach relative to another. While most NIRCI measurements are made in diffuse reflectance mode, it is also possible to perform transflectance and transmission measurements (Figure 14.2) with only minor modifications to the optical configuration. These approaches are particularly appealing for samples that are poor diffuse reflectors or

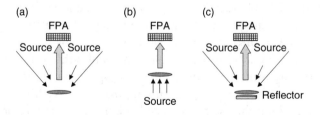
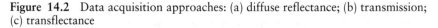

Figure 14.2 Data acquisition approaches: (a) diffuse reflectance; (b) transmission; (c) transflectance

very thin. Of course, the purpose of the experiment and nature of
the heterogeneity of the sample(s) should dictate the selection of the
imaging implementation. Samples with a degree of surface topology are
also much less of a concern in NIRCI than for other CI approaches.
NIRCI permits the use of optical components that provide a significantly
greater depth of field than is available for Raman or mid infrared CI
approaches. Samples with varying 'heights' of the surface remain within
the depth of field for a particular magnification. While a number of
parameters are involved and may be modified on certain instruments, a
system with lower magnification, i.e. larger field of view, is more permis-
sive of sample surface irregularities. Specular reflectance is a common
phenomenon in diffuse reflectance measurements and precautions must
be taken to minimize this effect. This effect is often seen in a CI experiment as localized areas of very high brightness in the image resulting in
poor quality data, nonlinearities and spectral saturation. One common
approach to deal with this issue is to polarize the radiation at the source
and use a second polarizer in front of the detector oriented at 90° to the
first, thereby eliminating the problem of specular reflectance. Figure 14.3
illustrates the effect of the use of polarizers on the image. The impact
of specular reflectance on the image should also be considered in the

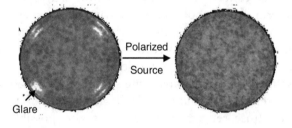

Figure 14.3 Specular reflectance on chemical images and the effect of cross-
polarizing the source in an acquisition using a liquid crystal tunable filter

choice of a support medium for the sample as this medium is often
present in the field of view. Mirror surfaces are a good selection because
they reflect the source radiation with the same polarization and appear
dark when viewed through the second polarizer. This in effect provides
'darkfield' illumination for the sample and enables the software to automatically distinguish the sample from the support media. When dealing
with thin or transparent samples, the opposite approach is often more
appropriate; a high-diffuse reflectance background is favorable because
it enables a transflectance (double pass transmission) measurement to
be made. White ceramic and Teflon^{TM} work well in this capacity.

14.4   WHAT, HOW MUCH AND WHERE

CI has established itself as a valuable and flexible tool with applications in a variety of industries. Its use in product development, quality control, quality assurance, root-cause analysis of manufacturing problems and process understanding are well documented (Lewis *et al.*, 2001; Hammond and Clarke, 2002; Koehler *et al.*, 2002; Lewis *et al.*, 2005 and references therein; 2004). Perhaps this is not so surprising since for many analytical chemists one of the most fundamental and generally useful 'tools of the trade' is the spectrometer, the measurement technology upon which CI is built. However, conventional spectrometers pose limitations in the analysis of complex, engineered manufactured products in that they cannot deliver an understanding of both the physical and spatial parameters designed to directly impact on product performance. Simultaneously, the biology and medical communities, who excel in image classification and at understanding structure/function relationships, now want to understand the biochemical basis of the highly complex images that are produced by their most common imaging tools, the human eye and the microscope. This simultaneous need for chemical specificity and image classification in both communities is the realm of CI. As a result CI offers interesting possibilities for understanding complex chemical systems whether they are manufactured or of biological origin. However, perhaps it is fair to say that after the initial experimental methods were developed and commercial systems entered the market, laboratory chemical (hyperspectral) imaging systems did not produce the number of enabling or critical measurement applications that might have been assumed. On the one hand the spectroscopy community was excited by the possibility of straightforwardly and quickly adding images to their spectral measurements, but the 'pictures' were not always perceived as having immediate analytical value. For biology and medicine, it was interesting and different because the contrast was intrinsic (no exogenous reagents) and ultimately based upon highly specific molecular or chemical markers, but the images were somewhat pedestrian, offering only low fidelity (limited numbers of pixels) and low resolution. Improvements in imaging instrumentation and the development of dedicated software that allows simultaneous spectroscopic and image analysis have helped address these issues.

In the chemical and pharmaceutical industry, problems that only involve measuring composition and concentration are well served with other instrumentation. However, as manufactured products evolve to more complex functionality, CI adds the capability to access the geometric, or spatial distribution information inherent in these types of materials. These data are available in addition to the more traditional metrics of composition and concentration. In essence the technology adds the 'where' to the 'what' and 'how much' data that are available from conventional spectroscopic instrumentation. The ability to access 'where' has little added value when products are relatively simple mixtures, but has significant implications for analyzing chemical and biochemical systems in which structure/function relationships define or drive functionality. For example the need to understand what determines or changes the dissolution characteristics of an extended release pharmaceutical product, or the behavior of a more complex drug delivery system, has brought about the need for a better understanding of product 'assembly' at the micro-chemical level (Hammond and Clarke, 2002; Lyon *et al.*, 2002; Clarke, 2003). Chemical images provide such insight.

However, the 'where' has always been of primary consideration in the biological community. In medical imaging for example, the interpretation of morphological images has traditionally been the primary factor guiding a diagnosis. Less emphasis was placed on understanding the fundamental chemical changes that underpin the observed contrast. As a result, biochemical testing and diagnostic imaging were not always directly connected. With more sophisticated CI instrumentation available to the medical community, integrated measurements and research is possible (Lee *et al.*, 2006). These tools can lead to better understanding of the biochemical processes that result in morphological changes, and ultimately may manifest and dictate the progress of a disease.

14.5 DATA ANALYSIS AND THE UNDERLYING STRUCTURE OF THE DATA

NIRCI is a massively parallel and spatially resolved implementation of NIR spectroscopy. As a result, the traditional function of NIR spectroscopy to determine sample composition and concentration is what ultimately provides the fundamental contrast in the images, and that also underpins the quantitative information that is extracted from CI data.

The basic assumption in conventional NIR spectroscopy is that samples are either spatially homogeneous or that the sampling optics of the spectrometer can be designed to make this approximation valid even for spatially heterogeneous samples. In such a measurement, a single spectrum is usually obtained from a sample, thereby producing a spectral signature that includes the chemical information from all chemical entities present in the sample, i.e. a mixture spectrum. However, in order for this spectrum to be useful for quantitative analysis, a calibration model must be produced that relates this mixture spectrum to the sample components. This is usually accomplished by recording spectra of multiple samples with the same chemical composition, but with varying concentrations. A prediction model may be developed based on these calibration spectra and the concentration matrix, using a variety of chemometrics tools. Finally, this model is applied to a spectrum acquired from a sample of the same chemical composition, but of unknown concentrations, with the goal of predicting the concentration of one or multiple components of interest.

In an entirely analogous manner, the spectra in a NIRCI dataset are usually treated as a series of individual measurements and used in the development of similar multivariate models to predict parameters such as concentration. In other words, each spectrum measured by individual pixels on the array detector can be analyzed quantitatively as a series of mixture spectra. The resulting images can be represented with intensities scored from '0' to '1', with '1' being the score for a 'pure pixel', i.e. one that contains a spectrum of only one of the individual chemical components comprising the mixture. Following from the example presented in Section 14.2 which addressed the change in perceived sample heterogeneity with magnification, it becomes apparent that the process of chemically imaging a heterogeneous sample results in a complex ensemble of mixture and pure component spectra, the characteristics of which depend on the size of the individual chemical domains within a sample as well as the magnification of the imaging optics. Further complications arise due to variations in spatial resolution as a result of differences in the wavelength of the absorption bands as well as factors associated with depth of penetration of the light into the sample. However, if the typical domain (or agglomeration) size of the chemical components present in the sample is larger than the pixel size imaged on the sample, then a spectrum from a single pixel will be very close to the spectrum of the pure chemical component. Spatially resolved pure components then no longer require an estimation of concentration and become good candidates for classification algorithms. In classification, a score of 1 indicates that the spectrum is very similar to the pure component and 0 means that it is not. The classification result for each pixel provides a 'picture' showing the spatial distribution of the component of interest. Furthermore, an estimation of abundance may be calculated by counting the number of pixels scoring close to 1 for a chemical component of interest. While the estimation of abundance is an example of novel valuable information obtained from NIRCI, this example more importantly introduces the concept of using the technique as a primary method, one that does not require *a priori* knowledge about the sample, but rather utilizes the intrinsic chemical contrast to characterize the system.

Simple geometrical shapes of various colors representing chemical moieties can be used to better describe the concept of spatial and spectral unmixing that are at the heart of using CI in both primary and secondary methods. Figures 14.4 and 14.5 show a schematic repre-

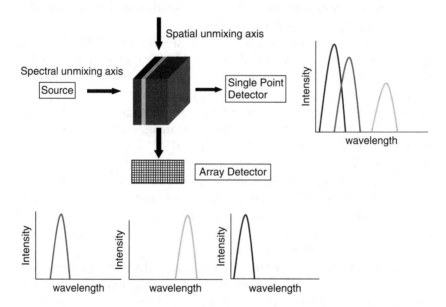

Figure 14.4   Spatial and spectral unmixing in a structurally simple multicomponent sample

sentation of the structure of the data and the resulting interpretation. The two axes available for analysis are identified at the top and on the left; they are the spectral and spatial unmixing axes. The spectral unmixing axis is common with single point NIR spectroscopy and provides a full spectrum amenable for use in a secondary method calibration. The component spectra combine to form a convoluted profile that can be resolved, or unmixed, with a calibration. When using the spectral unmixing axis, each detector of the array performs the same function a single detector placed in a conventional spectrometer does; the value of the array then resides in the large number of detectors available for

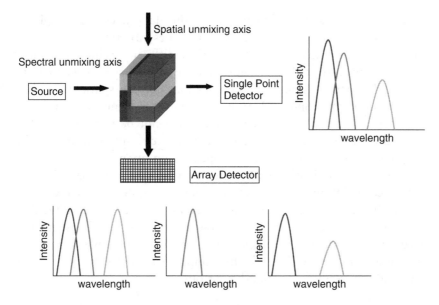

Figure 14.5 Spatial and spectral unmixing in a complex engineered sample

simultaneous acquisition. A commercial NIR imaging system can acquire approximately 82 000 spectra in approximately 2 min of data acquisition time. This tremendous number of spectra can be acquired from a single sample or from multiple samples positioned in the field of view of the instrument. The former is used to investigate heterogeneity in a single sample (intra-sample characterization) and the latter is the basis of high throughput applications (inter-sample characterization). Theoretically, ~82 000 samples could be positioned individually, each over a single detector. It is important to notice that for high throughput experiments, the spatial information in an image only serves to provide coordinates to identify the sample. In this scheme, the imaging hypercube can be unfolded into a 2-D matrix for analysis without any loss of information.

The spatial unmixing axis is specific to imaging data. Similarly to a photographic picture, the meaning of the data may reside in the arrangement of pixels and their associated tones. The image pixels can be 'unfolded', but the result is an incomprehensible mixture of tones; in order for the picture to meaningfully represent the subject, the spatial arrangement must be preserved. The same is true for some chemical images. The hypercube can be unfolded and yield the equivalent of a series of single point measurements (useful for high throughput applications as seen above), but the information content of the picture itself, i.e. the spatial arrangement of pixels, is not being used. The information contained in the spatial dimensions of the dataset can be employed to develop a primary method, one that does not rely on a calibration against parameters measured with a different technique. This is clearly a departure from the traditional use of NIR spectroscopic data. The development is simple for samples that contain spatially resolved pure chemical components (Figure 14.4). In other samples, the tone, or the spectral intensity, at a certain pixel may not be pure black (0) or pure white (1) but rather an intermediate value resulting from a multi-component mixture (Figure 14.5). Spectral unmixing may need to be considered in combination with spatial unmixing for these samples. The level of convolution of the spectral signature acquired by one detector pixel is directly related to the spatial resolution that can be attained by the imaging system and the morphology of the samples. Consequently, in order to obtain useful information, one fundamental experimental parameter must be addressed: the sample(s) must be analyzed using statistically meaningful spatial dimensions.

14.6 IMAGING WITH STATISTICALLY MEANINGFUL SPATIAL DIMENSIONS

The size of the field of view (system magnification) and the number of detectors in the array are at the heart of the relevance of an experiment with regards to a particular problem or sample type. Contaminant detection is a well-suited example to illustrate this concept. Figure 14.6 shows a schematic representation of a contaminant present at a level of $\sim$1–2 % in a sample and the measured area obtained from three modalities. In macro scale or single point spectroscopy, the signal arising from the contaminant is diluted by the average signal originating from the rest of the object. A proper calibration may be able to detect and even quantify the contaminant, but overall, this approach is plagued with the problem of dilution. While this is not an imaging situation per se, the same problem is observed when imaging is performed using a field of view that is too large for the sample. For example, imaging with a $100\,\mu m$ per pixel magnification, a range often designed for high throughput applications, would likely not easily permit the detection of a $10\,\mu m$ contaminant present in a $10\,000 \times 10\,000\,\mu m$ (1×1 cm) sample. The signal arising from the localized contaminant would appear in a single pixel and be diluted in the signal arising from the matrix in the $100 \times 100\,\mu m$ pixel spectrum.

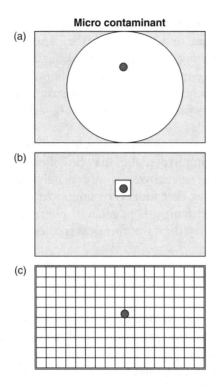

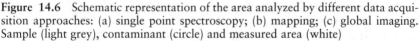

Figure 14.6 Schematic representation of the area analyzed by different data acquisition approaches: (a) single point spectroscopy; (b) mapping; (c) global imaging. Sample (light grey), contaminant (circle) and measured area (white)

Looking at this problem with an instrument using a different size scale would present a different challenge. If a magnification of 1 μm per pixel was used for example, it would either be necessary to image the sample multiple times, moving the sample or the imaging head between measurements, in order to acquire data from the whole sample in a 'searching' mode. As a result, the acquisition time would be multiplied manyfold and the amount of data generated would be proportionally larger. If the time budgeted for the analysis is not extensive, then it would be necessary to image only a portion of the sample, one that is representative of the whole, and hope that the contaminant is present in the sampled area. When the contaminant is detected using this approach, it usually presents itself in the image cube as numerous spectra of the chemical making up the contaminant.

The best approach should combine the benefits from the two schemes illustrated above. The objective is to tailor the size of the field of view to image as much of the sample as possible at once, to reduce the acquisition time, while ensuring that the pixel size is in scale with the contaminant or object of interest. This results in a few pixel spectra showing strong spectral contributions from the contaminant, in a data cube that contains information about most or all the sample surface. Luckily, a number of NIRCI instrument designs allow for such tailoring of the field of view and can be adapted for the problem at hand.

Once the data acquisition parameters are determined for a particular problem, NIR imaging hypercubes may be collected. They are generally rather large, tens of megabytes, and depending on the sample, may contain many spectra that look very similar to one another. The next task in method development is to generate contrast and numerical data from these images based on the chemical species that are present.

14.7   CHEMICAL CONTRAST

Spatially heterogeneous samples are well suited to the application of unsupervised classification procedures, and can be analyzed based solely on the information present in the data cube. Figure 14.7 illustrates an example where the intensity image at one specific wavelength shows the spatial distribution of various chemical components in the sample. The dashed spectrum displays a much higher intensity at 2080 nm; consequently, the pre-processed single channel image (i.e. the image plane) at 2080 nm principally represents the spatial distribution of the chemical component giving rise to the dashed spectrum. Low brightness in the image therefore indicates either absence of or very little contribution from this chemical species. Prior knowledge of the ingredients making up the sample is not necessary in this type of approach. This univariate analysis, i.e. monitoring one variable consisting in this case of the intensity at a selected wavelength, is highly intuitive and fast, but may not be sufficient for complicated samples containing chemical species with overlapping spectral bands. Multivariate analysis generally performs better in such conditions. Numerous multivariate analysis methods are available and the sample type and purpose of the analysis should guide the selection of an appropriate processing scheme.

In simple systems characterized by well resolved peaks arising from one or a few species present in the sample, a ratio of two peaks or the peak height above a 'baseline' may be used to derive qualitative and quantitative information. Similarly, the area under a peak can be used for quantitative measurement when components do not overlap.

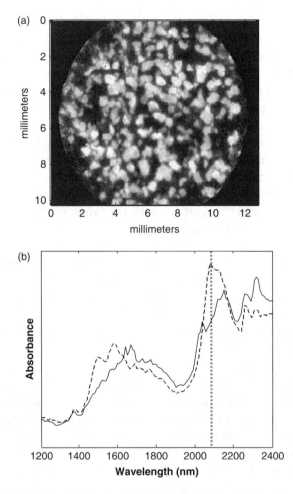

Figure 14.7 NIR chemical image highlighting the spatial location of chemical species (a) within a sample and (b) the corresponding spectra

Changes in band position, band width and band shape across a spectral image can also be used to generate good chemical contrast.

In cases where the interpretation of the spectra is not straightforward because the system contains chemical species with overlapping spectral bands, one of the most useful unsupervised multivariate analysis methods available is principal component analysis (PCA). Briefly, PCA is a means of describing the information contained in an image with a smaller number of characteristics called components (Geladi and Grahn, 1996). Briefly, the first component is computed to represent the maximum variance in the original image data cube. This component contribution is removed from the original data and subsequent components are calculated in a similar fashion. PCA can be used as a data reduction method, in preparation for the application of another multivariate analysis, or as a classification method by comparing the projection of a dataset in its original form onto loading vectors calculated for each component (i.e. calculation of scores). Projecting scores as a function of spatial position often provides quick access to interpretable chemical differences that characterize the sample. Notably, PCA loading vectors are not necessarily related solely to the chemical composition of the sample, and may represent variance that is due to physical characteristics or even artifacts. Careful examination of the loading vectors is necessary to correctly interpret both the origin of the loading vector and the meaning of the scores.

Other unsupervised classification techniques include cluster analysis, where the spectra are grouped into clusters that may be more or less predefined by the user. Some implementations of cluster analysis indeed allow the user to enter only the number of clusters that are represented in the dataset, while others require the input of centroid coordinates. Various correlation measurements (e.g. Euclidean, correlation coefficient) can also be used to generate chemical contrast without a need for a calibration. Pixels or areas in an image can be used to calculate correlation parameters with all other pixels in the image. This approach highlights domains of similar chemical composition, regardless of whether the actual identity of the chemical makeup is known or not.

When the ingredients or components of a sample are known and are available for analysis, they can be used in comparative approaches to yield a classification or concentration measurement for each pixel of the chemical image. For example, correlation coefficients can be calculated between the spectrum of a pure component (a reference library spectrum for example) and the spectra measured at all pixels in the image to generate a correlation coefficient image. Pixels displaying a high correlation coefficient would then be classified as 'similar' to the library spectrum. All the processing schemes mentioned so far produce a number, be it a correlation coefficient, a classification score, a distance from a centroid or peak height, which becomes the source of the image contrast. It is called chemical contrast because the numbers are directly related to the chemical composition of the individual pixel spectra.

In more complicated systems, where multiple ingredients produce overlapping bands in the spectra, data reduction or calibrations are often preferable to straightforward comparison of spectra. The results of an unsupervised multivariate analysis can still be used as input in a supervised approach, thereby benefiting from the reduced size of the dataset, but methods that perform their own data reduction, such as principal component regression (PCR), are also quite popular. PCR is useful for systems consisting of a few ingredients of known concentration. Briefly, a set of concentrations or class identities provided by the user are regressed against principal component (PC) scores of the spectra corresponding to the members of that set. It uses the coefficients from that regression to predict the concentrations or classification of unknown spectra based on the scores of those spectra with respect to the loading vectors from the original PC calculation. The process of building a model consists of determining the number and identity of the calculated factors required for robust prediction of concentration or classification.

More complicated systems composed of multiple ingredients may require more flexible multivariate tools such as partial least squares (PLS) analysis. PLS is a minimally restrictive extension of multiple linear regression models and its flexibility makes it appropriate where the use of traditional multivariate methods is severely limited, such as when there are fewer observations than predictor variables. The main advantage of PLS is that one does not need to know the identity of all species present in a system as long as the samples in the training set are a good representation of the system. PLS is an effective classifier and quantifier because both the target class and the spectra are factored in the calibration. PLS is particularly useful for imaging data in that a classification PLS is appropriate for heterogeneous samples, and the resulting classification image can be used to both identify ingredients and to generate quantitative information in the form of an estimation of abundance. A concentration PLS operates much like traditional NIR calibrations, where a concentration value is calculated for each pixel in the image. This type of approach requires a calibration matrix, while the classification PLS only requires the availability of pure component spectra. It should be borne in mind that a calibration set for a concentration calculation is not as easy to prepare in the context of imaging because the concentration must be homogeneous across all pixels of the detector encountering the concentration standard. In other words, a perfect dispersion or a perfect blend needs to be achieved in the calibration samples at the microscopic scale if this is the magnification setting used to acquire the images. This difficulty, in combination with the compelling results seen with classification PLS models, probably explain the popularity of the much simpler latter method. The classification PLS assigns a continuously varying score (0 to 1) to each pixel, which is usually related to the concentration of the component in each pixel ($1 = 100\%$). The mean of the scores over the entire sample typically corresponds to the mean abundance in the sample and each pixel has its own abundance estimation in the resulting image. Such a calibration requires a spectral library of pure components only, and no mixtures. It is possible to further simplify the result by setting threshold values for the value of scores that are considered high enough for the pixel to be classified as 'rich' in the particular component and hence produce binary images from the scores images. Figure 14.8 shows such an example of a PLS classification scores image using a binary result approach. This

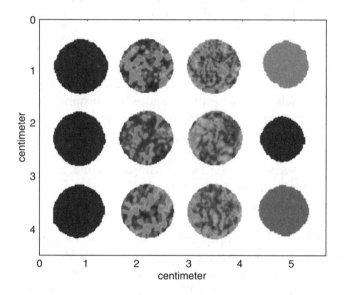

Figure 14.8 RGB image created from PLS scores transformed into binary values using a threshold method

image provides a qualitative assessment of the distribution of the chemical of interest in the sample. Here, the field of view was selected to analyze multiple samples at once and wells were reserved for three pure components found in the samples. The pure components serve as a real-time calibration set for the classification approach; the main advantage of this method is that the calibration and samples are imaged at once, eliminating the need for any calibration transfer procedures. Using the PLS classification, results are obtained for the three classes, i.e. the three pure ingredients, and displayed in the form of a composite red, green and blue image illustrating the quantity and spatial arrangement of the ingredients.

14.8   MEASURE, COUNT AND COMPARE

While calibration to an external spectral standard coupled with multivariate regression can be a useful way to characterize an imaging sample, external calibration is not the only way to extract quantitative information from spectral images. If contrast is developed in a consistent way across a series of related samples, then several kinds of quantitative treatments are possible. Spatially independent information is available in the form of histograms of pixel intensities at image planes (or channels) displaying contrast. These distributions can be based on single wavelengths or the results of more advanced supervised or unsupervised processing. Such distributions are not referenced to size, and if the field of view is suitably chosen, should be size independent. One may compare the statistics of the intensity distributions from image to image, looking for shifts in the mean as an indication of a concentration variation between samples, or one could compare the standard deviation, skew and kurtosis of the distributions as measures of the degree of uniformity of the samples. In many cases it may be appropriate to model the distributions as sums of Gaussians, in which case the histograms may be fit, and the variation of the fit parameters from sample to sample correlated against changes in physical characteristics or behavior.

Figure 14.9 shows a series of histogram plots and their corresponding images generated from the analysis of a blending process as a function of blending time (Lyon *et al.*, 2002). In this case the image contrast is the result of a classification PLS to the active pharmaceutical ingredient (API). Pixels with spectra very close to that of the pure API are mapped to values close to one in this image (white), while pixels containing mostly or only excipients are mapped to values close to zero (black). The histograms show a steady evolution from an apparently multimodal distribution corresponding to well-segregated macroscopic (on the spatial scale of the image) domains, through intermediate, single-mode distributions characterized by large skew and kurtosis, and finally, to an apparent Gaussian distribution, corresponding to the mixing end point. The mean value of the distribution also evolves with time, showing that there is a net change in concentration of the API over the field of view as mixing progresses. In this series it is obvious that one could construct a quantitative measure of the 'goodness of mixing', looking only at the statistical distribution of intensities in the derived images, without making any direct reference to those images.

The threshold approach can be used to access a variety of information. For example, comparing the number, size, and even spatial

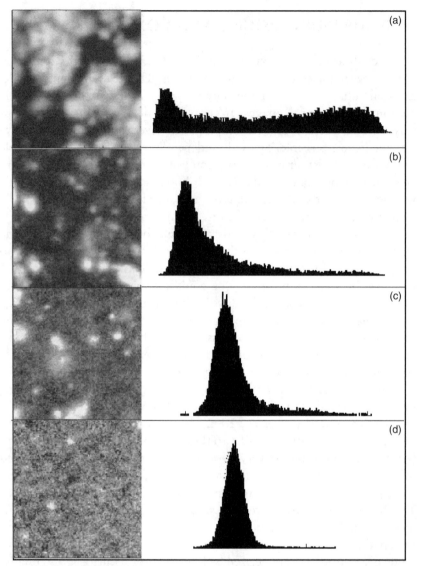

Figure 14.9 Histogram plots of the intensity measured at all points in the image. From (a) to (d), the shape of the plot is indicative of an increasingly more homogeneous blend.

relationships of the derived chemical domains can yield quantitative information possibly related to functional variations among the samples. When applied to pixel intensities, it can serve to isolate domains of a certain chemical composition, which in turn may be measured to compare distributions of domain sizes from sample to sample.

Statistics describing the spatial relationships of the domains may also be of interest; one might think of investigating the spatial relationships of hydrated excipients with highly hygroscopic ingredients for example. Such spatial correlation between domains corresponding to different chemical species is of interest for both pharmaceutical and physiological applications. Density variation of domains across individual samples is also potentially important in pharmaceutical quality control.

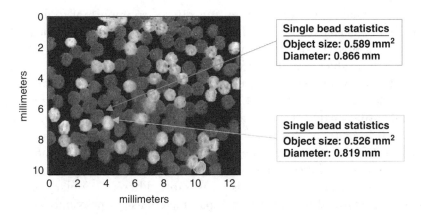

Figure 14.10 Chemical image of pharmaceutical beads that comprise a time-release product. The size and equivalent diameter can be calculated for any or all beads

Figure 14.10 shows pharmaceutical beads that comprise a time-release product in capsule form; statistics calculated from the chemical image are displayed in Table 14.1. The number and average size of beads from each of the two types present (immediate and extended release) are indicative of their relative abundance and thereby directly related to the profile of activity of the product. However, beyond the abundance calculation, it is useful to determine the individual size of each bead because significant departures from the average size may have an effect

Table 14.1 Statistics related to the composition and size characteristics of beads

	Type I beads	Type II beads	Overall
Number of beads	44	91	135
Mean bead size (mm^2)	0.477	0.536	0.517
Standard deviation of bead size (mm^2)	0.059	0.095	0.089
Mean diameter (mm)	0.779	0.826	0.811
Standard deriation of diameter (mm)	0.275	0.348	0.337

on the properties of the capsule content by skewing the time-release profile. In other words, even if only one bead was unusually large, it could have a large effect on the functional properties of the bead mixture. CI has a distinct advantage over a bulk or averaging chemical analysis technique for this type of sample in that chemical and physical data are available for perhaps thousands of individual granules at once. A variety of statistics can be derived characterizing variations within a particular bead type or between bead types in a single measurement on a finished product.

In pharmaceutical applications, it is also often important to measure the homogeneity of the API distribution within a sample, particularly when the dosage form may be segmented, e.g. if a patient is told to take half a tablet. Significant variations in the dosage provided by the two halves of tablets have been reported (Cook *et al.*, 2004) and are cause for concern. Figure 14.11 shows a pressed tablet displaying a number of domains of varied sizes. Statistics on the spatial relationships of API domains in this tablet indicate that the center of mass of the API is 2.2 mm from the center of mass of the tablet [as indicated by the green and red crosshairs in Figure 14.11(c)]. The dosage of the two halves of this tablet therefore is measurably different. Similar analysis can be performed to measure such parameters as the proximity of two

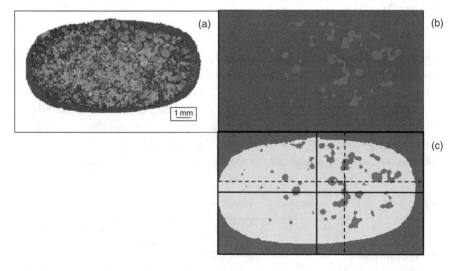

Figure 14.11 (a) RGB chemical image of various components of a pharmaceutical tablet. (b) Binary image showing the API only. (c) Projection of the API distribution on the total surface of the tablet and cross-hairs indicating the center of mass of the tablet (solid) and of the API (dashed)

chemical species for deriving measurements of component affinity or consequences of particular manufacturing processes. One might, for instance, be interested in the distribution of lubricants in a finished solid dosage form when the order of ingredients inclusion in the blender is modified, or when a lubricant is pre-mixed before introduction in the blender. Such a measurement does not require any calibration and is therefore a primary method measuring data that is informative about the process itself as well as the finished product. The same dataset can be used in a secondary method and correlated with relevant sample quality parameter measurements traditionally performed in the industry, such as dissolution testing.

Figure 14.10 illustrates one example of application that does not need pictures. In this image, the goal is to count the number of objects belonging to each of the two classes, their sizes and some product-specific statistics; in this case, the information of interest is the variability in the size of the coated beads. It is clear that the information is available, but for it to have any value, the metric that is measured must be related to a particular physical or chemical characteristic of the sample of interest.

14.9 CORRELATING DATA TO SAMPLE PERFORMANCE AND/OR BEHAVIOR: THE VALUE OF NIRCI DATA

Numerous pieces of information contained in CI datasets can be translated into numerical output. This is the point where spectroscopy and spatial analysis combine to provide robust analytical information on both the chemical and physical properties of a sample. The challenge for the user and ultimately the value of the technology lies in understanding the links between these measurable characteristics and product performance. Unfortunately, this is an area where little is published and those organizations, mainly pharmaceutical, that are routinely employing NIRCI have little incentive or inclination to publish the work for a variety of strategic reasons.

Beyond the strategic advantages conferred by the information on the finished product, many useful characteristics about physical processes or raw materials can be obtained. Such knowledge is particularly valuable in the new quality by design approach brought about by the process analytical technology (PAT) initiative of the USFDA (Arrivo, 2003; Lewis et al., 2005; http://www.fda.gov/cder/OPS/PAT.htm). For example, an interesting piece of information may be the measurement of the spatial distribution of a hydrated excipient in relation to the spatial distribution of a highly hygroscopic API whose anhydrous form exhibits different behavior or performance compared with the hydrated form. In this case, it is critical to be able to measure the spatial relationship between the two components, i.e. a metric that would reveal the efficacy of the manufacturing process that was probably tailored to keep them isolated from each other. In terms of the functionality of the product, knowledge about the distribution of a hygroscopic component, such as a disintegrant, could provide information about the solubility characteristics of the pressed tablet. Similarly, component distribution and stability in storage may be related in certain products and the distribution of lubricant carbohydrate may be associated with flow characteristics. As stated earlier, it is perhaps only the specialist in the product being analyzed that truly understands the value of information that can be acquired by CI.

14.10   CONCLUSIONS

NIRCI is a versatile analytical technique that simultaneously measures highly specific chemical and morphological information about samples with minimal or no sample preparation. Changes to the optical configuration of NIRCI instrumentation are simple and a change of objective is the only step necessary to go from high spatial resolution micro data collection to relatively large fields of view or high throughput spectroscopic approaches. The collected hyperspectral data contain a wealth of information about the chemical and physical characteristics of a particular sample and the end-user can mine the data in a variety of ways. The large number of individual spectra available from a single sample allows for robust statistical assessments of concentration distributions as well as the calculation of size and shape information for localized chemical domains or particles in a single sample. The rapidity of the measurement also enables multiple samples to be measured quickly, thereby enhancing the statistical robustness of inter- and intra-sample comparisons. As a result it is likely that the industrial applications of NIRCI will continue to grow and expand into new areas. While this chapter has drawn on numerous examples from the pharmaceutical industry, and in particular pharmaceutical formulation, applications in the food, cosmetics, polymer and other industries have been documented and provide equally compelling arguments that justify routine application.

REFERENCES

Arrivo, S. M. (2003) The role of PAT in pharmaceutical research and development, *Am. Pharmaceut. Rev.*, **6** (2), 46–53.

Clarke, F. C. (2003) Extracting process-related information from pharmaceutical dosage forms using near infrared microscopy, *Vib. Spectrosc.*, **34**, 25–35.

Clarke, F. C., Hammond, S. V., Jee, R. D. and Moffat, A. C. (2002) Determination of the information depth and sample size for the analysis of pharmaceutical materials using reflectance near-infrared microscopy, *Appl. Spectrosc.*, **56**, 1475–1483.

Colarusso, P., Kidder, L. H., Levin, I. W., Fraser, J. C. and Lewis, E. N. (1998) Infrared spectroscopic imaging: from planetary to cellular systems, *Appl. Spectrosc.*, **52**, 3, 106A.

Cook, T. J., Edwards, S., Gyemah, C., Shah, M., Shah, I. and Fox, T. (2004) Variability in tablet fragment weights, *J. Am. Pharm Assoc.*, **44**, 583–586.

Geladi, P. and Grahn, H. (1996) *Multivariate Image Analysis*, John Wiley & Sons, Ltd, Chichester, Chapters 6 and 7.

Hammond, S. V. and Clarke, F. C. (2002) Near-infrared microspectroscopy. In: *Handbook of Vibrational Spectroscopy*, Vol. 2, J. M. Chalmers and P. R. Griffiths, Eds, John Wiley & Sons, Ltd, Chichester, pp. 1405–1418.

Kidder, L. H., Haka, A. S., and Lewis, E. N. (2002) Instrumentation for FT-IR imaging. In: *Handbook of Vibrational Spectroscopy*, Vol. 2, J. M. Chalmers and P. R. Griffiths, Eds., John Wiley & Sons, Ltd, Chichester, pp. 1386–1404.

Koehler IV, F. W., Lee, E., Kidder, L. H. and Lewis, E. N. (2002) Near-infrared spectroscopy: the practical chemical imaging solution, *Spectrosc. Eur.*, **14**, 12–19.

Lee, E., Kidder, L. H., Kalasinsky, V. F., Schoppelrei, J. W. and Lewis, E. N. (2006) Forensic visualization of foreign matter in human tissue by near-infrared spectral imaging: Methodology and data mining strategies, *Cytometry*, **69**A(8), 888–896.

Lewis, E. N. (2002) *High-Throughput Infrared Spectroscopy*, US Patent 6 483 112, 11/01/2002

Lewis, E. N., Carroll, J. E. and Clarke, F. C. (2001) A near infrared view of pharmaceutical formulation analysis, *NIR News*, **12**, 16–18.

Lewis, E. N., Lee, E. and Kidder, L. H. (2004) Combining imaging and spectroscopy: solving problems with near infrared chemical imaging, *Microscopy Today*, Nov., 8–12.

Lewis, E. N., Schoppelrei, J., Lee, E. and Kidder, L. H. (2005) Near-infrared chemical imaging as a process analytical tool. In: *Process Analytical Technology*, K. A. Bakeev, Ed., Blackwell Publishing Ltd, Oxford, pp. 187–225.

Lyon, R. C., Lester, D. S., Lewis, E. N., Lee, E., Yu, L. X., Jefferson, E. H. and Hussain, A. (2002) Near-infrared spectral imaging for quality assurance of pharmaceutical products: analysis of tablets to assess powder blend homogeneity, *AAPS PharmSciTech*, **3**(3), article 17 [PAT].

Index

Note: page numbers in italics refer to figures.